VITAL STATISTICS

VITAL

STATISTICS

Michael Orkin
Richard Drogin

PROFESSORS OF STATISTICS
CALIFORNIA STATE UNIVERSITY, HAYWARD

McGRAW-HILL BOOK COMPANY

New York   St. Louis   San Francisco   Auckland
Düsseldorf   Johannesburg   Kuala Lumpur   London
Mexico   Montreal   New Delhi   Panama   Paris
São Paulo   Singapore   Sydney   Tokyo   Toronto

Library of Congress Cataloging Publication Data

Orkin, Michael.
    Vital statistics.

    1. Probabilities.   2. Mathematical statistics.
I. Drogin, Richard, joint author.   II. Title.
QA273.065   1975   519.2   74-10818
ISBN 0-07-047720-5

VITAL STATISTICS

456789 FGFG 7987

This book was set in Times Roman by Textbook Services, Inc.
The editors were A. Anthony Arthur and Andrea Stryker-Rodda;
the designer was Ben Kann;
the production supervisor was Charles Hess.
The drawings were done by B. Handelman Associates Inc.

to
Miles, Ruben, Ezra,
Leon, and Jesse

Contents

Preface

This text is the outgrowth of notes we have built up over the past few years while teaching the familiar introductory statistics course for students with minimal mathematics backgrounds. One of our main goals in developing this material has been to liven up the subject matter without destroying its usefulness. In doing this, we have tried to make the book a versatile one, appealing to many disciplines. We have included enough material, in a flexible format, to enable instructors with a wide range of tastes to select a suitable syllabus for a one- or two-semester course.

We describe now some of the main features and basic structure of the book. Chapters 1 to 6 compose a core for an introductory course. Chapters 7 to 13 add to this in a logical sequence, yet are written so that selections can be made at the instructor's discretion. Chapters 14 to 17 go more deeply into probabilistic concepts than the first 13 chapters and can be used, all or in part, any time after Chapter 4. Nowhere, however, is a knowledge of calculus needed. Chapter 18 is an interesting sidelight and could be used to liven up the last week of a course.

The many problems and examples are practical in nature, especially in the earlier chapters, yet are designed to make the book interesting for students who think statistics is too dry or abstract. A true-false test is included at the end of each chapter.

Some course outlines possible for this book are as follows.

Very thorough one-semester course: Chapters 1 to 8, 10, 13
Emphasis on nonparametric methods: Chapters 1 to 6, 9 to 11, 13
Emphasis on "classical" methods: Chapters 1 to 8, 10, 12, 13
Bayesian: Chapters 1 to 6, 14, 17
Emphasis on probability and decision theory: Chapters 1 to 5, 14 to 18 (could be used as a math course)
Two-semester course: 1st semester, Chapters 1 to 8; 2d semester, Chapters 9 to 16 (also 17 and 18 if time permits)

We wish to thank the typists at Case Western Reserve University, Columbia University, and California State University, Hayward, who helped in preparing this manuscript. Our special thanks go to Ann Cambra and Susanne Whittner. We thank Professors David Blackwell, Judah Rosenblatt, and Stephen Book for their valuable comments and suggestions. Also, we thank the numerous students, including Dexter Jung, Dave Robinson, and Dale Duke, who helped in proofreading. Finally, we wish to thank our wives Jane and Ellen for their patience during the long hours of preparation of this book.

Michael Orkin

Richard Drogin

VITAL
STATISTICS

Introduction 1

Statistics can be thought of as the quantification of methods for interpreting and understanding random phenomena. Most people use statistical ideas everyday without even realizing it. Whenever you fasten your seat belt, try to predict the weather, take vitamin C because you feel a cold coming on, or choose to wait in the shortest line in a crowded bank, you are using the ideas upon which statistics is based.

We have organized the book to appeal to your intuition and to demonstrate how to make some of your intuitive ideas more precise. The first step in our project is to illustrate how certain common situations can be described numerically. We shall be particularly concerned with situations called *random experiments,* that is, experiments with outcomes that cannot be predicted in advance. Next, we shall introduce the concept of *probability* to describe the likelihood of the various possible outcomes of a random experiment. In the later part of the book, we shall see how the description of a random experiment in terms of probabilities can be used to derive important information and make statistical inferences. Accordingly, three basic topics are covered:

Methods of describing data

Concepts of probability

Methods of making statistical inferences

We now present some sample problems from each of these three topics.

Describing data

1   You have just completed a large study concerning the effects of urban air pollution on respiratory disease. You have data from samples taken in five cities with varying degrees of pollution and also sample data from rural areas. The data include disease rates for various diseases for people of diverse age and economic status. You couldn't possibly list all these data in your final report. How could you summarize the data in an easy-to-read yet informative way?

2   You are interested in the ways in which plants react to their surroundings. You expose a variety of plants to a wide range of environments, including sudden changes in temperature, lighting, and noise level. You measure the plants' reactions with sensitive electronic devices which detect subtle changes in plant physiology. You also have records of growth rates of houseplants in the same house over a long period of time, some of the plants having been talked to and treated with personal affection by their owner and others being treated impersonally. How should you display your data?

Probability

1   You are going to toss a coin 10 times. What is the probability that you will get 5 heads? (The answer is not 1/2.)

2   Why has the proportion of males and females remained at about 1/2 throughout the history of mankind even though we cannot predict the sex of a baby before its birth?

3   Why do Las Vegas casinoes keep on winning year after year, with none of them ever going broke?

4   If you randomly select a jury of 12 people from a population containing 45% women, what is the probability that no women are selected?

5   Each time an underground nuclear test is conducted, there is a .0001 chance that radioactive dust will escape into the atmosphere. If 100 such tests are made, what is the probability that some dust will escape?

Statistical inference

1   You conduct a study to see if students who practice transcendental meditation do better in school, on the average, than people who don't meditate. You cannot ask everybody, and so you take a random sample from each group. How could you make your decision?

2   You wish to estimate the proportion of people in a large population who will vote for Mr. Smiley. You take a sample of size 200 and find that 112 will vote for Mr. Smiley. What should your estimate be and how accurate is it?

3   You want to see if an automobile manufacturer's claim about the gas mileage of a certain compact car is accurate. How would you go about this?

4   You want to see if a certain coin is evenly balanced, and so you toss it 100 times, observing 53 heads and 47 tails. What should you conclude, and how likely is it that your conclusion is correct?

5   You want to see if there is a difference in the effects of four different types of fertilizers. You plant four similar gardens, using one of the fertilizers for each garden. How should you make your decision?

In the following chapters, you will learn how to solve problems such as these. We ask you not to feel overwhelmed by the seemingly large amount of mathematical material which confronts you, but to remember the words of the wise Chinese philosopher Lao Tzu: "Even a journey of a thousand miles must begin with a single step."

Describing and summarizing data 2

Chills run down the spine and a cold sweat breaks out on many students when they hear words like *statistics* and *data*. They even think that the use of statistics makes their studies and reports dry and uninteresting. We feel that although sometimes the use (or misuse) of statistics may have this effect, this need not be the case. In fact, mastery of a few simple methods of describing data can often enable you to clarify and improve an otherwise vague and confusing study. In this chapter we shall present such methods.

Recall from Chap. 1 that our main goal is to study methods of statistical inference, that is, methods of using partial information derived from a random sample to infer things about the population as a whole. In this chapter we shall introduce some types of information often used and discuss techniques of summarizing and describing data. We shall not discuss methods of inferring facts about unknown quantities until later. In a sense, then, this chapter provides the setting for the rest of the book. Before proceeding, we give one piece of advice: In the future, when you prepare a report or study of your own, don't try to be fancy and overly mathematical. A few simple techniques learned well are far more useful than many techniques not clearly understood. If you study this chapter, you should have no trouble organizing data that arise in a wide variety of situations.

Consider, now, how you would handle a topic like one of these:

1   A study to determine if there is a relationship between pornography and crimes of violence

2   A study to determine if excessive eating of meat has a tendency to induce heart disease

3   The evaluation of a new method of teaching

4   A study comparing the incomes of men and women who are doing similar jobs

5   A study of the use of drugs by public school children in a certain large city

6   A study of the conditions of prisons in a certain state

A discussion of any one of these topics will require the presentation and analysis of data, that is, numerical information about the topic under study.

Let us now examine topic 5. Suppose you have to conduct a study of the use of drugs by public school children in a large city. As the researcher, you are immediately confronted with a major problem. You have neither the time nor the money to obtain information (especially detailed information) about every public school child in the city because there are simply too many students for you to do so. In fact, you have resources available to obtain information concerning only a small proportion of the total population of students. The first step, then, should be to decide how you will make your selection in a *representative* way, that is, in a way in which the information given by the sample gives an accurate picture of the entire population.

The standard method of such selection is to take what is known as a *random sample*. We shall call the group from which the random sample is drawn the *population*. In statistics, the word "population" has a more general meaning than the one in everyday use. A population does not have to consist of people but can merely signify the objects under study, whether they be students, machines, laboratory animals, or medicines. We say that an element of a population is *selected at random* if each member of the population has the same chance of being selected. Accordingly, we say that a collection of elements selected from a population is a *random sample* if every possible collection of the same number of elements has the same chance of being selected.

> **Definition.** A collection of elements of a population is said to constitute a *random sample* if the selection is done in such a way that every possible collection of the same number of elements has the same chance of being selected.

It is surprising how accurate a picture of the population is obtained from a random sample. (Later we shall be more explicit as to what "accurate" means.)

Suppose that you wish to select a random sample of 400 students from the population of all students in a certain city. If you had a list of the names of each student, you could put each name on a slip of paper and put the slips of paper in a large hat. Then you could mix the slips well and select 400 slips, letting the associated names constitute your sample.

Of course, this is a rather cumbersome procedure, and a more convenient method might be sought. In fact, you might decide that for many reasons it is inconvenient to select 400 students at random from the entire population. For example, you might not wish to spend the time it would take to travel from school to school throughout the city, which would be necessary if you selected your sample from the entire population. Perhaps you consider it important to become well acquainted with the students and personnel of the schools you work in. For these reasons you might decide to concentrate your study in one or two schools which you consider representative of the city as a whole and then perform an in-depth study of a random sample of students from each of these schools.

Suppose you decide to pick one representative school and choose a random sample from it. Again, the most straightforward method of doing this is to put all the names in a hat and choose at random from these, obtaining a random sample of students from this school. But this method is still rather cumbersome. Instead, a *table of random numbers* may be used to obtain a random sample. A table of random numbers is a table which contains lists of numbers that have been selected or generated according to some random mechanism and then have been listed in their order of selection. In other words, someone has selected numbers out of a hat for you and has even been nice enough to tabulate the results. All you have to do is read the table and use the numbers accordingly.

To utilize random-number tables in choosing 400 students, you must first decide to use some combination of numbers which identify the students. These might be telephone numbers, social security numbers, addresses, or any other convenient numbers that may serve as a label. For our example, let's suppose each student has a four-digit identification number. Then, to choose 400 students at random, you just choose 400 ID numbers at random and then select the students who have these ID numbers for your sample. To do this using the table of random numbers, you would start at an arbitrary place in the table and let the first four digits from the starting place be the identification number of the first student to be selected; the next four digits would be the second identification number; the third four digits would be the third number; etc. In the event that a selected number does not correspond to anyone's identification number, you would skip this number and proceed as before.

Sometimes, modifications might have to be made in your method owing to the nature of the identification numbers in a particular population; for example, they may all begin with a 1 or end in a 0 or have some other such feature. The required modification of the selection procedure is usually straightforward; for example, if every number begins with 1, you could select the next three digits from the table. (A more detailed discussion of random sampling will be presented in the next chapter.)

Example 2-1. Two rows from a table of random numbers are as follows: 04433 80674 24520 18222 10610 05794 37515 60298 47829 72648 37414 75755. We wish to select five 4-digit numbers at random, and so we use the first five groups of 4 digits: 0443, 3806, 7424, 5201, 8222.

Once you have decided upon your sample, you are ready to begin collecting data (sometimes, especially when doing such things as taking surveys, data are collected in the process of taking the sample). Of course, you must decide which types of information you consider to be relevant to the study. In the present situation, you may decide to find out the types of drugs (if any) used by each student. Also, you may wish to know the frequency and length of time that each drug has been used. Additional interests may be the drug source (e.g., from another student, parents' medicine chest, or an out-of-school contact), cost, and social attitudes concerning each drug. Information concerning the home environment of the student, degree of success achieved in school, IQ, and social status may also be useful. You would not be interested in the facts with regard to any particular student but only to get a general picture of the situation. It might even be wise to take some safeguards concerning the confidentiality of your information before gathering any data of a sensitive nature.

Once you have collected the data, you will have on your hands a mass of information which no one in his right mind would read. It is your job to summarize and display the data in an interesting and informative manner. Throughout the rest of this chapter we shall be concerned with the task of summarizing, describing, and displaying data in an interesting and informative way.

Grouping and displaying data

Suppose that you are interested in the relation between the use of marijuana and scores achieved on an aptitude test. Let's say you have collected the scores on an aptitude test given to 60 marijuana smokers, and suppose you also have scores on the same test from 60 non-marijuana smokers. You would probably want to compare the two sets of scores. For now, however, we shall confine our interest to how to display these data. Consider the test scores of the 60 marijuana smokers, which are displayed in Fig. 2-1.

Arranged in this way, the data are difficult to interpret. There are too many numbers to grasp at once. One way to organize these data is

Test scores for 60 marijuana smokers

84.1	88.9	90.2	80.4
87.5	82.3	83.6	83.7
86.9	84.4	88.8	81.5
85.8	87.7	85.0	83.3
87.2	87.7	86.0	83.9
85.2	85.5	84.2	83.9
80.1	82.8	81.3	86.7
82.9	79.1	88.2	84.7
87.0	77.4	89.8	83.6
92.0	80.7	87.8	86.3
82.1	80.6	82.6	83.1
82.6	86.9	89.1	81.8
81.9	86.4	86.1	85.5
83.9	84.5	90.8	88.3
86.7	83.4	91.2	84.8

Figure 2-1

to group the data by size, that is, to partition the data into evenly spaced intervals, which we shall call *group,* or *class, intervals.* Usually about ten intervals are a suitable number; however, the choice of numbers is subjective and will depend on the number of data values and how they are clustered. In general, we want a number of intervals small enough to make a clear display but large enough to furnish a reasonable amount of information.

Assuming we have made the decision to divide the data into 10 group intervals, we must decide what the interval lengths should be. Note that the largest score is 92 and the lowest score is 77.4; that is, the *range* of scores is $92 - 77.4 = 14.6$. Since $14.6/10 = 1.46$, it seems reasonable to have the interval lengths be close to 1.46. It is convenient to have the interval lengths have the same number of decimal places as the data. In the present case (looking at Fig. 2-1), we let the interval lengths be carried out to tenths and let each interval length be 1.5.

Our next task is to decide where to place the lowest interval. Since the intervals are to be adjacent, once we have positioned the lowest interval, we shall know immediately where to place the others. Also, since we want the 10 intervals to contain all the scores, it is appropriate to have the lowest interval contain the lowest score. With this in mind, we shall let the midpoint of the first interval be 78.0 (also having the same number of decimal places as the data values). Since the intervals have length 1.5, the midpoint of the next interval should be $78 + 1.5 = 79.5$; the midpoint of the next interval should be $79.5 + 1.5 = 81.0$; and the midpoint of the next interval should be 82.5; and so on. Once the interval midpoints have been determined, the boundaries follow immediately. Each midpoint is midway between the two adjacent boundaries, and so a distance of $1.5/2 = .75$ from each midpoint gives the corresponding interval boundaries. The lower boundary of the first interval, then, is $78.0 - .75 = 77.25$, and the upper boundary is $78.0 + .75 = 78.75$; the next interval goes from 78.75 to 80.25; the next from 80.25 to 81.75; the next from 81.75 to 83.25; and so on.

We can now construct a table, called a *frequency table,* listing the group intervals, their midpoints, and the frequency of scores in each interval. The easiest way to determine the frequencies is to have a tally column on the table and to go through the data, tallying the scores in the appropriate intervals as we come to them, as we have done in Fig. 2-2. In Fig. 2-2 we have also listed the *relative frequencies* of each interval. The relative frequency of an interval is obtained by dividing the frequency of that interval by the total number of scores (in this case,

60). The completed frequency table gives an organized and easy-to-read picture of the distribution of data. The only information which is lost is the exact placement of scores within each group interval, not much of a price to pay for such improved clarity. By looking at Fig. 2-2 we see, for example, that 13 scores are between 83.25 and 84.75, 11 scores (or a proportion of .1833 of the scores) are between 86.25 and 87.75, and 6 scores (or a proportion of .1 of the scores) are between 87.75 and 89.25.

Remember that when grouping data, there is no exact rule which tells you what to do. You should strive for clarity, numbers that are easy to work with, yet an informative and accurate display. The following is a review of the procedure for grouping data.

HOW TO GROUP DATA

1   Decide upon an appropriate number of intervals (10 is usually a reasonable amount). This decision should depend upon the amount and range of the data and the amount of information you wish to retain in the frequency table.

2   Divide the range of the data by the number of intervals you wish to have to obtain an appropriate interval length. Then decide the exact interval length.

3   Decide on an appropriate first interval midpoint. It should be

Frequency distribution for group intervals of test scores for 60 marijuana smokers

Midpoints	Boundaries	Tallies	_	Frequency	Relative frequency
78.0	77.25–78.75	1		1	.0167
79.5	78.75–80.25	11		2	.0333
81.0	80.25–81.75	₮₮₮		5	.0833
82.5	81.75–83.25	₮₮₮ 1111		9	.1500
84.0	83.25–84.75	₮₮₮ ₮₮₮ 111		13	.2167
85.5	84.75–86.25	₮₮₮ 111		8	.1333
87.0	86.25–87.75	₮₮₮ ₮₮₮ 1		11	.1833
88.5	87.75–89.25	₮₮₮ 1		6	.1000
90.0	89.25–90.75	11		2	.0333
91.5	90.75–92.25	111		3	.0500

Figure 2-2

such that the lowest score is in the first interval and all data will be included in the intervals you have decided upon.

4   Determine all the interval midpoints by successively adding the interval length to the midpoints. Then determine the interval boundaries by adding one-half the interval length to each midpoint. The lower boundary of the first interval is obtained by subtracting half the interval length from the first midpoint.

5   Make a frequency table showing the frequency and relative frequency of data points in each group interval and the boundaries and midpoints of each interval.

We note that grouping the data is necessary only when there are large amounts of data. If the range is extremely large and there are a large number of data values, you may wish to have many more than 10 intervals. If there are "gaps" in the data, you may even have intervals with nothing in them. We suggest that if the data are clustered in a certain region and there are one or two extreme scores nowhere near the rest of the data, then when you are grouping the data, you should forget about the extreme scores, mentioning them as a parenthetical remark in the margin of the frequency table.

HISTOGRAMS

A *histogram* is a graphic display of the group frequency table. The histogram is extremely useful as a descriptive method and has theoretical applications as well as descriptive ones. To construct a histogram, we mark off coordinate axes and carry out the following procedure.

1   On the horizontal axis mark off intervals corresponding to the group intervals on the group frequency table (in other words, the horizontal axis is scaled according to data values).

2   Scale the vertical axis to correspond to the interval frequencies (we can also include an extra vertical axis on the right side of the histogram scaled according to relative frequencies).

3   Make a bar over each interval, the height being the frequency of values in that interval. Because of this step, a histogram is often called a *bar graph*.

4   Label the histogram clearly.

In Fig. 2-3 the histogram is displayed corresponding to the table in Fig. 2-2. The histogram gives us a picture of the *distribution* of the data. We can see, for example, in Fig. 2-3 that $1 + 2 + 5 = 8$ of the 60 scores, or 13%, are less than or equal to 81.75. At a glance it seems that about one-half the scores (actually 31) are less than 84.75. In this way, we might use the histogram to plan for future experiments using the assumption that future data might yield the same patterns. We shall see later that there is strong theoretical justification for this idea.

Problems

2-1 Group the data and make a histogram for the following final-exam scores of statistics students at a certain university:

88	79	90	88	71
83	65	98	86	33
71	49	100	77	93
60	91	80	63	67
99	66	42	75	62
69	68	75	96	83
63	87	84	58	90
21	87	68	89	80

2-2 A random sample of 25 residents from a certain neighborhood were surveyed to determine the amount of natural gas they con-

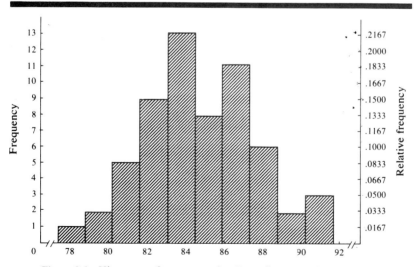

Figure 2-3 Histogram of test scores for 60 marijuana smokers.

sumed over a 1-month period. The results given below are in units of volume of gas in therms:

110	101	98	102	106
98	97	96	90	93
99	97	90	98	98
100	89	89	95	89
99	93	99	85	87

a. Group these data and construct an appropriate frequency table and histogram. Use 9 intervals, each of length 3, beginning the first interval at 84.5.

b. Looking at the frequency table, find the proportion of values between 84 and 96.5.

c. What percentage of residents use at least 90.5 therms a month?

CUMULATIVE FREQUENCIES, PERCENTILES, AND QUARTILES

Another useful description of grouped data is given by listing the *cumulative frequency* of each interval. As the name implies, the cumulative frequency describes how the distribution of data "accumulates."

> **Definition.** The cumulative frequency of a given group interval is the frequency of scores in that interval and in all intervals before it. Analogously, the cumulative frequency of a particular score is the number of scores which are less than or equal to that score.

In Fig. 2-4 we have expanded Fig. 2-2 to include a column giving cumulative frequencies and the corresponding *relative cumulative frequencies*. The relative cumulative frequency of an interval is the proportion of scores which are in that interval and all intervals before it. From Fig. 2-4 we read, for example, that 17 of the scores, or a proportion of .2833, are less than or equal to 83.25; likewise, 55 of the scores, or a proportion of .9167, are less than or equal to 89.25.

A pictorial display of cumulative frequencies can be given by the *cumulative frequency polygon.* To construct the corresponding cumulative frequency polygon, mark off coordinate axes and proceed in the following way.

1 Scale the horizontal axis to correspond to data values; on this axis mark off the interval boundaries.

2 Scale the vertical axis to correspond to cumulative frequencies.

Distribution of grouped data for test scores of
60 marijuana smokers with cumulative frequencies

Boundaries	Frequency	Relative frequency	Cumulative frequency	Cumulative relative frequency
77.25–78.75	1	.0167	1	.0167
78.75–80.25	2	.0333	3	.0500
80.25–81.75	5	.0833	8	.1333
81.75–83.25	9	.1500	17	.2833
83.25–84.75	13	.2167	30	.5000
84.75–86.25	8	.1333	38	.6333
86.25–87.75	11	.1833	49	.8167
87.75–89.25	6	.1000	55	.9167
89.25–90.75	2	.0333	57	.9500
90.75–92.25	3	.0500	60	1.0000

Figure 2-4

3   At the upper boundary of each interval, put a dot at the height of the cumulative frequency for that interval. Put a dot at height zero at the lower boundary of the first interval (no scores are less than this point).

4   Connect adjacent dots with straight lines, and you have the cumulative frequency polygon.

In Fig. 2-5 we display the cumulative frequency polygon resulting

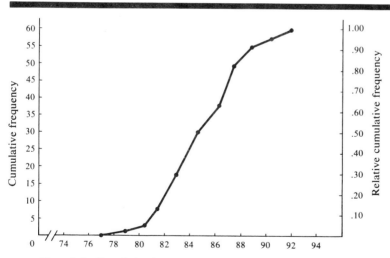

Figure 2-5   Cumulative frequency polygon. Test scores for marijuana smokers.

from Fig. 2-4. Notice that from the cumulative frequency polygon, it is easy to compute the frequency of scores in any interval by merely taking the difference of the heights of the boundaries for that interval. For example, the number of observations in the interval from 86.25 to 87.75 is the difference in corresponding heights; that is, $49 - 38 = 11$.

Relative cumulative frequencies provide an effective way of describing the relative placement of a particular score in the data. The percentage of scores which are less than or equal to a particular score, obtained by multiplying the relative cumulative frequency by 100, is called the *percentile rank* of the score.

> **Definition.** The percentile rank of a data value is the percentage (rounded off to the nearest whole percent) of values which are less than or equal to that value. In other words, the percentile rank of a particular value is the cumulative relative frequency of that value times 100:
>
> $$\text{Percentile rank of } x = \frac{\text{no. of scores} \le x}{\text{total no. of scores}}(100)$$

If we wish to compute the percentile rank of a score of, say, 85.8 on the test in our example, we would compute

$$\text{Percentile rank of } 85.8 = \frac{\text{no. of scores} \le 85.8}{\text{total no. of scores}}(100)$$
$$= \frac{36}{60}(100) = 60$$

Thus, 60% of the scores on the test are less than or equal to 85.8.

Conversely, a *percentile*, say, the 40th percentile, of a sample is the number such that 40% of the data values are less than or equal to that number. If we wish to find a score corresponding to the 40th percentile, we carry out the above procedure in reverse and obtain the score with 40% of the scores equal to or below it. Since there are 60 scores altogether, we wish to find the score k such that $k/60 = .4$. Solving for k yields $k = 24$. Looking at the raw data, we see that the twenty-fourth ordered score is 83.9. The score 83.9 is thus the score at the 40th percentile. We call the 25th and 75th percentiles the *first* and *third quartiles,* respectively. The 50th percentile, or *second quartile,* corresponds to a number called the *median* of the data.

Percentile ranks and percentiles thus give a measure of the relative placement of a value in a large collection of data. If we have only a small number of observations, percentile ranks and percentiles are not particularly useful since we can determine the relative placement of a particular score at a glance.

COMPARING TWO SETS OF DATA

Cumulative frequency polygons and histograms provide a quick method of comparing two sets of data. We can graph polygons and histograms for two (or more) sets of data and compare the corresponding distributions of data at a glance. In Fig. 2-6a and 2-6b we have graphed the polygon in Fig. 2-5 and the histogram in Fig. 2-3, with the corresponding polygon and histogram representing the test scores of 60 non-marijuana smokers. Figure 2-7 presents the corresponding group frequency table and histogram for 60 non-marijuana smokers. In later chapters we shall explain how to make *inferences* in such situations. For now, just using your intuition, what conclusions can you draw?

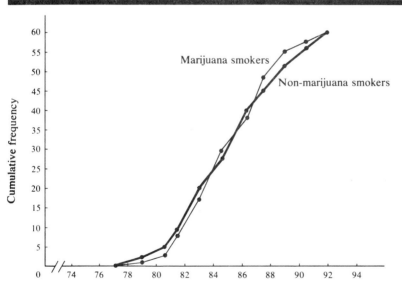

Figure 2-6 (*a*) Cumulative frequency polygons. Test scores for marijuana smokers versus test scores for non-marijuana smokers.

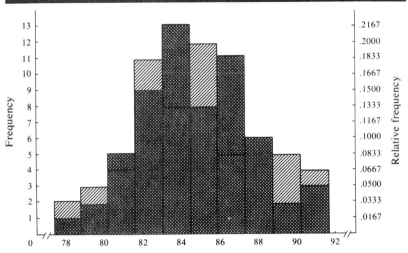

Figure 2-6 (*b*) Histograms. Test scores for marijuana smokers (dark) versus test scores for non-marijuana smokers (light).

Problems

2-3   Construct the cumulative frequency polygon for the data in Prob. 2-1. Find the percentile rank of the scores 90, 98, 87, 68.

2-4   Construct the cumulative frequency polygon for the data in Prob. 2-2. Find the percentile rank for 99, 89, and 87 therms. What value corresponds to the 58th percentile?

Descriptive measures

We leave the drug study now, but we shall still assume that our data have already been gathered from some population. If the data have arisen from a random sample, they are called *sample data*; if the data represent all the elements of a population under study, they are called *population data*. Usually, as in the drug study, we are unable to gather population data and must content ourselves with sample data. Therefore we shall ordinarily assume that we are working with sample data.

MEASURES OF CENTRAL TENDENCY

We are often interested in one number (or numbers) which describes the *center,* or *middle,* of the data. One such number, frequently used to represent the "typical" member of the population, is called the mean.

The Mean

Definition. The mean of a collection of numbers is the numerical average of these numbers. To compute the mean, add the numbers and divide by the number in the collection.

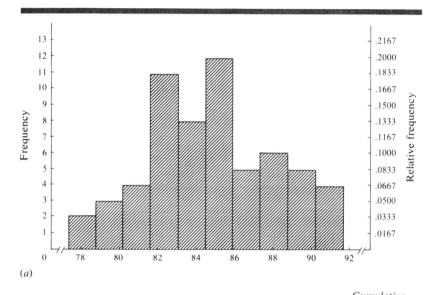

(a)

Boundaries	Frequency	Relative frequency	Cumulative frequency	Cumulative relative frequency
77.25–78.75	2	.0333	2	.0333
78.75–80.25	3	.0500	5	.0833
80.25–81.75	4	.0667	9	.1500
81.75–83.25	11	.1833	20	.3333
83.25–84.75	8	.1333	28	.4667
84.75–86.25	12	.2000	40	.6667
86.25–87.75	5	.0833	45	.7500
87.75–89.25	6	.1000	51	.8500
89.25–90.75	5	.0833	56	.9333
90.75–92.25	4	.0667	60	1.000

(b)

Figure 2-7 (a) Histogram for test scores of students who do not smoke marijuana; (b) frequency and cumulative frequency table for test scores of students who do not smoke marijuana.

Example 2-2. The mean of the numbers 1, 5, −3, 6 is $[1 + 5 + (-3) + 6]/4 = 2.25$.

If the data represent every member of the population, then the mean is called the *population mean* and is denoted by the Greek letter μ (pronounced "mew"). If the data represent the result of a random sample (i.e., are sample data), the corresponding mean is called the *sample mean* and is denoted by the symbol $\bar{X}$ (pronounced "ex-bar"). To avoid the confusion which sometimes arises when considering means from more than one set of data at a time, occasionally we shall denote sample means by other letters and subscripts, such as $\bar{Y}, \bar{Z}, \bar{X}_1,$ $\bar{X}_2,$ and $\bar{X}_3$.

Example 2-3. A sample is taken of scores on a certain test, yielding the following sample data:

100	85	60
100	65	60
90	60	10

We compute the sample mean:

$$\bar{X} = \frac{100 + 100 + 90 + 85 + 65 + 60 + 60 + 60 + 10}{9} = 70$$

An alternative method for computing the mean in cases where the same value appears many times is the following. List the numbers and in an adjacent column list the frequencies with which they occur. Then multiply each value by its frequency, add these products, and (as before) divide by the number of scores. For the data in Example 2-3, we have the following table:

Value	Frequency
100	2
90	1
85	1
65	1
60	3
10	1
	Total = 9

The necessary computations yield

$$\bar{X} = \frac{(100 \times 2) + (90 \times 1) + (85 \times 1) + (65 \times 1) + (60 \times 3) + (10 \times 1)}{9} = 70$$

Exercise 2-1. Compute $\bar{X}$ for the following sets of sample data.

(a) $-50, -40, -10, 10, 40, 50$ (*Ans.:* $\bar{X} = 0$)
(b) $3/2, 7/2, -1/4, 0, 0, 2$ (*Ans.:* $\bar{X} = 1\ 1/8$)
(c) $50, 50, 50, 50, -10,000$ (*Ans.:* $\bar{X} = -1,960$)

When computing the mean of a collection of data, you should always convert the numbers into the same units of measurement. For example, if your data consist of the heights of certain individuals but some of the heights are measured in inches and others in feet, you cannot compute the mean directly from the numbers. You would first have to convert all heights into inches or feet, whichever you preferred.

Example 2-4. The heights of five children sampled in a certain study are $4'3''$, $3'2''$, $39''$, $4'$, and $41''$. We convert all heights into inches and compute

$$\bar{X} = \frac{51'' + 38'' + 39'' + 48'' + 41''}{5} = 43.4''$$

When you have only sample data available, the sample mean often serves as an estimate of the population mean. If you are studying heights of children, you may be interested in knowing the population mean μ but have knowledge only of the sample mean $\bar{X}$. Observe that $\bar{X}$ is a *random* quantity; that is, its value varies and depends upon the particular sample you have selected. On the other hand, μ, the population mean, is not a random quantity; it is a number which describes a certain aspect of the population as a whole and does not vary unless the population itself changes. You may wish to use $\bar{X}$ to estimate the unknown quantity μ.

Numbers such as μ, which measure basic properties of the population under study and which do not vary unless the population itself changes, are called *parameters*. It is often the case that the statistician is concerned with obtaining information about a population parameter based on the results of a random sample.

Example 2-5. A population consists of eight tickets in a hat, numbered as follows: 1, 1, 3, 5, 4, 1, 7, 2. The population mean is

$$\mu = \frac{1 + 1 + 3 + 5 + 4 + 1 + 7 + 2}{8} = 3$$

Suppose you take a random sample of two tickets from this hat, selecting tickets that are marked 1 and 3. For this particular sample, $\bar{X} = (1 + 3)/2 = 2$. Suppose, now, that you replace these tickets and take another sample of two tickets, this time selecting tickets marked 5 and 1. For this particular sample, $\bar{X} = (5 + 1)/2 = 3$. Notice that the value of $\bar{X}$ may change according to the results of the sample, but that μ is a number intrinsic to the population and does not change unless the makeup of tickets in the hat is changed. Notice, also, that $\bar{X}$ may not be the same as μ although the average of $\bar{X}$'s for every possible sample of size 2 (or any other size) equals μ.

Coding Methods. *Coding* is a technique which simplifies calculations when computing the mean of data with very small or large (cumbersome to work with) numbers. From each data value a fixed number is subtracted (or multiplied) in order to obtain more manageable data. Then the necessary calculations are performed with the modified data, finally adding (or dividing) back the number which was previously subtracted (or multiplied). For example, if we label our data X and subtract 700 from each value, we would label our new data $X - 700$. The coding formula says that to compute $\bar{X}$, we would compute the mean of the modified data and then add 700 to the modified mean. If we multiply each X value by 10, we would label our new data $10X$. The coding formula says to compute $\bar{X}$, we would compute the mean of the modified data and then divide the modified mean by 10. The coding formulas are as follows:

Coding formulas.   For any number c, $(\overline{X + c}) = \bar{X} + c$. For any number c, $(\overline{cX}) = c\bar{X}$

Example 2-6. To demonstrate the use of coding, the mean is computed for the following data.

(a) X: 981, 980, 1,000, 978, 980. Without coding, verify $\bar{X} = 983.8$. Now, subtract 980 from each number to get $(X - 980)$: 1, 0, 20, -2, 0.   Compute the mean of the modified data to get $(\overline{X - 980}) = (1 + 0 + 20 - 2 + 0)/5 = 3.8$. Add back 980 to get $\bar{X} = 983.8$.

(b) Y: .02, .0201, .0202, .0190.   Without coding, verify $\bar{Y} = .019825$. Now, multiply each number by 10,000 and then subtract 201 to get $(10,000Y - 201)$: -1, 0, 1, -11. The modified mean is -2.75. Now add back 201 and divide by 10,000 to get $\bar{Y} = .019825$.

A word of caution: Don't memorize formulas without trying to understand them. Coding methods are conceptually quite simple, and you can obtain a feeling for how to use them by working problems. Remember that coding is only a time-saving tool and does not affect your results in any way. The numbers that you use for coding should be chosen to make your calculations as easy as possible. There is no predetermined way to choose them.

Combining Means. You may have occasion to combine means from different sets of measurements from the same population to calculate an overall mean. Suppose that you have, say, three different sets of data X_1, X_2, and X_3, with a total of n_1 values for X_1, n_2 values for X_2, and n_3 values for X_3, and that you wish to compute the overall mean $\bar{X}$. If you already know the individual means $\bar{X}_1$, $\bar{X}_2$, and $\bar{X}_3$, the following formula gives a quick method of computing the overall mean $\bar{X}$:

$$\bar{X} = \frac{n_1\bar{X}_1 + n_2\bar{X}_2 + n_3\bar{X}_3}{n_1 + n_2 + n_3}$$

Example 2-7. The average length (in inches) of rats from four different samples is found to be $\bar{X}_1 = 5$ inches (with sample size $n_1 = 40$), $\bar{X}_2 = 7$ inches ($n_2 = 20$), $\bar{X}_3 = 6.2$ inches ($n_3 = 25$), $\bar{X}_4 = 8$ inches ($n_4 = 15$). What is the overall average length $\bar{X}$?
  Using the above formula, compute $\bar{X}$:

$$\bar{X} = \frac{(40 \times 5'') + (20 \times 7'') + (25 \times 6.2'') + (15 \times 8'')}{40 + 20 + 25 + 15} = 6.15''$$

Exercise 2-2. Compute the overall mean $\bar{Y}$ if $\bar{Y}_1 = 20$ with sample size 10 and $\bar{Y}_2 = 30$ with sample size 10.   (Ans.: $\bar{Y} = 25$)

The Median. The mean is sensitive to extreme values and sometimes gives an inaccurate description of the "center" of the data, as the following example illustrates.
  A group of workers in a certain factory complained that they were underpaid. Their boss refuted this by saying that the mean salary in his factory was $10,000. The figures presented by the boss were not false although they were certainly misleading. His own salary was $91,000, and each of the nine workers received an annual wage of $1,000.
  In this example the salary of the boss has a substantial effect on the mean, even though it represents only 1 salary out of 10. A measure which is not so sensitive to extreme values is the *median*.

Definition. The median is the "middle" score if we arrange the data in increasing order. More precisely, if there are an odd number of scores, the median is the middle score, and if there are an even number of scores, the median is the average of the two middle scores. The median is the 50th percentile.

Example 2-8.

(a) *X:* 0, −2, −2, 1, 5. Compute the median for these data. Arranging the data in order, we obtain −2, −2, 0, 1, and 5 and the median is 0. (There are an odd number of scores, and so we take the middle score.)

(b) *Y:* 1.5, 1.5, 0, −6, 1, 4. Compute the median for these data. Arranging the data in order, we obtain −6, 0, 1, 1.5, 1.5, and 4 and the median is $(1 + 1.5)/2 = 1.25$. (There are an even number of scores, and so we take the average of the two middle scores.)

(c) *Z:* 40, 20, 18, 35, 50,000. Verify that the median is 35.

In the example about the boss and his workers, the median is $1,000. This and the result of Example 2-8(c) illustrates how the median is not affected by extreme data values. When describing data, it would be best, of course, to give both the mean and the median.

The Mode. The final measure of the center of the data is the *mode*. The mode, which is not as widely used as the mean and the median, is the score (or scores) that occurs most often. For example, the mode of the numbers 1, 19, 2, 19, and 7 is 19, and the modes of the numbers 1, 2, 2, 3, and 3 are 2 and 3. (We call the latter collection of data *bimodal* since there are two modes. Trimodal or multimodal data are also possible). If each value occurs the same number of times, there is no mode. The mode is a useful description in conducting inventory studies or preference surveys (the mode is the item which is most popular).

Problems

2-5 Compute the mean, median, and mode for the following sets of data:

a. *X:* 1, 1, 3, 3, 5, 60

b. *Y:* 1, 1, 3, 3, 5, 7, 7, 5

c. *Z:* −30, −20, 0, 0, 0, 20, 50

2-6 The weights at birth (in pounds) of 12 children in a certain large family are as follows: 8.1, 9.0, 7.0, 6.3, 5.9, 11.2, 6.6, 10.0, 5.0, 8.3, 9.1, 5.1. Compute the mean and median birth weight.

2-7 Workers in a certain factory were surveyed to determine the distance they commuted to work each day. The poll included 30 workers from the day shift, 20 workers from the swing shift, and 10 workers from the graveyard shift. The means for these shifts are given below. Find the overall mean.

Day shift: $\bar{X} = 9.8$ miles
Swing shift: $\bar{Y} = 10.7$ miles
Graveyard shift: $\bar{Z} = 21.0$ miles

2-8 The Easy-Way Loan Company has loaned \$10,000 at 5% interest, \$15,000 at 7% interest, and \$6,000 at 10% interest. Find the mean interest the company receives from its loans.

2-9 A survey was taken to determine the average selling price of a particular make of car. In a certain city 10 dealers were surveyed and the selling prices are listed below. Find the mean and median selling prices.

\$2,505, \$2,505, \$2,415, \$2,435, \$2,490, \$3,000, \$2,500, \$2,520, \$2,510, \$3,580

MEASURES OF VARIATION

Suppose you like nut-filled candy bars which have about 40 nuts per bar, and you don't like the taste if there are too few or too many nuts. If you wish to determine the brand most suitable for you, a natural thing to do would be to take a sample box of candy bars from several candy companies and count up the number of nuts in each bar. You choose three boxes X, Y, and Z, one from each of three different companies, with six bars in each box. Suppose the average nut content of a candy bar in box X is $\bar{X} = 40$, $\bar{Y} = 45$, and $\bar{Z} = 40$. Which box would be preferable to you?

If you were told nothing else, you would probably choose either X or Z since each box has an average nut content of 40 while Y does not. However, let's look at the nut contents of the individual candy bars in each box:

X: 5, 5, 1, 75, 75, 79

Y: 40, 40, 40, 40, 40, 70

Z: 40, 40, 40, 40, 39, 41

Observe that Y would be preferable to X, because five of the six candy bars in Y are exactly as you like them while none of the candy bars in X are as you like them (even though $\bar{X} = 40$ and $\bar{Y} = 45$). Also, Z seems to be the best of the three boxes (remember, you like bars with 40 nuts per bar). What is important is that you can't learn any of this information from the means, medians, or modes alone. More information is needed to describe the data adequately. Notice that the numbers in X vary widely while the numbers in Y and Z are clustered close to the mean. We need a measurement to describe the amount of variation in the data.

Range. A simple but not very informative measure of the spread or variation of data is the *range,* which is the difference between the highest score and the lowest. This measure of variation frequently does not tell us much since it involves only upper and lower extreme values and not the values in between. However, the range is useful in obtaining a quick, initial appraisal of the spread of a set of data values. If you discover that the range is relatively small, you know for sure that all the values are clustered together. But if the range is large, you know only that the highest and lowest values are far apart; the other values might be clustered tightly together. Thus, when the range is large, one must look at the data more closely to determine the amount of variation, as illustrated by the candy-bar example. Here, knowing the ranges of X, Y, and Z immediately makes one suspect X. The range of X is $79 - 1 = 78$; the range of Y is $70 - 40 = 30$; and the range of Z is 2. Notice that the range of Y is a bit misleading. In fact, all but one of the Y candy bars have exactly 40 nuts in them, but the range is 30. We need some measure of the spread of data which takes into account all values, not just the extreme ones.

Standard Deviation. At first, you might think of using the absolute difference between the data values and their mean as a measure of variation within the data. Unfortunately, using absolute values causes many mathematical difficulties. As a substitute for absolute differences, we shall use squared differences. This may seem rather arbitrary, but squared differences are much easier to work with mathematically than absolute differences. Also, in many problems involving random phenomena (which we shall study in later chapters), the squared differences between the mean and the data values arise as "natural" units of measurement. Accordingly, we shall introduce the common measure of variation known as the *standard deviation.*

The standard deviation for population data is denoted by σ (sigma) and is the square root of the average squared distance of the values from the mean. Letting X = the population values, μ = the population mean, and N = the number of elements in the population, and using the Greek letter Σ (capital sigma), the standard mathematical symbol for summation, we obtain the formula for the *population standard deviation:*

$$\sigma = \sqrt{\frac{\Sigma\,(X - \mu)^2}{N}}$$

The *sample standard deviation,* which we shall denote by s, is computed in a similar way. First, we compute the sample mean $\bar{X}$. Next, we compute the squared differences between the data values and $\bar{X}$. Then we sum these squared differences, divide by $n - 1$ (if n = sample size), and take the square root. Again using Σ to denote summation, we obtain

$$s = \sqrt{\frac{\Sigma(X - \bar{X})^2}{n - 1}}$$

As in the case of the sample mean, the sample standard deviation s is a random quantity which depends upon the particular sample we have selected. The population standard deviation σ is a population parameter and is a measure of the variation of the population we are studying. It is often the case that σ is an unknown quantity which we estimate by s. In fact, we use the factor $n - 1$ instead of n in the computation of s because statisticians have found that using this quantity makes s a better estimate of σ.

Notice from the formulas that the standard deviation is always positive (zero if there is no variation, i.e., if every value is the same). Data with much variation will have a large standard deviation, and clustered data will have a small standard deviation. As with the mean, the standard deviation should be computed only when all data values are expressed in the same units of measurement, in which case the standard deviation is also expressed in these units.

When computing the standard deviation (as well as when doing other statistical computations), it is extremely helpful to use an electronic calculator. Although we are not able to provide you with one, the table of square roots in the Appendix may lighten the work for some of the calculations you will have to make. In a moment, we shall discuss the use of the calculator to compute $\bar{X}$ and s.

Example 2-9. We compute the sample standard deviation s for the following sets of sample data.

(*a*) X: 3, -1, 5, 4, 4. $\bar{X} = 3$. (It is a good idea to make a table to keep track of the calculations, as we do in Figs. 2-8 and 2-9.)

X	$X - \bar{X}$	$(X - \bar{X})^2$
3	0	0
-1	-4	16
5	2	4
4	1	1
4	1	1

Figure 2-8

Therefore,

$$s = \sqrt{\frac{\Sigma(X - \bar{X})^2}{n - 1}} = \sqrt{\frac{0 + 16 + 4 + 1 + 1}{4}} = 2.35$$

(*b*) Y: $-2, -2, 0, 12$. $\bar{Y} = 2$.

Y	$Y - \bar{Y}$	$(Y - \bar{Y})^2$
-2	-4	16
-2	-4	16
0	-2	4
12	10	100

Figure 2-9

Therefore,

$$s = \sqrt{\frac{16 + 16 + 4 + 100}{3}} = 6.73$$

Exercise 2-3. Verify that in the candy-bar example $s_x = 39.85$ nuts, $s_y = 12.2$ nuts, and $s_z = .63$ nuts. (To avoid ambiguous notation, we attach subscripts to s to denote which set of data we are referring to.) Notice that box Z has a much smaller standard deviation than X or Y; therefore, we may infer that box Z is better than boxes X and Y. Y.

Coding Formulas for the Standard Deviation. We can simplify some standard deviation calculations by coding in a way similar to the coding methods we used for the mean. If we add or subtract the same number to every data value, notice that we are not changing the variation of the data but shifting all the data by the same amount; therefore, the standard deviation remains unchanged. Also, if we multiply each data value by the same amount, we multiply the standard deviation by the same absolute value. We obtain the following.

Coding formulas for s. For any number c and set of data X,

$$s_{x+c} = s_x$$
$$s_{cx} = |c| s_x$$

Example 2-10. Using the coding formulas, we compute s for the following sets of data.

(a) X: 375, 377, 373, 373. We subtract 373 from each value to get the modified data 2, 4, 0, 0. We compute s for the modified data and obtain $s_{x-373} = 1.92$. Therefore, $s_x = s_{x-373} = 1.92$.

(b) Y: 12,400, 12,800, 12,800, 12,400. First we divide each of the numbers by 100 (multiply by 1/100) to get 124, 128, 128, 124. Now we subtract 124 from each number to get $1/100 \, Y - 124$: 0, 4,4,0. We compute $s_{(1/100)Y-124} = 100(2.31) = 231$. Again, we mention that the choice of c in the coding formula is somewhat arbitrary and is chosen only to make computations easier. It will not affect the final answer.

Problems

2-10 Eight people weighed a 1-pound bag of flour and obtained the following weights (in pounds): 1.0, 1.1, .9, .9, 1.3, 1.0, .5, 1.1. Find the sample standard deviation s.

2-11 In a large city five dentists were surveyed to determine the price of a large filling: $15, $13, $19, $18, $15. Find s for these data.

2-12 The following data give the times (in minutes) it took a certain beleaguered husband to wash the dinner dishes over a 1-week period: 18.1, 18.8, 19.3, 18.0, 19.0, 19.1, 18.5. Compute the standard deviation for these sample data.

2-13 The morning school bus arrived at a certain corner at the following times over a 10-day period: 8:45, 8:47, 8:53, 8:45, 8:38, 8:55, 8:48, 8:50, 8:56, 8:44. Compute s for these times.

Using a Calculator to Compute $\bar{X}$ and s.   When we must describe a large set of data, computations of $\bar{X}$ and s can become very tedious. The use of an electronic calculator makes these computations much easier, however, especially when we use the following alternate formula.

Calculator formula for s.

$$s = \sqrt{\frac{\Sigma X^2 - (\Sigma X)^2/n}{n-1}}$$

This formula gives the same result as the original formula for s but makes computations much easier in most cases.

Exercise 2-4. Use a calculator and the calculator formula for s to compute s for the data in Examples 2-9 and 2-10.

Exercise 2-5. Use a calculator to compute $\bar{X}$ and s for the data in the drug study of Fig. 2-1 ($\bar{X} = 85.17$; $s = 3.31$).

We recommend using a calculator wherever possible, especially with large amounts of data. Computing relative frequencies, percentile ranks, and other measures becomes an easier task when a calculator is used.

Types of data

In many situations the data we obtain from our statistical experiment may be of a different nature than the data we have discussed so far. More specifically, we may encounter data in which the observations are not physical measurements as such, or we may wish to transform measurements into other forms. Consider the following three examples and see if you can detect differences in the types of data presented.

Example 2-11. Two groups of middle-aged businessmen, smokers and nonsmokers, are tested to see how long they can hold their breaths. The results are tabulated in Fig. 2-10.

Example 2-12. Two groups of students, smokers and nonsmokers, are given a series of physical fitness tests and then are ranked according to how well they perform. Rank 1 means the best overall performance in both groups, rank 2 the second best, etc. The results are tabulated in Fig. 2-11.

Time of holding breath (in minutes)

Smokers	Nonsmokers
.6	.9
1.2	1.6
1.1	2.1
.7	1.1
1.9	2.4
2.1	1.8
1.0	2.8

Figure 2-10

Rankings in physical fitness test

Smokers	Nonsmokers
4	1
7	5
10	6
11	2
12	8
9	3
14	13

Figure 2-11

Example 2-13. Two groups of elderly women, smokers and non-smokers, are asked if they have recurring dizzy spells. The number of women in each group having recurring dizzy spells is given in Fig. 2-12.

These examples illustrate the three types of data encountered most frequently by the statistical experimenter. The first and most common type, as is shown in Example 2-11, is called *measure data,* that is, data in which the observations are physical measurements. With measure data, such as heights, weights, scores on tests, and waiting times, we can perform the usual arithmetic operations and compute

Occurrence of dizzy spells

	No. having dizzy spells	No. not having dizzy spells
Smokers	4	3
Nonsmokers	1	6

Figure 2-12

sample means, medians, standard deviations and the common descriptive measures. Measure data are sometimes called *cardinal,* or *continuous, data* and are the type of data we shall encounter most frequently.

The second type of data encountered frequently, as illustrated in Example 2-12, is called *ordinal data.* With ordinal data, the observations may not be physical measurements, but they always can be compared or ranked; that is, with ordinal data we always can take two observations and say one is "better than" or "bigger than" the other. Sometimes, when comparing different sets of data, it is useful to transform physical measurements into ranks using the usual numerical order and then study the ranks themselves rather than the original measurements. Although our information is reduced, there are important theoretical reasons for doing this (as we shall see in Chap. 9). Also, transforming from measurements to ranks may decrease the biasing effect that one or two "wild" measurements may have on a set of data.

The third type of data encountered frequently, as illustrated in Example 2-13, is called *nominal,* or *count,* data. Count data are obtained by the classification of the data into different groups, listing how many observations are in each group. In Example 2-13, the categories are smokers and nonsmokers and those who have and those who don't have dizzy spells. We shall analyze count data in Chap. 10.

To aid your understanding of the three common types of data we summarize this discussion in Fig. 2-13.

Type of data	*Description*
Measure	Physical measurements
Ordinal	Rankings, or orderings
Count, or nominal	Classification of data into groups

Figure 2-13

Summary

We have discussed techniques of summarizing and describing data. The mean, median, and mode are measures of the center of the data; the standard deviation is a measure of the amount of variation in the data. We discussed coding for the mean and standard deviation, a technique which sometimes makes calculations much easier. We also discussed a method of grouping large amounts of data into group intervals and tabulating this information in a group frequency table. We have seen that a useful way of displaying this information is in a histogram. Another useful method of displaying the distribution of data is the cumulative frequency polygon.

We used the histogram and cumulative frequency polygon to compare two sets of data. We also used cumulative frequencies and percentiles to describe the distribution of data and the relative placement of particular values. Finally, we discussed three different types of data which occur most frequently in statistical situations.

Important terms

Count data

Cumulative frequency

Cumulative frequency polygon

Frequency table

Group intervals

Histogram

Mean

Measure data

Median

Mode

Ordinal data

Percentile

Percentile rank

Standard Deviation

Additional problems

2-14 Find the sample mean, standard deviation, median, and mode for the following sets of data.

    a. −6, 8, 4, 17, −6, 2, 0

b.   1.7, 1.5, 2.9, −2.1, 3.6, 4.2

c.   375,110; 375,100; 375,108; 375,108; 375,111

d.   4, 6, 6, 6, 6, 6, 6, 6, 6, 5, 6, 6, 989

2-15 The annual family incomes (in thousands of dollars) of 50 high school leaders are as follows:

Annual family incomes (in thousands of dollars)

8	14	21	9	20
7	18	26	17	38
10	23	24	52	19
22	37	12	48	41
28	25	15	49	41
34	11	13	39	12
11	12	18	20	33
11	9	18	48	36
9	17	21	20	29
15	29	28	44	45

a.   Group these data in an appropriate number of class intervals and make a frequency table, histogram, and cumulative frequency polygon of the grouped data.

b.   Use a calculator to compute $\bar{X}$ and s.

c.   Find the percentile ranks for incomes of $20,000 and $38,000.

d.   What income is at the 50th percentile? What is the median income?

2-16 Use the coding formulas to compute $\bar{X}$ and s for the following sets of data:

a.

86,259	86,258	86,268	86,260
86,263	86,265	86,265	86,266
86,269	86,271	86,259	86,255

b.

.0023	.0027	.0029	.00281
.0024	.0023	.0030	.0026
.0025	.0031	.0032	.0030

2-17 Group the following data in an appropriate number of class intervals and make a frequency table, histogram, cumulative frequency polygon for the grouped data.

12.3	6.4	18.3	15.3
8.4	7.3	20.0	8.6
3.1	11.7	26.1	6.7
15.9	21.5	25.1	3.9
17.2	23.8	21.5	10.0
11.8	25.2	14.3	10.0
13.7	9.6	6.8	11.1
8.8	28.4	28.6	12.3
2.9	20.0	19.1	18.9
3.8	14.7	18.7	22.3

2-18 Using a calculator, compute $\bar{X}$ and s for the data in Prob. 2-17.

2-19 a.   Using the data of Prob. 2-17, find the percentile ranks of the numbers 19.1, 25.1, 22.3, 10.0.

   b.   Find the observations corresponding to the 20th, 50th, and 70th percentiles.

True-false test

Circle the correct letter.[1]

T F   1. Exactly one-half of the data values are bigger than the sample mean.

T F   2. Whenever you are grouping data, you should use 10 class intervals.

T F   3. The standard deviation is a measure of the amount of variation in data.

T F   4. The mean is always an accurate indicator of the center of data.

T F   5. The main purpose of grouping data is to display it in an informative, yet easy-to-interpret way.

[1]Answers to all True-False tests are given at the back of the book.

T F   6. The percentile rank of your score on a test indicates your relative placement.

T F   7. A value with a relative cumulative frequency of .95 is one of the smallest values in the data.

T F   8. The histogram is a pictorial version of the frequency table of grouped data.

T F   9. Any characteristic of the population under study can be completely determined by appropriate study of sample data.

T F   10. Any sample taken from a large enough population can be considered to be random.

Probability 3

When making statistical inferences, we use information from a random sample to infer properties of the parent population. In order to do this, we must know the types of samples to which different parent populations are likely to give rise. In particular, we must be able to compute the probabilities that a sample from a given population will have various attributes, e.g., that a sample of size 2 from a population of 20 good and 30 bad units will contain only good units. For this reason, we must develop some knowlege of *probability*.

Probability has been formally studied since about the seventeenth century when members of the European aristocracy hired some of the best mathematicians of the day to figure out strategies and compute odds in gambling games for which they had a passion. Since then, particularly since the early 1900's, probability theory has undergone extensive growth and development. It now stands as an area of major interest in pure mathematics; in addition, it has provided the mathematical foundation for modern statistics.

In this chapter we shall introduce a few basic concepts of probability and relate them to problems of random sampling. We shall not go into depth or inundate you with details but shall try to argue in an intuitive way wherever possible.

Basic probability model

In any problem involving probability or statistics there is always a *random experiment* which takes place or may be envisioned as taking place. By random experiment we mean any experiment in which the result cannot be predicted beforehand. We shall be concerned primarily with a particular type of random experiment, namely, one in which a random sample is taken from a population under study.

Although we cannot predict for certain the result of a random experiment, we may be able to make a statement regarding how likely a particular result is to occur. This is exactly what probability means. To use a clear notation, from now on, let us call a result of a random experiment an *outcome*. We shall call any collection of outcomes an *event*. The probability of a specific event represents how likely it is that that event will occur. We express this as a proportion:

> *The probability of an event represents the proportion of times we expect the event to occur if we perform the experiment over and over a large number of times.*

For experiments with equally likely outcomes, there is a simple formula for computing the probability of an event:

> **Probability formula.** When selecting at random from a population consisting of N elements, the probability that a specific element is selected equals $1/N$. If an event consists of K outcomes, the probability of this event equals K/N.

Examples 3-1 to 3-6 contain some simple experiments that illustrate the use of this formula.

Example 3-1. A random experiment consists of spinning a spinner with numbers 1, 2, 3, 4, 5 (see Fig. 3-1) and then observing the number on which the spinner lands. Notice that there are 5 possible outcomes, and they are all equally likely. Since there is only 1 way the spinner can point to 1, the probability that the spinner points to 1 is 1/5. It is convenient to write this fact as P(spinner lands on 1) = 1/5, or simply $P(1) = 1/5$. Likewise $P(1 \text{ or } 2) = 2/5$, and $P(1, 2, \text{ or } 5) = 3/5$.

Example 3-2. In a certain bag of apples, 10% are rotten. An apple is selected at random from this bag. Thus, each apple has the same chance of being selected. Since 10% are rotten, no. of rotten apples/total no. of apples = .1; therefore, P(selected apple is rotten) = .1.

Example 3-3. A "fair" coin is tossed. The outcomes are heads and tails. $P(\text{heads}) = 1/2$; $P(\text{tails}) = 1/2$.

Example 3-4. A ball is selected from a hat which contains 20 black balls and 10 white balls. Since there are 30 possible equally likely outcomes, 20 resulting in a black ball, P(black ball is selected) = 20/30 = 2/3; $P(\text{white ball}) = 1/3$.

Example 3-5. A card is selected at random from an ordinary deck of playing cards. Since there are 52 cards altogether, and since only 1 of them is the queen of spades, P(queen of spades) = 1/52. Also, since there are 13 spades, $P(\text{spade}) = 13/52 = 1/4$. Similarly, $P(\text{queen}) = 4/52 = 1/13$.

Example 3-6. In a certain city .6 of the registered voters favor legalization of marijuana. If a voter is selected at random from the city, P(favors legalization) = .6.

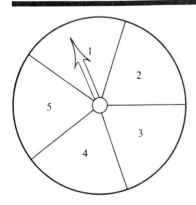

Figure 3-1 Selecting a number at random.

COMPLEMENTARY EVENTS

For any event in a probability experiment (for simplicity of conversation, let's call the event *C*), we can speak of the *complementary event* (not *C*), which is merely the collection of all outcomes that are not in *C*. In other words, (not *C*) occurs whenever *C* doesn't occur. From the probability formula we obtain *P*(not *C*) = no. of outcomes not in *C*/total no. of outcomes. It follows that $P(C) + P(\text{not } C) = 1$. Rearranging, we obtain:

For any event *C*, $P(C) = 1 - P(\text{not } C)$

Example 3-7. A group of 10 birds consists of 2 gray sparrows, 1 blue sparrow, 4 gray doves, and 3 red cardinals. One bird is selected at random. Let the events *G* = (a gray bird is selected) and *S* = (a sparrow is selected). Then $P(G) = 6/10$, $P(\text{not } G) = 1 - P(G) = 4/10$; $P(S) = 3/10$, $P(\text{not } S) = 1 - P(S) = 7/10$.

COMBINING EVENTS

If we have two events *A*, *B*, we can define new events (*A* or *B*), (*A* and *B*) as follows:

(*A* or *B*) = all outcomes that are in at least one of the events *A*, *B*

(*A* and *B*) = all outcomes that are in both *A* and *B*

We shall sometimes abbreviate (*A* and *B*) by writing (*AB*). Figure 3-2 describes these combinations of events.

Example 3-8. Consider the birds in Example 3-7. The event (*G* or *S*) occurs if either *G* or *S* occurs, for example, if either a sparrow or a gray bird is selected; $P(G \text{ or } S) = 7/10$. Also, the event (*G* and *S*) occurs if the bird selected is gray *and* a sparrow; $P(G \text{ and } S) = 2/10$.

Example 3-9. An encounter group consists of 2 male jugglers, 3 female jugglers, 5 male puppeteers, 4 female puppeteers, and 1 male speed-reading champion. Of this group, 1 member is selected at random. Let *M* = (the selected person is a male), *J* = (the selected person is a juggler), and *S* = (the selected person is the speed-reading champion). Then $P(M) = 8/15$, $P(J) = 5/15$, $P(S) = 1/15$, $P(M \text{ or } J) = 11/15$, $P(M \text{ and } J) = 2/15$, $P(J \text{ or } S) = 6/15$, and $P(J \text{ and } S) = 0$. In the last case, notice that *J* and *S* cannot occur at the same time. When this is the case, we say that the events in question are *mutually exclusive*.

Problems

3-1 In a packet of flower seeds, there are 5 type-*A* seeds, 4 type-*B* seeds, and 7 type-*C* seeds. If you randomly select a seed to plant, what is *P*(it is type *B*)? What is *P*(it is not type *B*)?

3-2 A bag contains 6 crisp apples and 5 noncrisp apples. Of the crisp apples, 4 are big; and of the noncrisp apples, 2 are big. An apple is selected at random from the bag. Find the following.

a. *P*(apple selected is crisp) = 6/11

b. *P*(apple selected is big) = 6/11

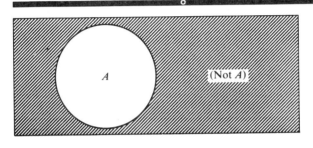

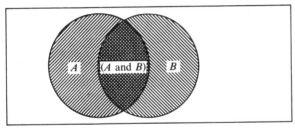

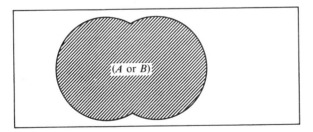

Figure 3-2 Combinations of events.

    c. *P*(apple selected is crisp and big)

    d. *P*(apple selected is neither crisp nor big)

    e. *P*(apple selected is crisp or big)

3-3 A three-digit number is randomly selected from a random-number table; in other words, the number selected can range from 000 to 999. Find the following.

    a. *P*(number selected is odd)

    b. *P*(second digit is 6)

    c. *P*(first digit is 8 and second digit is 6)

    d. *P*(sum of the first two digits is less than 4)

    e. *P*(number selected is bigger than 800)

    f. *P*(there are no 6s)

Sampling with replacement; independent events

Consider the following experiment. A gambler has a bag which contains 3 red balls and 2 green balls. In order to play a certain game, 2 balls are selected at random from this bag, the first ball selected being replaced in the bag before the next ball is drawn. We shall call this procedure *sampling with replacement*. Suppose we wish to compute the probability that a green ball will be selected first and a red ball selected second; i.e., we wish to find $P(GR)$. If we tried to use the probability formula, we would run into difficulty since the *outcomes GR, GG, RR, RG* are not equally likely. We would have to consider the balls individually, perhaps numbering them, rather than distinguishing only their color. Fortunately, however, there is a quick method for computing such probabilities, providing the events in question are independent.

> **Definition.** We say that two events are *independent* if the occurrence of one event does not affect the likelihood of the occurrence of the other.

The most common type of independent events we shall encounter results from independent sampling, or sampling with replacement. In our gambling experiment, the result of selecting the first ball in no way affects the selection of the second ball since the first ball is replaced before the second ball is drawn. In fact, we can reason intuitively and compute $P(GR)$ as follows: The probability of selecting a G on the first selection is 2/5. In other words, if we perform this experiment many times, we would expect to obtain a green ball on the first selection about 2/5 of the time. Likewise, we would expect to obtain a red ball on the second selection about 3/5 of the time independently of the outcome of the first selection. Thus, we would expect to select a green ball first and then a red ball about 2/5 of 3/5, or 6/25 of the time. For example, if we performed the experiment 100 times, we would expect to obtain a green ball on the first selection about $2/5 \times 100 = 40$ times, and of these 40 times we would expect also to obtain a red ball on the second selection about $3/5 \times 40 = 24$ times. We would expect to obtain (GR) about $2/5 \times 3/5 \times 100 = 6/25 \times 100 = 24$ times, or 6/25 of the time.

This reasoning leads us to a general rule for independent events which we call the *multiplication rule:*

> **Multiplication rule for independent events.** If two events A, B, are independent, then $P(AB) = P(A) \times P(B)$. Likewise, if many events A, B, C, ..., are independent, then $P(ABC ...) = P(A)P(B)P(C)....$

Using the multiplication rule, we obtain: $P(RR) = P(R) \times P(R) = 3/5 \times 3/5 = 9/25$; $P(GG) = 2/5 \times 2/5 = 4/25$; $P(RG) = 3/5 \times 2/5 = 6/25$.

We now let the gambler select 3 balls from the bag instead of 2. Remember that since the gambler is selecting with replacement, the result of a draw is independent of the results of any other draw. Also, on any draw, $P(R) = 3/5$ and $P(G) = 2/5$. We use the multiplication rule for independent events and obtain $P(RRR) = 3/5 \times 3/5 \times 3/5 = 27/125$; $P(RGR) = 3/5 \times 2/5 \times 3/5 = 18/125$; and so on. Thus we can compute probabilities for all outcomes in the experiment, and these are tabulated in Fig. 3-3.

In order to compute probabilities for events in this experiment, we introduce a basic probability axiom, which we shall call the addition rule.

> **The addition rule.** The probability of an event is the sum of the probabilities of its outcomes.

The probability formula for experiments with equally likely outcomes is

Outcome	Probability
RRR	$3/5 \times 3/5 \times 3/5 = 27/125$
RRG	$3/5 \times 3/5 \times 2/5 = 18/125$
RGR	$18/125$
GRR	$18/125$
RGG	$3/5 \times 2/5 \times 2/5 = 12/125$
GRG	$12/125$
GGR	$12/125$
GGG	$2/5 \times 2/5 \times 2/5 = 8/125$

Figure 3-3

a special case of the addition rule. For example, if an experiment has N equally likely outcomes (each having probability $1/N$) the addition rule states that an event with K outcomes has probability $1/N + 1/N + \cdots + 1/N$ (K times) $= K(1/N) = K/N$, as the earlier formula states. Also, the addition rule implies that if events A, B are mutually exclusive, then $P(A \text{ or } B) = P(A) + P(B)$. (Why is this true?)

To find the probability of an event using the addition rule, then, we compute the probabilities of all the outcomes in the event and add them up. In the discussion which follows, we shall help make this clear by listing the outcomes of the events we analyze. For example, in the gambling experiment, we denote the event that the gambler selects exactly one red ball out of 3 by (RGG, GRG, GGR), since these are the outcomes which make up (exactly 1 R).

Using the addition rule, we now compute the probabilities of some events in this experiment. We see, for example, that $P(\text{exactly 1 } R) = P(RGG, GRG, GGR) = 12/125 + 12/125 + 12/125 = 36/125$; $P(\text{exactly 1 } G) = P(RRG, RGR, GRR) = 18/125 + 18/125 + 18/125 = 54/125$; $P(\text{at least 1 } G) = P(GGG, GGR, GRG, RGG, RRG, RGR, GRR) = 98/125$. Notice that by using the rule $P(A) = 1 - P(\text{not } A)$, we can compute $P(\text{at least 1 } G)$ in an easy way: $[\text{not (at least 1 } G)] = (0 \text{ } G\text{'s}) = (RRR)$, and so $P(\text{at least 1 } G) = 1 - P(RRR) = 1 - 27/125 = 98/125$. In the same way, we see that $P(\text{at least 1 } R) = 1 - P(GGG) = 1 - 8/125 = 117/125$.

It is important to realize that in computing probabilities for this experiment we used the fact that the selections are made *with* replacement. For each draw the distribution of balls in the bag is the same as for previous draws; in other words the possible results of one draw are independent of the possible results of any other draw. The same is true for coin tossing: The result of one toss is independent of the outcomes of any other tosses. (This is true also for roulette-wheel spinning, dice rolling, etc.) We can compute probabilities for coin tossing in a fashion similar to the previous experiment: $P(HT) = 1/2 \times 1/2 = 1/4$; $P(HTHT) = 1/2 \times 1/2 \times 1/2 \times 1/2 = 1/16$; $P(HHHHT) = 1/2 \times 1/2 \times 1/2 \times 1/2 \times 1/2 = (1/2)^5 = 1/32$.

In the next four examples we illustrate the multiplication and addition rules.

Example 3-10. An astronaut's oxygen supply comes from two independent sources. Source A has probability .9 of working, and source B has probability .8 of working. What is $P(A$ or B is working)?

> SOLUTION: If we let $\bar{A} = (A$ is not working), $\bar{B} = (B$ is not working), then $(A$ or $B) = (AB, A\bar{B}, \bar{A}B)$. Since $P(A) = .9$, $P(\bar{A}) = .1$. Likewise we see $P(\bar{B}) = .2$. We use the multiplication rule to compute: $P(AB) = .9 \times .8 = .72$; $P(A\bar{B}) = .9 \times .2 = .18$; $P(\bar{A}B) = .1 \times .8 = .08$. Then we use the addition rule and compute $P(A$ or $B) = .72 + .18 + .08 = .98$. Another way to compute this probability is $P(A$ or $B) = 1 - P(\text{not } (A \text{ or } B)) = 1 - P(\bar{A}\bar{B}) = 1 - .1 \times .2 = .98$.

Example 3-11. If you wander into a certain protest demonstration, your chances of getting tear-gassed are 2/3. Independently of that, your chances of being hit by a rock are 1/4. Compute $P(G$ or $R)$, where G means gassed and R means hit by a rock (also $\bar{G}$ means not gassed and $\bar{R}$ means not hit by a rock).

> SOLUTION: We use the same reasoning as before: $P(G$ or $R) = P(GR, \bar{G}R, G\bar{R})$. $P(GR) = 2/3 \times 1/4 = 1/6$; $P(\bar{G}R) = 1/3 \times 1/4 = 1/12$; $P(G\bar{R}) = 2/3 \times 3/4 = 1/2$. Now, by the addition rule, $P(G$ or $R) = 1/6 + 1/12 + 1/2 = 3/4$.

Example 3-12. In a large population of voters, of which 60% are Democrats, 3 voters are selected at random. Find $P(1$ of the voters selected is a Democrat).

> SOLUTION: Whenever a relatively small sample is taken from a large population, we assume that the selections are independent because a small number of selections has no significant effect on the distribution of the population as a whole. Thus, in such a situation, we compute probabilities as if the selections were made with replacement.
>
> The event $(1\ D$ is selected) is composed of the outcomes DNN, NDN, NND. Since we are assuming independent selec-

tions, we can use the multiplication rule. Using this rule and the fact that for one selection $P(D) = .6$, and $P(N) = .4$, we have $P(DNN) = P(NDN) = P(NND) = .6 \times .4 \times .4$ (in some order) $= .096$. Thus $P(1\ D$ is selected$) = .096 + .096 + .096 = .288$.

Example 3-13. The commissioner of public safety wishes to select 3 people at random from a large group containing a proportion of .2 reckless fools. What is P(at least 1 reckless fool is selected)?

SOLUTION: Note that P(at least 1 reckless fool is selected) $= 1 - P(0$ reckless fools are selected). Since the population contains .2 reckless fools, it contains .8 who aren't reckless fools and so the multiplication rule yields $P(0$ reckless fools are selected$) = .8 \times .8 \times .8 = .512$. Thus, we have P(at least 1 reckless fool is selected$) = 1 - .512 = .488$.

Before proceeding to the next section, we observe that the multiplication rule for independent events cannot be used for dependent events. Consider a population of shoes in which a proportion of .2 are rubber-soled and .1 are tennis shoes. If we select a shoe at random from this population, then P(rubber-soled shoe) $= .2$, P(tennis shoe) $= .1$; but it is *not* true that P(rubber-soled shoe and tennis shoe) $= (.2)(.1) = .02$. In fact, every tennis shoe has rubber soles, and so P(rubber-soled shoe and tennis shoe) $= P$(tennis shoe) $= .1$.

Problems

3-4 A certain vaccine has probability .9 of being effective. Suppose 4 people are given this vaccine.

    a. What is P(it is effective for all 4 people)?

    b. What is P(it is effective for 3 people)?

3-5 From a large collection of cameras, 40% of which contain built-in light meters, 3 cameras are selected at random. What is P(2 cameras have built-in light meters)?

Note: Whenever a relatively small sample is taken from a large population, we assume the sampling is done with replacement. Why is this the case?

3-6 The governor of a certain state vetoes legislation with probability .2 (independent of the type of bill). What is P(3 bills are signed before the governor vetoes 1 bill)?

3-7   From a box containing 3 tickets marked "tall," 2 marked "medium," and 4 marked "short," 4 tickets are selected at random with replacement. Find the following.

a. P(2 tickets are marked "tall")

b. P(2 tickets are marked "tall" and 2 tickets are marked "short")

c. P(3 tickets are marked "medium")

Dependent selections; tree diagrams

We now consider the case in which sampling is made *without replacement* from a small population; that is, a member that is selected is not replaced before the next selection is made. Suppose, then, that we have a population consisting of 3 women and 2 men from which 2 people are to be selected at random to serve on a certain committee. We wish to compute probabilities for this experiment.

First observe that the possible outcomes of the experiment are the same as if we replaced the first selection before making the second one: *WW, WM, MW, MM*. The probabilities of these outcomes are different, however, than in the case of independent events. Notice that at the time of the first selection the population contains 3 women and 2 men, and so $P(W$ on the first selection$) = 3/5$. Whenever a woman is selected first, there remain 2 women and 2 men in the population. Thus, in the situation where the first selection is a woman, $P(W$ on the second selection$) = 2/4 = 1/2$.

If we perform this experiment a large number of times, we would expect to obtain a woman on the first selection about 3/5 of the time; and of the times when a woman is selected first, we would expect to obtain a woman on the second selection about 1/2 of the time. We would expect to obtain a woman on both selections about 1/2 of 3/5, or 3/10 of the time. Thus $P(WW) = 3/5 \times 1/2 = 3/10$. Continuing with the same line of reasoning, we compute: $P(WM) = 3/5 \times 1/2 = 3/10$; $P(MW) = 2/5 \times 3/4 = 3/10$; $P(MM) = 2/5 \times 1/4 = 1/10$. We cannot use the multiplication rule in this situation because the selections are not independent, as they would be if we were selecting with replacement.

Here are probabilities of some events in this experiment: P(exactly 1 $M) = P(MW, WM) = (3/5 \times 1/2) + (2/5 \times 3/4) = 3/5$; $P(M$ on the second selection$) = P(WM, MM) = 3/10 + 1/10 = 2/5$; $P(W$ on the second selection$) = 1 - P(M$ on the second selection$) = 3/5$. The meaning of an event such as (M on the second selection) may be confusing. Remember, a sure way to describe an event is to list the outcomes which make it up. Thus, as listed above, the event (*M* on the second selection) is the event composed of the outcomes *WM, MM,* which takes into account all possible results of the first selection.

TREE DIAGRAMS

An extremely useful method for representing information from an experiment which takes place in dependent stages is a *tree diagram.* A tree diagram is a sequential chart listing the possible outcomes at each stage of an experiment, along with the probabilities for each of these outcomes, given whatever sequence of outcomes has occurred.

In Fig. 3-4 we present the tree diagram for the previous sampling experiment. Each outcome of the experiment is represented by an appropriate path of branches, starting at the top and continuing along the various stages of the experiment. To compute the probability of an outcome, you merely multiply the probabilities along the path of branches which represents it. You should check now that the tree diagram in Fig. 3-4 yields the correct probabilities for the previous experiment. Notice that the main value of a tree diagram is to keep track of probabilities at the various stages of an experiment. In the independent case (sampling with replacement) a tree diagram is not often necessary because the multiplication rule can be applied.

Example 3-14. When you travel on a certain freeway during rush hour, you have a .7 chance of getting off at the correct exit (exit *A*). If you get off at exit *A*, you have a .9 chance of not getting lost. If, however, you miss exit *A* and have to get off at exit *B*, your chances of getting lost are .8. What is *P*(you don't get lost)?

> SOLUTION: Study the appropriate tree diagram in Fig. 3-5 (*A* means you get off at exit *A; N* means you don't get lost; etc.) and compute $P(AN) = .7 \times .9 = .63$, $P(BN) = .3 \times .2 = .06$. Since the event (you don't get lost) = (*AN, BN*), *P*(you don't get lost) = $.63 + .06 = .69$.

Example 3-15. In a certain community 1/3 of the residents are angry. It is known that an angry person will give you directions with probability 1/4 while a person who is not angry will give you directions with probability 9/10. If you are traveling through this community and ask someone for directions, what is the probability that you get them?

> SOLUTION: The tree diagram in Fig. 3-6 displays the relevant in-

formation. We see that (you get directions) = (angry and directions, not angry and directions), and so *P*(you get directions) = 1/3 × 1/4 + 2/3 × 9/10 = 41/60.

Example 3-16. Vaccine *A* has a .9 chance of being effective if a person is type X and only a .5 chance if a person is not type X. Vaccine *B*, on the other hand, has a .8 chance of being effective if a person is type X but a .6 chance if a person is not type X. Also, 40% of the pop-

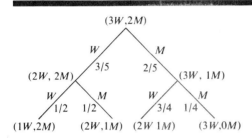

Figure 3-4

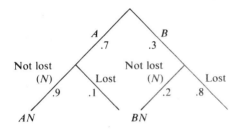

Figure 3-5

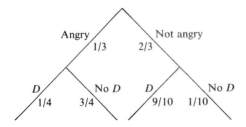

Figure 3-6

ulation is type X. If a person is selected at random and the vaccine is administered, which vaccine has a greater probability of being effective?

SOLUTION: In Fig. 3-7 we show tree diagrams for vaccines A and B, where E means effective and N means not effective. From the tree diagrams we obtain: $P(A$ is effective$) = .4 \times .9 + .6 \times .5 = .66$; $P(B$ is effective$) = .4 \times .8 + .6 \times .6 = .68$. Thus, vaccine B has a greater probability of being effective.

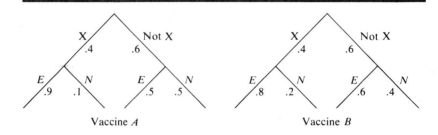

Vaccine A Vaccine B

Figure 3-7

Problems

3-8 A room contains 4 males and 3 females. A person is selected at random, and then this person and another person of the same sex are replaced in the room. This procedure is then repeated. What is P(there are now 5 males and 4 females in the room)?

3-9 You select at random a committee of size 3 from a group of 4 men and 4 women.

    a. What is P(committee contains 2 women)?

    b. What is P(committee contains at least 2 women)?

3-10 A box contains 1 ticket marked "winner" and 4 marked "loser." You select 2 tickets at random without replacing the first ticket before selecting the second. Find the following:

    a. P(the first ticket selected is a winner)

    b. P(the second ticket selected is a winner)

    c. P(both tickets selected are losers)

    d. P(1 ticket selected is a winner)

3-11 From a room containing 5 communists and 5 capitalists, 4 people are selected at random without replacement. Find the following:

    a. P(the second person selected is a communist)

    b. P(the third person selected is a communist)

    c. P(the fourth person selected is a communist)

    d. P(at least two people selected are capitalists)

Random variables

To make notation easier we sometimes denote possible numerical outcomes of a random experiment by a capital letter such as X or Y. We call this labeling process a *random variable*. Here are some examples of random variables.

    Example 3-17. A coin is tossed three times. Let $S = $ the total number of heads in the three tosses. Then S is a random variable; it can have values of either 0, 1, 2, or 3. This provides a convenient way of writing probabilities: Instead of writing P(obtaining 3 heads) $= 1/8$, we need write only $P(S = 3) = 1/8$. Likewise, $P(S = 2)$ means the probability of obtaining 2 heads in 3 tosses; $P(S = 2) = 3/8$. In fact, we can list this information in a table (as in Fig. 3-8), called the *distribution of S*, which lists the probability of the occurrence of each possible value of S.

Distribution of S

S value	Probability
0	1/8
1	3/8
2	3/8
3	1/8

Figure 3-8

    Example 3-18. A random sample of size 2 is taken from a group of businessmen, 1/3 of whom make \$20,000 a year and 2/3 of whom make \$40,000 a year. Let $\bar{X} = $ the average salary of the 2 men selected. (Notice that $\bar{X}$ is the sample mean, as defined in Chap. 2.) We see now that $\bar{X}$ is a random variable. The possible values for $\bar{X}$ are (\$20,000 +

$40,000)/2 = \$30,000$; $(\$20,000 + \$20,000)/2 = \$20,000$; $(\$40,000 + \$40,000)/2 = \$40,000$. We also see that $\bar{X} = \$20,000$ if both business-men selected make \$20,000 per year; $\bar{X} = \$30,000$ if one makes \$20,000 and the other makes \$40,000; and $\bar{X} = \$40,000$ if both businessmen selected make \$40,000. Thus, $P(\bar{X} = \$20,000) = 1/3 \times 1/3 = 1/9$; $P(\bar{X} = \$30,000) = 2 \times (1/3 \times 2/3) = 4/9$; and $P(\bar{X} = \$40,000) = 2/3 \times 2/3 = 4/9$.

Example 3-18 blends ideas from Chap. 2 with ideas from this chapter. We see how sample descriptive statistics can be described in a probabilistic way. The other descriptive measures we discussed, the sample standard deviation and the sample median, are also random variables. Study of the probability distributions of these sampling variables plays an important role in statistical theory. In Chap. 4 we shall discuss two important distributions which arise from sampling variables, binomial and normal distributions.

MEAN AND STANDARD DEVIATION OF RANDOM VARIABLES

Just as populations and collections of data have means and standard deviations, so do random variables. The means and standard deviations of random variables have great theoretical importance and are valuable measures of the variables under study. In Chaps. 14 and 15 we shall discuss some of their basic properties. For now, we mention only that the mean of a random variable (also called the *expectation*) measures its average with respect to the probabilities of the various values it assumes. The standard deviation of a random variable measures the average amount of variation of the variable. Both measures are closely related to the types of means and standard deviations discussed in Chap. 2. When we have occasion to mention them, we shall use the notation μ and σ as we did for the corresponding population parameters.

Summary

We have discussed some basic concepts of probability theory with applications to two types of sampling situations, independent and dependent cases. The multiplication rule has been used to compute probabilities for sequences of independent events. Tree diagrams have been employed as an aid in sequential experiments where the sequence of events are dependent. Also, we have discussed random (or sampling) variables and their distributions, mentioning how the descriptive measures of Chap. 2 can be discussed in a probabilistic way.

Important terms

Addition rule of probability

Event

Independent events

Multiplication rule for independent events

Outcome

Probability formula

Random experiment

Random variable

Tree diagram

Additional problems

(*Starred problems are more difficult.*)

3-12 From a bag containing 2 balls marked A and 1 ball marked B, 2 balls are selected at random with replacement. Find P(both balls selected are A), $P(AB)$, $P(BA)$, and $P(BB)$.

3-13 From a room containing 3 students and 2 deans, 3 people are selected at random with replacement. Find P(exactly 1 student is selected) and P(at least 1 student is selected).

3-14 In a chicken coop 50% of the chickens are speckled and 50% are plain. A sample of 3 chickens are selected at random with replacement. Find P(the first and third chickens selected are speckled); P(at least 2 chickens selected are speckled).

3-15 A certain restaurant has good food with probability 2/3 and, independently of that, good service with probability 1/2.

   a. If you go to this restuarant, what is P(you get good food and good service)?

    b. What is *P*(you get good food or good service)?

   **c.* If you go to this restaurant twice, what is *P*(you get good food once and bad service once)?

3-16 If you go on a certain diet, there is a 1/5 chance that you will lose more than 15 pounds, a 2/5 chance that you will lose more than 10 pounds, and a 3/5 chance that you will lose more than 5 pounds.

    a. What is *P*(you will lose more than 5 pounds but no more than 10 pounds)?

    b. What is *P*(you will lose either no more than 5 pounds or no more than 10 pounds)?

3-17 From a box containing 50 tickets marked "yes" and 50 tickets marked "no," 10 tickets are selected at random with replacement. Find the following:

    a. *P*(all 10 tickets are marked "no")

    b. *P*(at least 1 ticket is marked "no")

    c. *P*(the first 5 tickets are marked "no" and the second 5 are marked "yes")

   **d.* *P*(exactly 5 tickets are marked "no") (*Hint:* The answer is *not* 1/2. This problem should be difficult for you. We shall discuss such problems in Chap. 4.)

3-18 A gambler has a weighted coin with *P*(heads) = 2/3. He bets you that in 3 tosses the coin will land heads twice. What is *P*(he wins)?

3-19 You have a 1/3 chance of winning a certain game if you use strategy *A*. If you use this strategy and play the game 3 times, what is *P*(you win the game at least once)?

3-20 From a box containing 10 balls numbered from 1 to 10, 4 balls are selected at random without replacement. Find the following:

    a. *P*(first ball selected has an odd number)

    b. *P*(first 2 balls selected have odd numbers)

    c. *P*(3 balls have odd numbers)

    d. *P*(2 balls selected have odd numbers and one is bigger than 5)

3-21 If you roll a pair of dice three times, what is *P*(you will get 7 two times)?

3-22 A magician claims to have ESP. If you place 3 cards numbered from1 to 3 face down, he claims to be able to guess the numbers. Suppose he is really guessing at random. Find the following:

   a.   *P*(he guesses 0 cards correctly)

   b.   *P*(he guesses all 3 cards correctly)

   c.   *P*(he guesses exactly 1 card correctly)

   d.   *P*(he guesses exactly 2 cards correctly)

3-23 You are in a group of 30 people, 5 of whom will be given free tickets to a concert. What is *P*(you are 1 of the 5 selected)? Justify your answer using the basic rules of probability theory.

3-24 You are on a trip and reach a fork in the road. You know that one route leads you to city A, where the chance of being robbed is 1/3. However, the city A police department is quite efficient, and if you get robbed, there is a 9/10 chance that your money will be recovered. The other route leads you to city B, where the chance of being robbed is only 1/5. However, in city B the police department is inept and the chance of recovering your money is 1/2.

   a.   In which city do you have a better chance of keeping your money?

   b.   Suppose you do not know which route leads to which city and so you toss a coin to decide which one to take. What is *P*(you keep your money)?

3-25 You are on a trip. If the weather is bad, you will stop at a motel, where the probability of getting a good night's sleep is 1/2 and the probability of spending a restless night is 1/2. If the weather is good, you will stop at a campground, where the probability of getting a good night's sleep is 1/3 and the probability of spending a restless night is 2/3. If there is a 3/4 chance of good weather, what is *P*(you get a good night's sleep)?

3-26 You are a traveling salesman, selling "Magic Bubble Bath." You know a member of the Mystic Bath cult will buy your product with probability .9 and that a nonmember will buy your product with probability .1. In a certain city, .005 of the people belong to the Mystic Bath cult. If you select someone at random from this city, what is *P*(the person buys your product)?

3-27 From a room containing 8 sick people and 12 healthy people, 6 people are selected without replacement. Find the following:

  a. P(the sixth person selected is healthy)

  b. P(the selections occur in an alternating pattern, i.e., no 2 people of the same type are selected in a row)

3-28 A certain particle has a .2 chance of splitting (into 2 particles) after 1 year. If you start with 1 particle, what is P(you will have 3 particles after the third year)?

3-29 You randomly place 5 identical balls into 3 boxes.

  a. How many different arrangements are there?

  b. What is P(all balls are placed into the same box)?

  c. What is P(no box remains empty)?

* *3-30 Confidence intervals for the median.* The median of a collection of numbers is the "middle" number. More precisely, if there are an odd number in our collection, the median is the middle number if we arrange the numbers in order; if there are an even number in our collection, the median is the average of the two middle numbers. (For example, the median of the numbers $-1, 2, 1, 4, 5$ is 2, and the median of the numbers 4, 8, 12, 4 is 6.) Suppose you are interested in estimating the median weight of a large number of people. Suppose also that there are an even number of people altogether and that no two people weigh exactly the same. To estimate the median, take a random sample of size 6 (with replacement) and weigh the people in your sample, arranging the weights in order.

  a. What is P(the median is somewhere between the smallest and largest weights in your sample)?

  b. What is P(the median is between the second smallest and second largest weights in your sample)?

* *3-31* The Atomic Energy Commission claims that in a certain underground nuclear test the probability of releasing radioactive dust into the atmosphere is .0000001. If similar tests are performed over and over again, what is P(eventually dust will be released)? Try to justify your answer using the laws of probability and intuition.

* *3-32* In city A P(temperature $= 70°) = .4$; in city B P(temperature $= 70°) = .6$. Also, P(maximum temperature in A and $B = 70°) = .3$. What is P(minimum temperature in A and $B = 70°$)? (Hint: Draw a picture.)

True-false test

Circle the correct letter.

T F   1. If a small sample is drawn from a large population, in computing probabilities it makes little difference whether or not the sample is drawn with or without replacement.

T F   2. A thorough knowledge of probability theory will enable us to determine the outcome of random experiments in advance.

T F   3. If two events are independent, they cannot possibly be mutually exclusive.

T F   4. It is possible to have negative probabilities for extremely unlikely events.

T F   5. Tree diagrams are useful in computing probabilities for experiments which take place in dependent stages.

T F   6. If the probability of an event is 1, and if we perform the associated experiment many times, the event in question will occur once.

T F   7. If you toss a fair coin and obtain 50 heads in a row, the chance of obtaining tails on the fifty-first toss is greater than 1/2.

T F   8. If $P(A) = 1/2$ and if $P(B) = 1/2$, then $P(AB) = 1/4$.

T F   9. If 1 person is selected at random from a population of 24 people, and if the probability of selecting a person from group A is .25, then there are 6 people in group A.

T F   10. If you toss a coin 1,000 times and obtain 300 heads, then you would suspect that the coin is not evenly balanced.

Sampling distributions: binomial and normal **4**

In Chap. 2 various quantities helpful in summarizing data, such as the mean and standard deviation, were considered. Recall that when data are based on a random sample from a population, then these quantities, called *sample statistics,* will be random; that is, their values will depend on which members of the population are included in the randomly selected sample. In order to make inferences about a population on the basis of sample statistics, it is important to understand their random behavior, especially what is called their *probability*, or *frequency, distribution.* Experience shows that a few distributions arise again and again in a variety of sampling situations. In this chapter we shall discuss two of the most common distributions, the *binomial* and *normal* distributions.

The binomial distribution

The *binomial distribution* arises whenever a sample is drawn with replacement from a population divided into two groups, such as

Democrat–Republican, male–female, long–short, or heads–tails. In these situations we are interested in the number of sample members in each of the two groups, or their percentages. For example, public opinion polls often report the results of a sample of voters by giving the percentage in favor of some government action. In reporting the effectiveness of a new treatment for a disease a doctor might give the number of patients treated and the number cured. Television commercials for pain relievers often give the results of studies comparing two brands by specifying the percentage in a sample who favored one brand over another. (Unfortunately, they rarely give the sample size, nor do they mention whether the sample was randomly selected.) The results of a coin-tossing experiment may be given by specifying the number of tosses and the number of times the coin landed heads. These examples have three properties in common:

1 There is a sequence of independent trials.

2 On each trial there are only two possible outcomes.

3 The probability of a particular outcome is the same for each trial.

Any experiment with these properties is called a *binomial* experiment.

In a binomial experiment the two possible outcomes on each trial are commonly referred to as *success* and *failure*. It doesn't really matter which outcome you call "success" and which outcome you call "failure" as long as you are consistent in a given problem. The letter p is used to denote the probability of success on a particular trial (we assume it's the same for each trial), and n refers to the number of trials in the experiment. The number of successes in a binomial experiment is denoted by S. The following examples will illustrate these terms.

Example 4-1. A fair coin is tossed 10 times. This is a binomial experiment. The tosses correspond to trials, and the two possible outcomes on each trial are heads and tails. We may associate heads with success and tails with failure. The number of trials is $n = 10$. The probability of success is $p = 1/2$ since the coin is fair.

Example 4-2. On a certain civil service exam given over many years it has been found that 70% of the examinees pass. Suppose the latest exam were taken by 15 people. This is a binomial experiment. The trials consist of people taking the test. There are 15 trials; hence $n = 15$. We say a success occurred on a trial if the person passed the test. The probability of passing is $p = .7$.

Example 4-3. An elementary school class consists of 11 boys and 9 girls. Suppose 5 children are chosen at random to be on a special committee. This is *not* a binomial experiment because the outcomes are not independent. For example, if a boy is chosen first, the probability of selecting a boy on the second choice is decreased; thus, condition 1 is not satisfied.

In any binomial experiment the quantity S, or number of successes, is random because the outcome of any trial is random and S is determined by the outcomes. Accordingly, we may speak of the *probability distribution of S*, which gives the probabilities of S for various values. Since binomial experiments arise so often in statistics, the probability distribution of S has been given a special name: the *binomial distribution*. Actually, there is a whole collection of binomial distributions. A particular distribution may be specified by giving the two numbers $n =$ sample size and $p =$ probability of success, these being *parameters* of the binomial distribution. Thus, in Example 4-2, we say S has a binomial distribution with parameters $n = 15$ and $p = .7$. The fact that $n = 15$ tells us the possible values of S are 0, 1, 2, ..., 15; that is, after the exam has been given, the number of people who pass could be 0, 1, 2, ..., or 15. The probabilities that S takes on these values are computed under the assumption $p = .7$.

Although there is a formula for computing probabilities of a binomial distribution, it involves tedious arithmetic computations. So that you will not have to make these computations, tables of probabilities for some binomial distributions are provided in the Appendix. The tables give probabilities of the form $P(S \leq k)$ for various values of the parameters n and p. The symbol $\leq$ means *less than or equal to*, and $P(S \leq k)$ means *the probability that S is less than or equal to k*. From the addition rule it follows that

$$P(S \leq k) = P(S = 0) + P(S = 1) + \cdots + P(S = k)$$

To illustrate the use of the tables of probabilities, consider the binomial distribution with parameters $n = 5$ and $p = .3$, which is shown in Fig. 4-1. Thus, for example, $P(S \leq 2) = .8369$ and $P(S \leq 4) = .9976$. From this table we can compute each of the probabilities $P(S = 0)$, $P(S = 1)$, $P(S = 2)$, ..., $P(S = 5)$. To obtain $P(S = 2)$, take the difference between $P(S \leq 2)$ and $P(S \leq 1)$; that is,

$$P(S = 2) = P(S \leq 2) - P(S \leq 1)$$

$$= .8369 - .5282 = .3087$$

k	$P(S \leq k)$
0	.1681
1	.5282
2	.8369
3	.9692
4	.9976
5	1.0000

Figure 4-1

Similarly,

$$P(S = 4) = P(S \leq 4) - P(S \leq 3)$$

$$= .9976 - .9692 = .0284$$

This manipulation follows from the addition rule of probability.

Example 4-4. In a certain community where 60% of the voters will vote for Mr. Smiley, 10 voters are selected at random with replacement. What is the probability that 4 of the 10 voters will vote for Mr. Smiley?

> SOLUTION: This is a binomial experiment. A trial consists of selecting a voter. Call a vote for Mr. Smiley a success. Then S = number of successes = number of votes for Smiley. S has a binomial distribution with parameters $n = 10$ and $p = .6$. From the binomial tables in the Appendix we find

$$P(S = 4) = P(S \leq 4) - P(S \leq 3)$$

$$= .1662 - .0548 = .1114$$

Example 4-5. Two baseball teams A and B will play each other 10 times this season. (Assume they are of equal ability.)

(a) Find the probability that team A wins 7 games and loses 3.

(b) What is the probability that one of the teams will win at least 8 games?

> SOLUTION: Again we have a binomial experiment. Here, a trial is a game. Let's say that a trial is a success if team A wins the game. Then S = number of successes = number of games won by team A. The quantity S has a binomial distribution with parameters $n = 10$ and $p = .5$.

(*a*) The desired probability is

$$P(S = 7) = P(S \le 7) - P(S \le 6)$$
$$= .9453 - .8281 = .1172$$

(*b*) The event that one of the teams wins at least 8 games occurs if team *A* wins at least 8 games or if team *B* wins at least 8 games. But the event that team *B* wins at least 8 games is the same as the event that team *A* wins 2 games or fewer. Hence,

$$P(\text{one team wins at least 8 games}) = P(\text{team } A \text{ wins at least 8 games})$$
$$+ P(\text{team } A \text{ wins 2 games or fewer})$$
$$= P(S \ge 8) + P(S \le 2)$$

From the binomial tables with $n = 10$ and $p = .5$, we find $P(S \le 2) = .0547$. To find $P(S \ge 8)$ notice that

$$P(S \ge 8) = 1 - P(S \le 7)$$
$$= 1 - .9453 = .0547$$

Therefore,

$$P(\text{one team wins at least 8 games}) = P(S \le 2) + P(S \ge 8)$$
$$= .0547 + .0547 = .1094$$

Example 4-6. A newly married couple plans to have 10 children. An astrologist tells them that based on their astrological reading, the probability of bearing a boy on any particular birth is .9. Since they like boys and girls equally well, the couple wishes to know the chances of having 5 boys and 5 girls. Find the probability of this event.

SOLUTION: Under the assumption that the probability of bearing a boy on a particular birth is .9, independent of the sex on other births, this is a binomial experiment. Let $S =$ number of boys. Then S has a binomial distribution with parameters $n = 10$ and $p = .9$. From the binomial tables we obtain

$$P(S = 5) = P(S \le 5) - P(S \le 4)$$
$$= .0016 - .0001 = .0015$$

So, there are only 15 chances in 10,000 that the hopes of the couple will be fulfilled.

Example 4-7. Suppose new tires are checked by a federal agency using the following sampling procedure: Out of each 1,000 tires, 10 are selected at random and tested for defects. If 2 or more defective tires are found among the 10 tires sampled, the entire lot of 1,000 is checked; otherwise, the lot is passed. If there are 200 defective tires among 1,000 in a particular lot, what is the probability that the lot will pass?

> SOLUTION: Even though sampling would be done without re-placement, since only 10 items are drawn from 1,000, sampling with or without replacement would be essentially the same. Thus, testing 10 tires can be considered to be a binomial experi-ment. Let S = number of defective tires in a sample of 10. Then S has a binomial distribution with parameters $n = 10$ and $p = .2$. From the binomial tables we find

> P(lot passes) = P(fewer than 2 defectives are found in a sample of 10)

> $$= P(S < 2)$$

> $$= P(S \leq 1)$$

> $$= .3758$$

Example 4-8. In a certain heart operation the probability of suc-cess is .4. If 5 patients undergo the operation, what is P(obtaining at least 4 successes)?

> SOLUTION: Let S = number of successful operations. Then S has a binomial distribution with $n = 5$ and $p = .4$. From the tables,

> $$P(S \geq 4) = 1 - P(S < 4)$$

> $$= 1 - P(S \leq 3)$$

> $$= 1 - .9130 = .0870$$

Problems

4-1 From a population of voters containing .4 Republicans, 10 voters are selected at random. Let S = no. of Republicans in the sample. Find:

   a. $P(S \leq 4)$

   b. $P(S \geq 5)$

   c. $P(S = 5)$

4-2 A fair coin is tossed 8 times. Let S = no. of heads. Find:

   a. $P(S = 5)$

   b. $P(S \geq 6)$

   c. $P(S \leq 3)$

4-3 From a large collection of cameras with .1 defectives, 15 cameras are selected at random. If S = no. of defectives in the sample, find:

   a. $P(S = 2)$

   b. $P(S \leq 3)$

4-4 On the average, 2 out of 10 people will respond favorably to a certain telephone sales pitch. If 20 people are called, letting S = the number who respond favorably, find:

   a. $P(S \geq 3)$

   b. $P(S < 6)$

The normal distribution

The most important frequency distribution in statistics is the *normal distribution.* It arises surprisingly often in natural phenomena and scientific experiments. Some quantities that often have normal distributions are test scores, heights of males in a large group, annual wage incomes of people doing similar work in a large city, velocities of particles in a volume, and errors in measurement.

    The basic properties of the normal distribution can be easily seen in the case of errors in measurement. A familiar example is the procedure often used to time races. Several judges stand at the finish line, each one timing the contestants as they cross the line. Everyone recognizes that there is usually variation among the times obtained by different judges, and so the results of a particular judge are not necessarily reliable. A similar phenomena occurs in the measurement of the size or weight of objects. Even when the number of objects in a given space are simply counted (as in the case when cells that are tightly packed on a microscope slide are counted), the results obtained by different observers may vary. This variation among observations can be thought of as being random since it cannot be predicted beforehand. Such random variation arising when many independent measurements are taken of the same quantity is often accurately described by the normal distribution.

To be specific, suppose 30 spectators are selected at random from people attending a championship track and field meet, and each person is given a stopwatch to time the winner of the 100-yard dash. If the stopwatches are finely calibrated, say, to one-hundredth of a second, then there will certainly be some discrepancies among their times. The data in Fig. 4-2 are plausible.

These data are summarized by the frequency chart in Fig. 4-3 and

Times observed by 30 spectators for the
winner of the 100-yard dash (in seconds)

9.41	9.46	9.39
9.46	9.44	9.42
9.39	9.49	9.41
9.37	9.43	9.47
9.51	9.43	9.50
9.45	9.36	9.48
9.48	9.39	9.43
9.48	9.41	9.35
9.41	9.41	9.43
9.38	9.38	9.40

Figure 4-2

Frequency chart for raw data in Fig. 4-2

Class intervals	Frequency
`9.345–9.365	2
9.365–9.385	3
9.385–9.405	4
9.405–9.425	6
9.425–9.445	5
9.445–9.465	3
9.465–9.485	4
9.485–9.505	2
9.505–9.525	1

Figure 4-3

by the resulting histogram (Fig. 4-4). We have grouped the data into class intervals of length .02 seconds, beginning at 9.345 seconds. Notice that the histogram (Fig. 4-4) resembles the bell-shaped curve in Fig. 4-5.

If many more people timed the race and the class intervals were made smaller, the resulting histogram would look more like the curve in Fig. 4-5. This means that the frequency distribution of times observed for the winner of the 100-yard dash may be represented by a *bell-shaped curve*. The probability that a judge or group of judges will obtain times within a certain specified range may be computed from this curve.

In statistics a bell-shaped curve is called a *normal curve*. A random quantity has a *normal distribution* if its probability, or frequency,

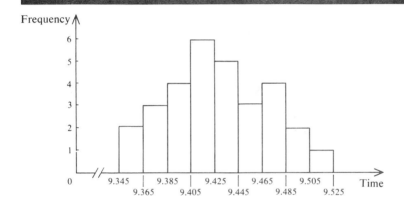

Figure 4-4

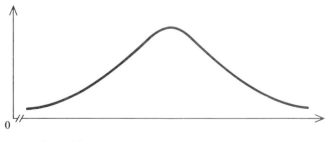

Figure 4-5

distribution is represented by a normal curve. Actually there are many normal curves, and each is characterized by two numbers. The *mean* gives the central value, or point of symmetry (normal curves are always symmetric). As we did for the population mean, we shall use the Greek letter μ to denote the mean of a normal curve. The *standard deviation* tells us the amount of spread of the curve, and it is denoted by σ. We use the terms mean and standard deviation because a sampling variable which has a distribution represented by a normal curve will have a corresponding mean and standard deviation.

Some examples of normal curves with a common mean and different standard deviations are shown in Fig. 4-6. Curve 1 has the smallest standard deviation, curve 2 has the next smallest standard deviation, and curve 3 has the largest standard deviation. Figure 4-7 illustrates three normal curves which have the same standard deviations but different means. As you can see, all normal curves are symmetric about their means.

OBTAINING NORMAL PROBABILITIES

Normal curves are used to obtain the probabilities of events involving variables that have a normal distribution. If X is a random quantity with a distribution that can be represented by a normal curve, then

> *The probability that the value of X is within a particular interval is approximately the area beneath the corresponding normal curve in that interval.*   (4-1)

For example, if X has a distribution represented by the normal curve in Fig. 4-8, then $P(7 \leq X \leq 13) = .68$. Moreover,

> *The area under a normal curve between the mean and any other value depends only on the difference between this value and the mean, expressed in units of standard deviation.*   (4-2)

This fact is illustrated by the three curves in Fig. 4-9.

Notice that even though the three normal curves in Fig. 4-9 have various means and standard deviations,

> *Approximately 68% of the area lies within one standard deviation of the mean.*   (4-3)

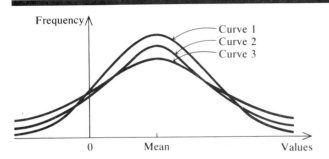

Figure 4-6

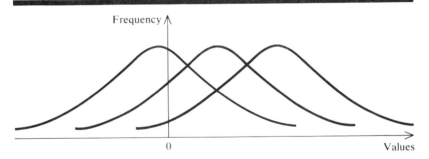

Figure 4-7

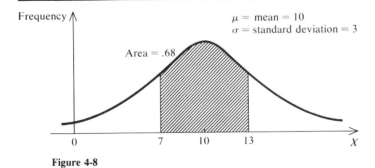

Figure 4-8

As illustrated in Fig. 4-10, it turns out that

> *Approximately 95% of the area lies within two standard devia-*
> *tions of the mean.* (4-4)

It will be helpful to memorize these two facts. They give a rule of thumb which may be applied to obtain a rough idea about the shape of a particular normal curve just by knowing the mean and standard deviation.

As we have stated, the area under a normal curve between two values depends only on the number of standard deviations between these values and the mean. This fact permits the use of one normal curve for computing the areas under any normal curve. For convenience, the normal curve with mean 0 and standard deviation 1 is used as the standard, and it is called the *standard normal curve*. Since we shall need to know areas under the standard normal curve to compute probabilities for all normal variables, we have provided tables in the Appendix which give these areas. We shall use the letter Z to represent a variable with a standard normal distribution.

The normal tables give probabilities of the form $P(Z \leq z)$ for various values of z (refer to Fig. 4-11). Any other probabilities that are needed can be computed from the table by using the facts that normal curves are symmetric and have a total area of 1. For example,

$$P(Z \geq .85) = 1 - P(Z \leq .85)$$

since the area under the curve to the right of .85 equals the total area minus the area to the left of .85. From the normal tables we find

$$P(Z \leq .85) = .8023$$

and hence

$$P(Z \geq .85) = 1 - .8023 = .1977$$

This is illustrated by Fig. 4-12.

An alternative method of finding $P(Z \geq .85)$ is to observe that since the standard normal curve is symmetric about 0,

$$P(Z \geq .85) = P(Z \leq -.85) = .1977$$

the latter probability being obtained from the normal tables. As for computing the probabilities that a standard normal variable Z falls in an interval, we observe, for example, that

$$P(1 \leq Z \leq 2) = P(Z \leq 2) - P(Z \leq 1)$$

$$= .9772 - .8413 = .1359$$

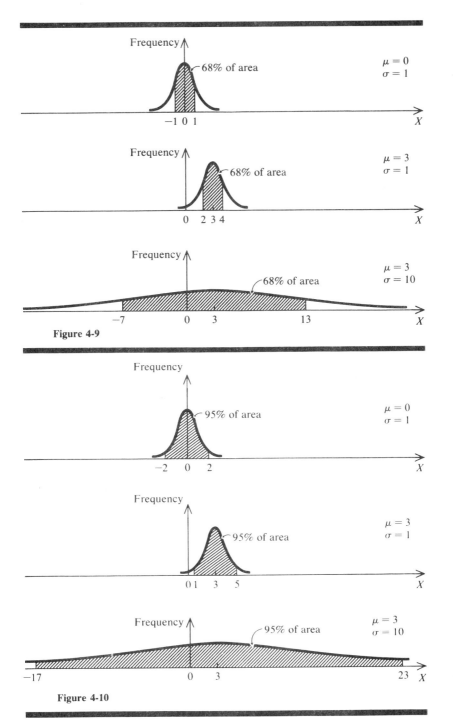

Figure 4-9

Figure 4-10

This is because the area between 1 and 2 is the same as the area to the left of 2 minus the area to the left of 1 (see Fig. 4-13). The same argument shows that

$$P(-3 \leq Z \leq 1.3) = P(Z \leq 1.3) - P(Z \leq -3)$$

$$= .9032 - .0013 = .9019$$

and

$$P(-2.2 \leq Z \leq .05) = P(Z \leq .05) - P(Z \leq -2.2)$$

$$= .5199 - .0139 = .5060$$

When using the normal tables you should always draw a sketch of the area you are trying to compute. You will find that this cuts down on careless mistakes.

Next consider the general case when the data of a variable X are approximately normal but not necessarily with mean 0 and standard deviation 1. To obtain the area under the normal curve for such variables, it is necessary to find the difference between the endpoints of the interval of interest and the mean in units of standard deviations and then use (4-2). This procedure is summarized as follows:

If X is a variable with a normal distribution having mean μ and standard deviation σ, and if a, b are two numbers with $a < b$, then

$$P(a \leq X \leq b) = P\left(\frac{a - \mu}{\sigma} \leq Z \leq \frac{b - \mu}{\sigma}\right) \qquad (4\text{-}5a)$$

For a picture, refer to Fig. 4-14. Notice that the quantity $(a - \mu)/\sigma$ gives simply the number of standard deviations between a and μ; the same is true for $(b - \mu)/\sigma$. Of course, the probability involving Z on the right side of the equation in (4-5a) can be obtained from the normal tables. Two special cases of (4-5a) are

$$P(X \geq a) = P\left(Z \geq \frac{a - \mu}{\sigma}\right) \qquad (4\text{-}5b)$$

$$P(X \leq b) = P\left(Z \leq \frac{b - \mu}{\sigma}\right) \qquad (4\text{-}5c)$$

Using formulas (4-5) and the facts that all normal curves are symmetric about their mean and have total area 1, any probability involving a normal variable may be determined. You should study carefully the computations in the following examples to make sure you understand them. It is essential that you be able to compute various normal

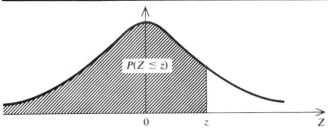

Figure 4-11

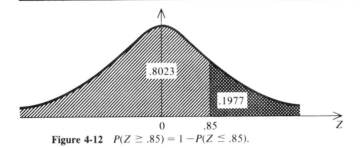

Figure 4-12 $P(Z \geq .85) = 1 - P(Z \leq .85)$.

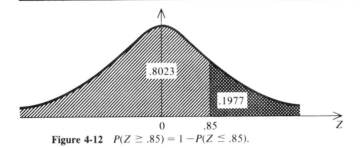

Figure 4-13 $A = A_1 - A_2$.

probabilities easily and without confusion before proceeding to the applications in later chapters.

Example 4-9. Suppose X is a normal variable having mean $\mu = 3$ and standard deviation $\sigma = 8$. Find the following:

(a) $P(3 \leq X \leq 10)$

(b) $P(X \leq 0)$

(c) $P(X \geq 0)$

(d) $P(-5 \leq X \leq -1)$

SOLUTION: We use (4-5).

(a) $P(3 \leq X \leq 10) = P\left(\dfrac{3-3}{8} \leq Z \leq \dfrac{10-3}{8}\right)$

$$= P(0 \leq Z \leq .88)$$

$$= .8106 - .5000$$

$$= .3106$$

(b) $P(X \leq 0) = P\left(Z \leq \dfrac{0-3}{8}\right)$

$$= P(Z \leq -.38)$$

$$= .3520$$

(c) $P(X \geq 0) = 1 - P(X \leq 0)$

$$= 1 - .3520$$

$$= .6480$$

(d) $P(-5 \leq X \leq -1) = P\left(\dfrac{-5-3}{8} \leq Z \leq \dfrac{-1-3}{8}\right)$

$$= P(-1 \leq Z \leq -.5)$$

$$= .3085 - .1587$$

$$= .1498$$

Example 4-10. Certain IQ test scores are approximately normally distributed with mean $\mu = 100$ and standard deviation $\sigma = 10$. In other words, if a person is chosen at random from the population and given an IQ test, the score can be considered as a normal random variable with mean $\mu = 100$ and standard deviation $\sigma = 10$. Find the following probabilities and interpret the results:

(a) $P(90 \leq X \leq 110)$

(b) $P(80 \le X \le 120)$

(c) $P(X \ge 125)$

SOLUTION:

(a) $P(90 \le X \le 110) = P\left(\dfrac{90 - 100}{10} \le Z \le \dfrac{110 - 100}{10}\right)$

$\qquad\qquad\qquad\qquad = P(-1 \le Z \le 1)$

$\qquad\qquad\qquad\qquad = .8413 - .1587$

$\qquad\qquad\qquad\qquad = .6826$

(b) $P(80 \le X \le 120) = P\left(\dfrac{80 - 100}{10} \le Z \le \dfrac{120 - 100}{10}\right)$

$\qquad\qquad\qquad\qquad = P(-2 \le Z \le 2)$

$\qquad\qquad\qquad\qquad = .9772 - .0228$

$\qquad\qquad\qquad\qquad = .9544$

Observe that the probabilities in (*a*) and (*b*) could have been obtained using the "rule of thumb" given in (4-3) and (4-4) since 90

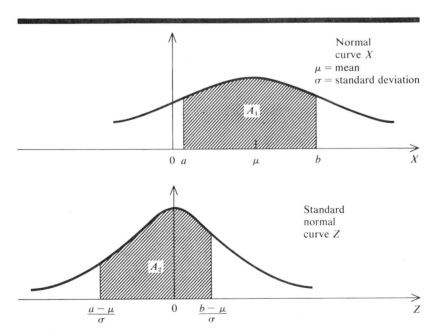

Figure 4-14 $A_1 = A_2$.

and 110 are exactly 1 standard deviation on each side of the mean, and 80 and 120 are 2 standard deviations.

(c) $P(X \geq 125) = P\left(Z \geq \dfrac{125 - 100}{10}\right)$

$$= P(Z \geq 2.5)$$

$$= 1 - .9938$$

$$= .0062$$

The probability statements in (*a*), (*b*), and (*c*) are equivalent to: "68.26% of the population have IQ's between 90 and 110." "95.44% of the population have IQ's between 80 and 120." ".62% of the population have IQ's of at least 125."

Example 4-11. Suppose all fourth graders in a certain school are taught to read by the same teaching method and at the end of the year they are tested for reading speed. If the $\mu = 200$ words per minute and $\sigma = 50$, what percentage of the students read more than 300 words per minute?

SOLUTION: It is commonly assumed that reading scores are normally distributed. Let $X =$ score of person chosen at random. Then we must find $P(X \geq 300)$:

$$P(X \geq 300) = P\left(Z \geq \dfrac{300 - 200}{50}\right)$$

$$= P(Z \geq 2)$$

$$= .0228$$

In other words, roughly 2 out of every 100 students will read more than 300 words per minute.

Problems

4-5   Let Z have a standard normal distribution. Compute each of the following probablilities:

a. $P(Z \geq 2)$

b. $P(-1.5 \leq Z \leq .7)$

c. $P(-2 \leq Z \leq -1)$

d. $P(Z \leq .08)$

4-6 Suppose X is normally distributed with $\mu = 5$ and $\sigma = 4$. Compute the following probabilities:

    a. $P(1 \leq X \leq 7.2)$

    b. $P(X \geq 10)$

    c. $P(X \leq 0)$

4-7 The incomes of industrial workers in a certain region are normally distributed with a mean of \$12,500 and a standard deviation of \$1,000. What proportion of the incomes are between \$11,000 and \$14,000?

4-8 A certain scale makes measurement errors that are normally distributed with a mean of 0 and a standard deviation of .1 ounce. If you weigh an object on this scale, what is the probability that the weight will be correct to within .3 ounce?

USING THE NORMAL TABLES IN REVERSE

A common problem in statistics is to find the value of a normal variable that will be exceeded a certain given fraction of the time. To solve such a problem we simply use the normal tables in reverse. For example, if Z is a standard normal variable, what is the number c such that $P(Z \geq c) = .05$? In other words, what is the value c on the horizontal axis for which the area to the right under the normal curve is .05?

In solving such a problem, always draw a picture like the one in Fig. 4-15, so that you know what you're looking for. When drawing this picture, even though we don't yet know the exact value of c, we know that it is a positive number because .5 of the area under the curve

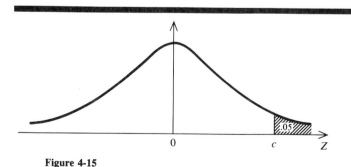

Figure 4-15

is to the right of 0 and so any number that has only .05 of the area to its right must be to the right of 0. Now, if the area to the right of c is .05, then the area to the left of c must be .95, as indicated in Fig. 4-16. From the normal tables in the Appendix we find

$$P(Z \leq 1.64) = .9495$$

$$P(Z \leq 1.65) = .9505$$

Thus, c must be between 1.64 and 1.65. If we had more extensive tables, the value of c could be determined more accurately; but, for our purposes we take simply $c = 1.65$.

If X is normal with mean μ and standard deviation σ, we can solve this kind of problem by using (4-5). As an illustration, suppose X is normal with $\mu = 6$ and $\sigma = 3$. What is the number c such that $P(X \geq c) = .1$? Using Eq. (4-5*b*),

$$P(X \geq c) = P\left(Z \geq \frac{c - 6}{3}\right)$$

and thus, c must satisfy $P(Z \geq (c - 6)/3) = .1$, or equivalently

$$P\left(Z \leq \frac{c - 6}{3}\right) = 1 - .1 = .9$$

From the table, $P(Z \leq 1.28) \approx .9$, and so $(c - 6)/3 = 1.28$. Solving for c, we have $c = 9.84$. For a picture, refer to Fig. 4-17.

Example 4-12. A certain club is having a meeting. There are 1,000 members who will attend but only 100 seats. It is decided that the oldest members should have seats. If the average age of members is 40, with standard deviation of 5, then what is the age of the youngest person who should have a seat? (Assume the ages are normally distributed.)

SOLUTION: Let $X =$ age of person chosen at random. We must find c such that

$$P(X \geq c) = \frac{100}{1,000} = .1$$

Since X is normal with $\mu = 40$ and $\sigma = 5$, we obtain

$$.1 = P(X \geq c) = P\left(Z \geq \frac{c - 40}{5}\right)$$

and hence $(c - 40)/5 = 1.28$. Solving, we obtain $c = 46.4$. Thus, the youngest person to have a seat will be approximately 46.4 years old. For an appropriate picture, refer to Fig. 4-18.

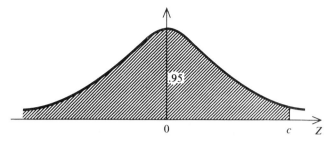

Figure 4-16

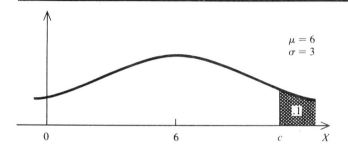

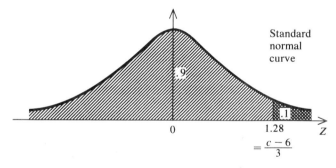

Figure 4-17

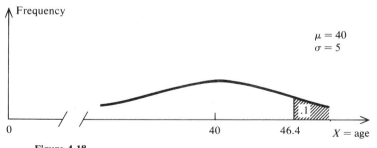

Figure 4-18

Example 4-13. Suppose that annual family incomes in a certain region are normally distributed with a mean of $7,500 and a standard deviation of $2,000. If the government wants to give welfare to those whose incomes are in the lowest 5%, what is the maximum annual income of those eligible for welfare?

SOLUTION: Let $X =$ income of person chosen at random. Then X is normal with $\mu = 7,500$ and $\sigma = 2,000$. We must find the number c such that $P(X \leq c) = .05$. So,

$$P(X \leq c) = P\left(Z \leq \frac{c - 7,500}{2,000}\right) = .05$$

implies

$$\frac{c - 7,500}{2,000} = -1.65$$

Solving, we obtain $c = \$4,200$. (Refer to Fig. 4-19 for the corresponding picture.)

Problems

4-9 If Z is a standard normal variable, find the number c such that $P(Z \leq c) = .3821$.

4-10 The scores on a certain test are normally distributed with a mean of 76 and a standard deviation of 5. How high a score must a student have to be in the top 10%?

4-11 The lengths of trout in a certain lake are normally distributed with a mean of 7 inches and a standard deviation of 2 inches. If the fish and game department wishes only the biggest 20% of the trout to be caught, what should the minimum size for "keepers" be?

Normal approximation to the binomial distribution

Binomial experiments are extremely common in sampling procedures, but owing to the unwieldiness of exact formulas for the binomial distribution when n is large, computational problems sometimes arise. Fortunately, binomial histograms (for large n) are shaped approximately like normal curves. Hence, binomial variables behave approximately like normal variables when n is large. This is made precise by the *central limit theorem for binomial variables:*

> **Central limit theorem for binomial variables.** Let S be a binomial variable with parameters n and p. Then the distribution of S is approximately normal with $\mu = np$ and $\sigma = \sqrt{np(1 - p)}$. This approximation is valid when $np(1 - p) > 5$.

Hence, binomial probabilities can be computed from the normal table by the following formula:

$$P(a \le S \le b) \approx P\left(\frac{a - np}{\sqrt{np(1 - p)}} \le Z \le \frac{b - np}{\sqrt{np(1 - p)}}\right) \qquad (4\text{-}6)$$

where Z is standard normal and a and b are given numbers. Also, $\approx$ means *approximately equal.*

Example 4-14. A random sample of size 500 is taken from a population which contains .1 who are "mentally deranged." What is the probability that at least 60 mentally deranged people are selected for our sample?

> SOLUTION: If we let $S = $ the number of mentally deranged people in the sample, then S has a binomial distribution with parameters $n = 500$ and $p = .1$. Thus, $np = 50$, $\sqrt{np(1 - p)} = 6.7$. According to the central limit theorem for binomial variables,
>
> $$P(S \ge 60) = P\left(Z \ge \frac{60 - 50}{6.7}\right)$$
>
> $$= P(Z \ge 1.49)$$
>
> $$= .0681$$

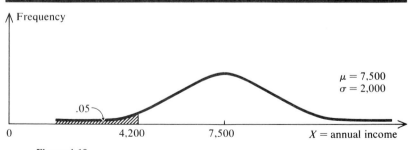

Figure 4-19

Example 4-15. A certain factory has 1,000 employees. It can operate at full efficiency if at least 935 employees show up for work. Assuming that the probability that an employee comes to work is .95, what is the probability that the factory will operate at full efficiency on any given day?

SOLUTION: Let S = number of employees who show up for work. Then S is binomial with $n = 1,000$ and $p = .95$. We obtain $np = 950$, and $\sqrt{np(1-p)} = \sqrt{1,000(.05)(.95)} = 6.9$. The desired probability is

$$P(S \geq 935) = P\left(Z \geq \frac{935 - 950}{6.9}\right)$$

$$= P(Z \geq -2.17)$$

$$= 1 - .015 = .985$$

Example 4-16. In a large city, approximately 40% of all fire alarms are false alarms. But the fire department cannot function properly if more than 50 out of 100 consecutive alarms are false. What is the probability that the fire department will not function properly?

SOLUTION: Consider a group of 100 consecutive alarms. Let S = number of false alarms among the 100 alarms. Then S is a binomial variable with $n = 100$ and $p = .4$. So, $np = 100(.4) = 40$, and $\sqrt{np(1-p)} = \sqrt{100(.4)(.6)} = 4.9$. Finally,

P(fire department does not function properly) $= P(S \geq 50)$

$$= P\left(Z \geq = \frac{50 - 40}{4.9}\right)$$

$$= P(Z \geq 2.04)$$

$$= .0207$$

Example 4-17. A professor gives a 100-question multiple-choice final exam. Each question has 4 choices. In order to pass, a student has to obtain at least 30 correct answers. A lazy student decides to guess at random on each question. What is the probability that the student passes the exam?

SOLUTION: Let S = number of correct answers achieved by the lazy student. S is binomial with $n = 100$ and $p = 1/4$ since there are 100 questions and 4 choices for each one. Thus, $np =$

$100(1/4) = 25$ and $\sqrt{np(1 - p)} = \sqrt{100(1/4)(3/4)} = 4.4$. Therefore

$P(\text{student passes}) = P(S \geq 30)$

$$= P\left(Z \geq \frac{30 - 25}{4.4}\right)$$

$$= P(Z \geq 1.14)$$

$$= .1271$$

Problems

4-12 A statistics professor believes that the students in his recent introductory course were exceptional. In the past about half the students scored higher than 70 on his final exam, but in his most recent class of 50 students, 35 students achieved scores above 70. If the students were actually similar to students in previous classes, what would be the probability that at least 35 students would achieve scores above 70?

4-13 A student works a 75-question multiple-choice test by rolling a die to determine which of 6 possible choices to make on each question. What is the probability that the student will answer at least 12 questions correctly?

4-14 A cold remedy relieves symptoms in 80% of the cases treated. If it is given to 1,000 cold sufferers chosen at random, what is the probability that at least 750 persons will notice relief? (Incidentally, how would you choose 1,000 cold sufferers at random?)

4-15 In a certain lottery, the winning number has ended in the digits 0, 1, or 2 in 20 out of the last 100 weeks. What is the probability that this event will occur in at least 20 out of 100 weeks, assuming each digit $(0, 1, \ldots, 9)$ is equally likely?

Normal approximation for sample proportions

In many problems we are not especially interested in the binomial variable $S = $ no. of successes. Rather, the *sample proportion* of successes $S/n = \hat{p}$ is important. The notation $\hat{p}$ is the standard way of denoting the sample proportion. Whenever we speak of proportion of voters for a certain proposition, percentage of bad apples in a barrel, probability of heads, etc., it is the random quantity $\hat{p}$ that is of interest. The histogram of $\hat{p}$ can be approximated by a normal curve, just as the histogram of S can be, when the sample size is large. This fact is called the *central limit theorem for sample proportions* and can be expressed by

$$P(a \le \hat{p} \le b) \approx P\left(\frac{a - p}{\sqrt{p(1 - p)/n}} \le Z \le \frac{b - p}{\sqrt{p(1 - p)/n}}\right) \quad (4\text{-}7)$$

where a and b are given numbers (that is, possible values of $\hat{p}$) and Z is a standard normal variable. This formula uses the fact that the mean of $\hat{p}$ is p, and the standard deviation of $\hat{p}$ is $\sqrt{p(1 - p)/n}$. As before, this approximation is valid if $np(1 - p) > 5$.

Example 4-18. Suppose that during the past few years 15% of all accidents on a certain freeway were serious, resulting in death for someone involved. Because of the energy crisis, the maximum speed limit was lowered from 65 to 55 miles per hour. The traffic commissioner claims that reducing the speed limit has reduced the probability that an accident will be serious, citing the fact that out of the past 150 accidents only 12 were serious. Does this evidence offer convincing support for the commissioner's claim?

SOLUTION: To evaluate the commissioner's claim, we wish to determine how likely the observed result is, if it has merely occurred by chance. That is, we compute the probability that so few accidents are serious, under the assumption that the probability that an accident is serious is still .15. If this probability is small, then we may infer that the observed decrease is not merely a coincidence for the 150 accidents observed, but is due to the reduced speed limit.

To find out how unlikely the evidence is, assuming $p = $ true probability that an accident is serious $= .15$, we need to compute the probability that a sample proportion $\hat{p}$ is as small as the observed value, $\hat{p} = 12/150 = .08$; that is, we want to find $P(\hat{p} \le .08)$. Using Eq. (4-7), with $n = 150$ and $p = .15$, we have

$$P(\hat{p} \le .08) \approx P\left(Z \le \frac{.08 - .15}{\sqrt{(.15)(.85)/150}}\right)$$

$$= P(Z \le -2.40) = .0082$$

Thus, we see that observing a sample proportion as small as .08 is very unlikely, if the probability that an accident is serious is still

.15. In fact, it would happen only about 8 times in 1,000 by chance. Therefore, the sample result is strong evidence that the probability that an accident will be serious has actually decreased.

Normal approximation for sample means

The most important sampling variable is the *sample mean* $\bar{X}$. Unfortunately, it is often quite difficult or simply impossible to find the exact distribution of this quantity. Because of the *central limit theorem for sample means,* it turns out that the distribution of $\bar{X}$ will be approximately normal. This theorem can be stated as follows:

Central limit theorem for sample means. Suppose a random sample of size n is drawn from a population with mean μ and standard deviation σ. Then the sample mean $\bar{X}$ is approximately normal with mean μ and standard deviation $\sigma/\sqrt{n}$ when n is large.

$$P(a \le \bar{X} \le b) \approx P\left(\frac{a - \mu}{\sigma/\sqrt{n}} \le Z \le \frac{b - \mu}{\sigma/\sqrt{n}}\right) \qquad (4\text{-}8)$$

where Z is standard normal, and a and b are any numbers.

Example 4-19. A grain dealer wants to buy rice from a grower. The rice comes in 50-pound bags, at least that's what it says on the label. To obtain some idea of the mean weight of all bags of rice, the grain dealer chooses 100 bags at random and then weighs them. He takes the sample mean $\bar{X} = 49$ pounds as an estimate of the mean weight μ of all bags. Assume the standard deviation of weights σ is 1 pound. What is the probability that $\bar{X}$ is less than 49 pounds if the mean weight μ of all bags actually is 50 pounds?

SOLUTION: According to the central limit theorem for sample means, $\bar{X}$ is approximately normal with a mean $\mu = 50$ pounds and a standard deviation $\sigma/\sqrt{n} = 1/\sqrt{100} = .1$. Thus,

$$P(\bar{X} \le 49) \approx P\left(Z \le \frac{(49 - 50)}{.1}\right)$$

$$= P(Z \le -10)$$

$$\approx 0$$

This means an observed value of $\bar{X}$ as low as 49 pounds is almost impossible under the assumption that $\mu = 50$. Hence, if the grain dealer observes such a low value of $\bar{X}$, he should seriously question the grower's claim.

Example 4-20. An anthropologist wants to estimate the age of ancient artifacts found in a certain ruin. He takes a sample of 30 artifacts and measures their age by radio-carbon dating procedures. (Assume that all the artifacts are actually 2,800 years old.) Due to random variations in the dating procedure, however, there is some deviation among the observed ages and the standard deviation of a measurement is 100 years. What is the probability that the sample average $\bar{X}$ is at least 2,830 years?

SOLUTION: According to the central limit theorem for sample means, $\bar{X}$ is approximately normal with a mean of 2,800 years and a standard deviation of $\sigma/\sqrt{n} = 100/\sqrt{30} = 18.2$. Thus,

$$P(\bar{X} \geq 2,830) \approx P\left(Z \geq \frac{2,830 - 2,800}{18.2}\right)$$

$$= P\left(Z \geq \frac{30}{18.2}\right)$$

$$= P(Z \geq 1.65)$$

$$= .05$$

Summary

In order to make inferences about a population on the basis of a sample statistic we need to know its probability, or frequency, distribution. The two most important distributions are the binomial and normal distributions. The binomial distribution arises when there is a sequence of independent trials having only two possible outcomes, called success and failure. If there are n trials and the probability of success on any trial is p, then the variable S = number of successes has a binomial distribution with parameters n and p. Tables of this distribution are provided in the Appendix.

A normal distribution is represented by a bell-shaped curve, called a normal curve. Each normal curve is characterized by two numbers, the mean μ and standard deviation σ. The normal distribution Z with $\mu = 0$ and $\sigma = 1$, is known as the standard normal distribution. The probabilities of any normal distribution may be computed from the tables of the standard normal distribution. If X is a variable with a normal distribution, then

$$P(a \leq X \leq b) = P\left(\frac{a - \mu}{\sigma} \leq Z \leq \frac{b - \mu}{\sigma}\right)$$

where a and b are numbers and Z has a standard normal distribution.

Many sampling variables have an approximately normal distribution. The binomial distribution with parameters n and p may be approximated by the normal distribution with $\mu = np$ and $\sigma = \sqrt{np\,(1 - p)}$ provided that $np\,(1 - p) > 5$. If a population contains a proportion p within a certain category, then the sample proportion $\hat{p}$, based on a sample of size n, will have approximately a normal distribution with $\mu = p$ and $\sigma = \sqrt{p\,(1 - p)/n}$. If $\bar{X}$ is the sample mean based on n observations from a population with a mean μ and standard deviation σ, then $\bar{X}$ has an approximately normal distribution with a mean μ and standard deviation $\sigma/\sqrt{n}$.

Important terms

Binomial distribution

Central limit theorem for sample means

Central limit theorem for sample proportions

Normal approximation to the binomial

Normal distribution

Additional Problems

4-16 A vaccine is 70% effective in preventing a certain type of dysentery. If 18 people who have been exposed to dysentery are given the vaccine, what is $P(3$ or fewer people get dysentery)?

4-17 On a certain dangerous stretch of freeway there is at least one serious accident on about 30% of the days.

    a. During a period of 20 consecutive days, what is $P($you observe a serious accident on at least 5 of the days)?

    b. If you observe 20 successive days with no accidents, what is $P($you observe a serious accident on the twenty-first day)?

4-18 Let Z have a standard normal distribution. Compute each of the following probabilities:

$P(Z \geq -1)$	$P(-.1 \leq Z \leq .2)$
$P(-2.2 \leq Z \leq .8)$	$P(.3 \leq Z \leq 1.5)$
$P(-5 \leq Z \leq 0)$	$P(Z \leq .6)$

4-19 Let X have a normal distribution.

a. If X has a mean of 10 and a standard deviation of 5, find each of the following probabilities:

$P(X \geq 11)$	$P(X \leq 4)$	$P(X \geq 16)$
$P(7 \leq X \leq 17)$	$P(X \geq 8)$	$P(X \geq 0)$

b. If X has a mean of 50 and a standard deviation of 100, find the following:

$P(X \geq 150)$ $P(X \leq 0)$ $P(-100 \leq X \leq 100)$

c. If X has a mean of -3 and a standard deviation of 8, find the following:

$P(X \geq 5)$ $P(X \leq -7)$ $P(-15 \leq X \leq 1)$

4-20 Let Z be standard normal. In each of the following cases find the number c which satisfies the equation:

$P(Z \leq c) = .95$	$P(Z \geq c) = .8$
$P(-c \leq Z \leq c) = .9$	$P(-2c \leq Z \leq 2c) = .9$

4-21 A certain company produces glass jars. On the average the jars contain 1 quart, but there is some variation among them. Assume that the sizes are normally distributed with standard deviation .01 quart.

a. What is the probability that a jar chosen at random will contain between .98 and 1.01 quarts?

b. What proportion of the jars will contain more than 1.05 quarts?

4-22 A hot-dog manufacturing company produces hot dogs having lengths that are normally distributed, with a mean length of 4 inches and a standard deviation .5 inch.

a. What proportion of hot dogs will exceed 4.1 inches?

b. In a sample of 100 hot dogs, what is the probability that their average length is at least 4.1 inches?

4-23 A certain automobile manufacturer claims that new cars get 30 miles to the gallon on the open road. Assume that the numbers of miles per gallon for various cars made by this manufacturer are normally distributed with a standard deviation of 5 miles per gallon.

a. What is the probability that a car chosen at random will get at least 28 miles per gallon?

b. If you buy two cars, what is the probability that both of them get at least 25 miles per gallon?

4-24 A supermarket sells about 1,000 loaves of bread per day with a standard deviation of 100 loaves.

a. If the market stocks 1,200 loaves per day, what is the probability that the loaves are sold out before the day is over?

b. How many loaves should the market stock in order to have a probability of .99 that it will not run out?

4-25 A certain carrot farm is famous for growing carrots of almost identical length. Suppose the mean length is 5 inches. What must the standard deviation be if it is known that 99% of the carrots are between 4.9 and 5.1 inches.

4-26 Suppose the price of an apple in a certain city varies slightly from day to day with a mean of 10 cents and a standard deviation of 2 cents.

a. On a day chosen at random, what is the probability that the price is at least 12 cents?

b. What is the probability that the average price over a 100-day period lies between 9.75 and 10.3 cents?

4-27 A game consists of tossing a fair coin 25 times. If at least 17 heads occur, the player wins a prize.

a. What is the probability of winning?

b. If the coin is unfair, with the probability of heads being .6, what is the probability of winning?

4-28 Suppose that 40% of the voters in a city favor a certain candidate for mayor. If a group of 75 voters chosen at random from the population are polled, what is the probability that the candidate will receive a majority?

4-29 A parachute which will break apart when used is worthless. Suppose a manufacturer can guarantee that 99% of its parachutes will work. If 10,000 parachutes are tested, what is the probability that at least 9,990 will work?

4-30 The scores on a certain college entrance examination are known to be normally distributed with a mean of 75 and a standard deviation of 10.

   a.   If anyone obtaining a score of at least 60 is passed, then what is the probability that a person chosen at random will pass?

   b.   If the board of examiners wants to pass 70% of those taking the test, what should the lowest passing score be?

*4-31 Transmitting by Morse code consists of sending a sequence of electronic signals of two types: short (represented by ·) and long (—). These signals are received by another station, pieced together, and translated into a message. Suppose the receiver correctly interprets a signal with probability .95.

   a.   What is the probability that in the transmission of 900 signals there are fewer than 100 errors?

   b.   Suppose the system is modified so that each signal is transmitted three times. Then the receiver interprets any of the sequences — — — or — — · or — · — or · — — as —. It interprets any of the signals · · · or · · — or · — · or — · · as ·. Under the modified system what is the probability of incorrectly interpreting the signal sent? What is the probability that the number of errors in 900 signals transmitted is fewer than 100?

*4-32 An airplane manufacturer needs washers which have a thickness of between .18 and .22 inch; any other thickness is unusable. One machine shop will sell washers for $1.00 per 1,000 washers; the thickness of their washers is normally distributed with a mean of .2 inch and a standard deviation of .1 inch. Another machine shop will sell washers for $.90 per 1,000; the thickness of their washers is normally distributed with a mean of .2 inch and a standard deviation of .11 inch. Which shop offers a better deal, using the price per usable washer as a criterion?

*4-33 For a certain brand of peanuts 2/3 of the bags contain at least 110 peanuts and 1/4 contain less than 100. Assuming the number of peanuts per bag is normally distributed, what is the mean number per bag and the standard deviation?

*4-34 As a promotional feature, suppose the Army offers a custom-length bed to each enlistee, giving him a choice of three different bed lengths. Assume that the enlistee will choose the bed which is closest in length to his height. The army has chosen the bed lengths so that equal proportions of each length will be used. What are the three lengths, assuming the heights of enlistees are normally distributed with a mean of 70 inches and a standard deviation of 2 inches?

True-false test

Circle the correct letter.

T F   1. Coin-tossing experiments can be described by the binomial distribution.

T F   2. If you toss a coin 20 times, the probability of obtaining 10 heads is 1/2.

T F   3. The parameters n and p completely determine a particular binomial distribution.

T F   4. Data arising from physical measurements are often thought to be normally distributed.

T F   5. If the sample size is large, the central limit theorem asserts that $\bar{X}$ is approximately normally distributed.

T F   6. The normal approximation to the binomial distribution allows you to make quick calculations of binomial probabilities.

T F   7. Measuring the weights of people who have gone on a certain diet may be considered a binomial experiment.

T F   8. The normal distribution cannot be used in experiments in abnormal psychology.

T F   9. In sample data from a population that is normally distributed with $\mu = 10$, every value is larger than the values of data from a population that is normally distributed with $\mu = 6$.

T F   10. If you know μ and σ for a particular normal distribution, then you can compute probabilities from the standard normal table.

Estimation 5

This chapter introduces statistical inference. You will play the role of statistical detective and learn how to make inferences about unknown quantities on the basis of limited information. Maybe you wish to know the average amount of time Americans spend watching television each day, or whether a civil service exam discriminates against some ethnic group. In such situations the quantities you are looking for cannot be found exactly because it's too time-consuming, too expensive, or just plain impossible to obtain them. The difficulty is that the unknown quantities reflect some numerical aspect of a whole population and the population is too large for a survey of every member. To bypass this difficulty, a random sample is chosen from the population, and the sample is then used as a representative model of the entire population. Of course, such a procedure has some chance of leading to false conclusions. If you're unlucky, the sample chosen may not be representative. Statistical inference gives us a quantitative method for describing these errors and techniques for minimizing them.

The vocabulary necessary for our discussion was given in Chap. 2. Recall that the term *population* refers to the collection of all items under study. A population may be adults in America, students attending a certain university, or hospital patients, but it doesn't necessarily have to be people. A population might consist of hot dogs produced by a certain machine, automobiles in some city, or lightbulbs sold by a certain hardware store. When a collection of data contains information about an entire population, with one number for each member, then it is called *population data*. Remember that when the mean is computed from population data, it is called the *population mean*; the median is called the *population median*; etc. Such population quantities are called *parameters*; they summarize some quantitative aspect of the population. When a random sample is selected from a population, the data associated with this subgroup are called *sample data*. In these terms, statistics is the science of making inferences about unknown parameters on the basis of sample data.

Statistics is a double-edged tool. It gives techniques for estimating the value of unknown parameters and tells how to decide between two possible descriptions of a population on the basis of sample data. For example, suppose the police department of a city wants to know the percentage of unpaid parking tickets but there are too many tickets to survey everyone. The police could estimate the unknown proportion using data obtained from a sample of tickets selected at random. This is a problem of estimation. But determining whether the incidence of cancer is higher among smokers than among nonsmokers is a problem of deciding between two descriptions of a population. We must choose one of the two descriptions "Smokers have a higher incidence of cancer than nonsmokers" or "Smokers do not have a higher incidence of cancer than nonsmokers." Often the decision of whether to accept one description of a population rather than another is based on a sample estimate of some unknown parameter. For this reason, our study of statistical inference begins with techniques for the estimation of parameters. Problems of decision making are discussed in later chapters.

Methods for estimating parameters

We shall consider methods for estimating three population parameters: the *population mean*, the *population proportion*, and the *population standard deviation*. We take a random sample from the population and compute the desired quantity from these sample data as if they were population data. This is summarized in Fig. 5-1.

Unknown population parameter	Quantity computed from sample to estimate population parameter
μ = population mean	$\bar{X}$ = sample mean
p = population proportion	$\hat{p}$ = sample proportion
σ = population standard deviation	s = sample standard deviation

Figure 5-1

You should know how to compute each of these quantities from the appropriate data. If you have forgotten, refer to Chap. 2.

Example 5-1. To find the average height of adults in America μ we would have to measure every adult, add the heights, and divide by the number of adults. This is much too hard, and so we settle for an estimate of this population mean by taking the average height $\bar{X}$ computed from a random sample of adults.

Example 5-2. To estimate the population proportion of New Yorkers who prefer gold to silver p, we take a random sample and compute the proportion in the sample who prefer gold $\hat{p}$.

Example 5-3. Suppose a psychologist wants to know the standard deviation σ of scores on a certain reading test for tenth graders in the United States. This quantity is estimated on the basis of a random sample, using the sample standard deviation s as the estimate.

The law of large numbers

It is often surprising to people that parameters associated with a large population can be accurately estimated on the basis of only a relatively small sample. In public opinion polls, for example, estimates of the percentage of voters favoring a certain presidential candidate or position are usually based on a sample of only about 1,000. Assuming there are 50 million voters, a sample of size 1,000 constitutes only 1/500 of 1% of the population. Yet, these estimates are usually quite accurate (when the sample is randomly selected).

To help understand why an estimate of some quantity associated with a large population based on only a relatively small sample can be accurate, consider the following experiment. A sheet of paper is torn into four equal pieces. The number 1 is written on two pieces, the number 2 on one piece, and the number 5 on the last piece. The four pieces are put in a hat and mixed well. One piece of paper is drawn out at random and the number on it is recorded; the paper is then replaced, and another draw is made. Suppose this procedure is repeated 60 times and the number drawn is recorded each time. We could obtain the data in Fig. 5-2.

	2	5	1	1	1	5
	5	1	1	1	2	2
	2	5	1	1	2	1
	2	5	1	2	2	5
	2	5	2	1	2	1
	5	1	1	5	1	5
	1	5	1	1	1	1
	5	1	1	5	2	2
	2	1	2	1	2	1
	5	1	1	1	1	1
Cumulative	31	61	73	92	108	132
totals						

Figure 5-2

The numbers in the first column were obtained on the first 10 draws; the numbers in the second column were obtained on the eleventh through twentieth draws; etc. Taking the averages of the first 10, 20, 30, 40, 50, and 60 draws, we would obtain the following:

Average based on

$$10 : \frac{2 + 5 + 2 + 2 + 2 + 5 + 1 + 5 + 2 + 5}{10} = \frac{31}{10} = 3.1$$

$$20 : \frac{61}{20} = 3.05$$

$$30 : \frac{73}{30} = 2.43$$

$$40 : \frac{92}{40} = 2.30$$

$$50 : \frac{108}{50} = 2.16$$

$$60 : \frac{132}{60} = 2.20$$

Plotting these averages on a graph, we would obtain the results in Fig. 5-3. Notice how the averages tend to stabilize around 2 1/4 for 30, 40, 50, and 60 draws. If the drawing were continued, the averages would get closer and closer to 2 1/4. (If you do not believe it, try it!)

What is the significance of 2 1/4? This number is just the average of the four numbers in the hat $(1 + 1 + 2 + 5)/4 = 2$ 1/4. It is a *population mean,* the population being the numbers in the hat. Why should the average of the numbers drawn randomly with replacement, as in the 60 draws, be close to the average of the four numbers in the hat? We reason this way: Since there are 2 pieces of paper with a 1 out of 4 pieces altogether, we have 2 chances out of 4 of drawing a 1 on any draw. Similarly, the chances of drawing a 2 or 5 are each 1 out of 4. So, in a large number of draws we would obtain a 1 about 1/2 the time, a 2 about 1/4 of the time, and a 5 about 1/4 of the time. This means that in 60 draws, about 30 of them would be 1, 15 would be 2, and 15 would be 5. Hence, the sum of numbers drawn would be about $30(1) + 15(2) + 15(5) = 135$, and the average would be roughly $135/60 = 2$ 1/4.

Now, pretend our hat has 2 million pieces of paper with a 1, 1

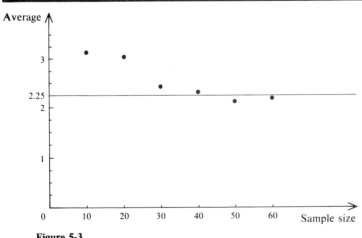

Figure 5-3

million pieces with a 2, and 1 million pieces with a 5. In other words, the proportions of 1s, 2s, and 5s are the same, but we have 1 million times as many pieces. (Notice that the population mean is the same as before, 2 1/4.) If we made 60 draws with replacement, we would still expect to draw a 1 about 1/2 of the time, a 2 about 1/4 of the time and a 5 about 1/4 of the time. So, by the reasoning just given, our average would be near 2 1/4 even though we made only 60 draws from 4 million pieces of paper instead of 4 pieces. This is an example of a very general law governing random phenomena, called the *law of averages*, or *law of large numbers*. Simply stated, it asserts the following:

> *Suppose a random experiment is repeated independently n times. When n is large, the sample mean $\bar{X}$ will be close to the population mean μ.*

Of course, the words "large" and "close" are vague, but it is surprising how small n can be and still have $\bar{X}$ extremely close to μ.

The law of large numbers justifies using the sample mean to estimate a population mean. For similar reasons, it also justifies using a sample proportion to estimate a population proportion, or a sample standard deviation to estimate a population standard deviation.

Accuracy

So far, methods for estimating three parameters have been proposed. These estimates will be good, as the law of large numbers asserts, if the random sample is large enough. However, two questions come to mind: First of all, exactly how large should the sample be? Second, how accurate is the estimate? These two questions are interrelated: The accuracy depends on the sample size.

At this point, a way to describe the accuracy of an estimate is needed. A widely used technique for specifying the accuracy of an estimate is the *confidence interval*. We shall discuss confidence intervals for the population mean and population proportion. To begin this discussion, consider the following example.

Suppose a judicial review board is making a study of the amount of time defendants who do not post bail spend in jail before their case is tried. An important quantity to know would be the average time in jail before trial. Call this number μ; it is a population mean, the population being all people who do not post bail. To estimate μ, the review board could select a random sample of 100 cases and then use the sample mean $\bar{X}$ as an estimate. Remember $\bar{X}$ is a random quantity; it depends on which 100 cases are selected for the sample. For some samples of 100 cases, $\bar{X}$ may be greater than μ; for others, it may be less than μ. In any case, some error is expected when using $\bar{X}$ to estimate μ. The error may be represented in symbols by $\bar{X} - \mu$, the difference between the estimate and the quantity being estimated.

Ideally, what's needed is a statement like "The error will be less than x days in all possible samples of 100 cases." The number x would then be the largest possible error, and therefore $\bar{X} - x < \mu < \bar{X} + x$ would give exact bounds on μ. (See Fig. 5.4.) Unfortunately, such a statement is seldom possible. Because of the random behavior of $\bar{X}$, no useful statements about error can be made with certainty. Therefore, as a compromise, a statistician will specify a bound on the error which will hold most of the time, but not always. A typical statement would be "The error will be less than x days in 95% of the possible samples of 100 cases." In other words, the error $\bar{X} - \mu$ will be less than x with probability .95 since the value of $\bar{X}$ is determined from a sample chosen at random. In symbols, this assertion is stated by the equation

$$P(-x < \bar{X} - \mu < x) = .95$$

We now explain how the number x is found. After our derivation, we shall provide an easy-to-use formula.

In order to determine the number x in a particular case we need to know the probability distribution of $\bar{X}$. According to the central limit theorem for sample means discussed in Chap. 4, if the sample size n is large enough, then the random quantity $\bar{X}$ will have approximately a normal distribution with a mean μ and a standard deviation $\sigma/\sqrt{n}$, where μ is the population mean and σ is the population standard deviation. Thus, for large samples, we can use the normal tables in the Appendix to compute probabilities for $\bar{X}$. From the tables, we see that a normal variable will lie within 1.96 standard deviations of the mean with probability .95; equivalently, $\bar{X}$ will be within $1.96\sigma/\sqrt{n}$ units of μ with probability .95. Suppose σ is known to be .2 days. Referring, then, to the judicial review board example, we find

$$1.96 \frac{\sigma}{\sqrt{n}} = \frac{1.96(.2)}{\sqrt{100}} = .04$$

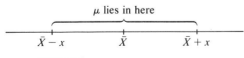

μ lies in here

$\bar{X} - x$ $\bar{X}$ $\bar{X} + x$

Figure 5-4

and hence

$$P(\bar{X} - .04 \leq \mu \leq \bar{X} + .04) = .95$$

The interval $(\bar{X} - .04, \bar{X} + .04)$ is called the *95% confidence interval* since the true value of μ has a probability of .95 of being in this interval. As you can see, this interval is random; that is, it depends on $\bar{X}$, a random quantity. If $\bar{X}$ turns out to be 10 days, then the 95% confidence interval for μ would be (9.96, 10.04); if $\bar{X}$ is 11.2 days, the 95% confidence interval would be (11.16, 11.24). The number .04 should be thought of as the *maximum error,* or a magnitude of error which *probably* will not be exceeded. The meaning of *probably* is specified by .95, the probability that a random error incurred while using $\bar{X}$ to estimate μ will not exceed .04. Such a probability is referred to as a *confidence coefficient.*

If a 90% confidence interval is desired, then the number y satisfying

$$P(-y \leq \bar{X} - \mu \leq y) = .9$$

must be determined. The number y is found in the same way as x, except .9 takes the place of .95. Since a normal variable will be within 1.65 standard deviations of the mean with probability .9, $\bar{X}$ and μ will be less than $1.65\sigma/\sqrt{n} = 1.65(.02) = .033$ units apart with probability .9. Thus, the interval $(\bar{X} - .033, \bar{X} + .033)$ is the 90% confidence interval for μ.

Notice that in order to obtain the 95% or 90% confidence intervals for μ, we needed to know the value of σ, the population standard deviation. Unfortunately, the value of σ is known only in rare situations. However, when the sample size n is greater than 30, the sample standard deviation s is usually an accurate estimate of σ. Thus, in most situations we are likely to encounter, s may safely be used in place of σ.

Large-sample confidence intervals for means

We can now present a general formula for computing a confidence interval for an unknown population mean μ, based on the sample mean $\bar{X}$. Assume a random sample of size n is drawn from a population with mean μ. Further, assume n is large enough so that the distribution of $\bar{X}$ will be approximately normal and so that the population standard deviation σ can be accurately estimated by the sample standard deviation s. As we have mentioned, $n > 30$ will usually suffice. The most commonly used confidence coefficients are .90, .95, and .99. The corresponding confidence intervals are computed according to the formulas below:

The 90% confidence interval for μ is

$$\bar{X} \pm 1.65 \, \frac{s}{\sqrt{n}} \qquad\qquad (5.2a)$$

The 95% confidence interval for μ is

$$\bar{X} \pm 1.96 \, \frac{s}{\sqrt{n}} \qquad\qquad (5.2b)$$

The 99% confidence interval for μ is

$$\bar{X} \pm 2.58 \, \frac{s}{\sqrt{n}} \qquad\qquad (5.2c)$$

The following is a program for computing a confidence interval for μ. Suppose a 95% confidence interval is desired:

1 Determine the value n = sample size.

2 Find the sample mean $\bar{X}$ and the sample standard deviation s.

3 Plug these values into the formula $\bar{X} \pm 1.96s/\sqrt{n}$.

Example 5-4. Assume $\bar{X} = 5$, $n = 100$, and $s = 3$. Find the 90% confidence interval for μ.

SOLUTION:

$$1.65\frac{s}{\sqrt{n}} = 1.65\left(\frac{3}{\sqrt{100}}\right) = .495$$

and so the 90% confidence interval for the mean is (4.505,5.495), or, equivalently, $5 \pm .495$. We are 90% confident that μ is somewhere between 4.505 and 5.495.

Example 5-5. Assume $\bar{X} = 11.2$, $n = 10,000$, and $s = 5$. Find the 99% confidence interval for the mean.

SOLUTION:

$$2.58\frac{s}{\sqrt{n}} = 2.58\left(\frac{5}{\sqrt{10,000}}\right) = .13$$

and so the 99% confidence interval for the mean is $11.20 \pm .13$.

In other words, we are 99% confident that μ is somewhere between 11.07 and 11.33.

Example 5-6. A fire department wants to know the average amount of time needed to put out fires. Over a 1-year period they choose 75 fires at random and observe the amount of time required to put out these fires. Suppose the average time needed to put out these 75 fires is $\bar{X} = 1.5$ hours, with a standard deviation of .60 hour. What is the 90% confidence interval for the mean time μ?

SOLUTION: $n = 75$, $s = .60$, and $\bar{X} = 1.5$; and so

$$1.65 \frac{s}{\sqrt{n}} = 1.65\left(\frac{.6}{\sqrt{75}}\right) = .11$$

Thus, the 90% confidence interval for the mean is $1.5 \pm .11$ hours. The fire department is 90% confident that μ is between 1.39 and 1.61 hours.

Example 5-7. Suppose the two countries, Academiland and Studin, are engaged in a fierce war. The leaders of Academiland want to know the average IQ for the population of Studinites in order to determine a strategy according to the intelligence of their opponents. The average IQ for 50 Studinites chosen at random is $\bar{X} = 112$, with a standard deviation of 10. What is the 95% confidence interval for μ, the average IQ for all Studinites?

SOLUTION: $n = 50$, $s = 10$, and $\bar{X} = 112$. Therefore,

$$1.96 \frac{s}{\sqrt{n}} = 1.96\left(\frac{10}{\sqrt{50}}\right) = 2.77$$

Thus, the 95% confidence interval for μ is 112 ± 2.77. The Academilanders are 95% confident that μ is between 109.23 and 114.77.

Problems

5-1 A sample of 100 boxes of candy are taken from a large shipment. Upon weighing them, the sample average is found to be $\bar{X} = .78$ pound with a standard deviation of $s = .01$. What is the 95% confidence interval for the mean weight of the shipment?

5-2 For a sample of 2,500 automobile insurance claims paid out during a 1-year period, it is found that the average is $400 per claim with a standard deviation of $100. What are the 90 and 99% confidence interval· for the mean claim payment?

5-3 In 100 horse races the average distance between the winner and runner-up horses is found to be 8.2 feet with a standard deviation of 2 feet.

   a. Find the 90 and 99% confidence intervals for the mean distance between the winner and runner-up horses.

   b. Compare the widths of these two intervals.

5-4 A machine which fills cans with tomato sauce is checked by weighing a sample of 100 cans out of every 2,000. The lot of 2,000 is rejected if the sample mean is less than 10 ounces and the 90% confidence interval does not include 10 ounces. If a sample has an average weight of 9.7 ounces with a standard deviation of .2 ounces, would the lot be rejected?

Large-sample confidence intervals for proportions

Sometimes a population is divided into two categories, for example, Democrat and Republican, male and female, or success and failure. In these cases it is often important to know the proportion of the population in a certain category. This number is called a *population proportion* and is denoted by p; it is estimated by the corresponding sample proportion $\hat{p}$. The central limit theorem for sample proportions tells us that if the sample size n is large, the random quantity $\hat{p}$ will have approximately a normal distribution with mean p and standard deviation $\sqrt{p(1-p)/n}$. Using this theorem, we may derive approximate confidence intervals for p for commonly used confidence coefficients as follows:

The 90% confidence interval for p is

$$\hat{p} \pm 1.65 \sqrt{\frac{\hat{p}(1-\hat{p})}{n}} \tag{5.3a}$$

The 95% confidence interval for p is

$$\hat{p} \pm 1.96 \sqrt{\frac{\hat{p}(1-\hat{p})}{n}} \tag{5.3b}$$

The 99% confidence interval for p is

$$\hat{p} \pm 2.58\sqrt{\frac{\hat{p}(1-\hat{p})}{n}} \qquad\qquad (5.3c)$$

Statisticians generally agree that these confidence intervals are accurate enough for most purposes if $np(1-p) > 5$. Since p is unknown, this condition may be checked by observing whether $n\hat{p}(1-\hat{p}) > 5$.

Example 5-8. An anthropologist found that in a sample of 100 people chosen at random from a certain remote region, 65% were males. Find the 90% confidence interval for the population proportion p of males in the region.

SOLUTION: We may use the appropriate formula in (5.3) with $n = 100$ and $\hat{p} = .65$ since $n\hat{p}(1-\hat{p}) = 100(.65)(.35) = 22.75 > 5$. Using the formula, we obtain

$$1.65\sqrt{\frac{\hat{p}(1-\hat{p})}{n}} = 1.65\sqrt{\frac{.65(.35)}{100}} = .08$$

Thus, the 90% confidence interval for the proportion of males is given by $.65 \pm .08$. We are 90% confident that the proportion of males is between .57 and .73.

People who haven't studied statistics often criticize the results of preelection polls by saying something like, "The sample size is only a fraction of the total population of voters, and so how can the estimate be accurate?" After reading this chapter, you should reply, "As long as the sample is randomly selected from the population, only the sample size will affect the accuracy of the estimate. The ratio of sample size to population size is irrelevant." To illustrate this idea, consider the following example.

Example 5-9. A random sample of 2,500 eligible voters was taken, and the proportion of those voting for Mr. Smiley was 44%. Find the 95% confidence interval for the population proportion p.

SOLUTION: Use the appropriate formula in (5-3), $\hat{p} = .44$, $n = 2,500$, and

$$1.96\sqrt{\frac{\hat{p}(1-\hat{p})}{n}} = 1.96\sqrt{\frac{.44(.56)}{2,500}} = .02$$

Hence, the 95% confidence interval for p is $.44 \pm .02$. In other words, we are 95% confident that p is between .42 and .46. But a sample size of 2,500 is only a minute fraction of the total population of eligible voters. This number is about 50 million and so the sample size is only .005% of the total population since

$$\frac{2,500}{50,000,000} = \frac{1}{20,000} = .00005$$

Formulas (5-3) can be used to obtain confidence intervals for probabilities also. Just treat them like proportions.

Example 5-10. In the moon shots there are several possible errors. Work is done perfecting the rocket until the possibility of error is reduced to an "acceptable" level. To estimate the true probability p of success (i.e., no errors) simulations of the takeoff and trip in space are made. Assume each simulation has p as its probability of success, where p is unknown. Suppose 10,000 simulations are made and the proportion of success in these is .99. What is the 95% confidence interval for p?

SOLUTION: We apply the appropriate formula in (5-3), with $\hat{p} = .99$ and $n = 10,000$:

$$1.96\sqrt{\frac{\hat{p}(1-\hat{p})}{n}} = 1.96\sqrt{\frac{.99(.01)}{10,000}} = .002$$

Thus, the 95% confidence interval for the probability p of a successful flight is $.99 \pm .002$. We are 95% certain that the probability of a successful flight is between .988 and .992.

Problems

5-5 In a certain city, a sample of 200 houses is taken to estimate the proportion of houses with faulty wiring. The sample data yielded $\hat{p} = .1$. Find the 95% confidence level for p, the true proportion of houses in this city with faulty wiring.

5-6 A department of transportation wishes to estimate the proportion of commuters in a certain city who drive to work with at least one other person. A sample of 500 workers yielded the sample proportion $\hat{p} = .4$.

   a. Find the 90% confidence interval for p.

   b. If you were conducting this study, how would you take the sample?

5-7   A study is done to estimate the proportion of prisoners p in a certain penitentiary who receive visitors each week. A sample of 100 prisoners yielded $\hat{p} = .2$. Find the 95% confidence interval for p.

Controlling the accuracy of estimates; computing the sample size

In many situations it is necessary that an estimate possess at least a certain minimum level of accuracy. The accuracy desired is specified by giving the maximum error in the width of the confidence interval to be obtained and the confidence coefficient. Ideally, narrow confidence intervals with a large confidence coefficient are best. From formulas (5.2) and (5.3) we see that the width of the confidence interval can be decreased by increasing n, and so for given accuracy the problem is to find the necessary sample size n.

For example, suppose the doctor is interested in the average time μ it takes for a certain dosage of penicillin to cure ear infections. The doctor might take a random sample of patients, administer the dosage, and measure the cure time, letting $\bar{X}$ be the sample mean. If a 95% confidence interval for μ of $\bar{X} \pm .5$ days is desired, then according to the appropriate formula in (5-2), the quantity $1.96 \, s/\sqrt{n}$ should be .5. If a value of $s = 5$ days has been obtained, say, on the basis of a preliminary study, then the sample size n can be determined by solving the equation

$$1.96 \, \frac{s}{\sqrt{n}} = .5$$

Squaring both sides of the equation gives $3.85(25)/n = .25$ and $n = 385$. If the doctor wants the estimate to have an error margin of no more than .5 days, a sample of size at least 385 should be taken. If a different confidence coefficient is desired, the computations with the appropriate number found in formula (5-2) should be repeated. Notice that solving for n yielded $n = 384.16$. We rounded off to 385 because 384.16 is the minimum value for n which gives the required accuracy.

In general, if a confidence interval $\bar{X} \pm e$ is desired, with error margin e given, then the necessary sample size is determined from the following formulas for the common confidence coefficients .9, .95, and .99:

For a 90% confidence interval for μ, take

$n = (1.65s/e)^2$

For a 95% confidence interval for μ, take

$n = (1.96s/e)^2$

For a 99% confidence interval for μ, take

$n = (2.58s/e)^2$

Unfortunately, the sample standard deviation s will not be known before the sample is taken. If the population standard deviation σ is known, then it should be used in place of s. If σ is unknown, it is common practice to do a pilot study first to obtain a value for s and then compute the necessary sample size n to be used in the actual study.

To obtain a confidence interval for a population proportion p with specified accuracy, say, $\hat{p} \pm e$, then use the appropriate formula below:

For a 90% confidence interval for p, take

$n = (1.65/2e)^2$

For a 95% confidence interval for p, take

$n = (1.96/2e)^2$

For a 99% confidence interval for p, take

$n = (2.58/2e)^2$

These formulas are derived from formulas (5-3) by using the fact that $\hat{p}(1 - \hat{p}) \leq 1/4$, no matter what the value of $\hat{p}$ *is*.

Example 5-11. Suppose the standard deviation of hot dogs produced by a certain machine is known to be .1 inch. To obtain an estimate of the average length of hot dogs which is accurate to within .01 inch with probability .95, how many observations should be taken?

SOLUTION: We use the sample average $\bar{X}$ to estimate μ, the average length of all hot dogs produced by the machine. The accuracy required is $e = .01$ inch with confidence coefficient .95. Using the given value of $\sigma = .1$ inch in place of s in the formula, we obtain the necessary sample size:

$$n = \left[\frac{(1.96)(.1)}{.01}\right]^2 = 385$$

Example 5-12. Suppose that in order to plan your gambling strategy properly, you need to know the probability of heads for a certain coin to within .001 with confidence coefficient .99. How many times would you have to toss the coin to obtain this level of accuracy for your estimate $\hat{p} =$ no. heads/no. tosses?

SOLUTION: The accuracy required is $e = .001$. Hence,

$$n = \left[\frac{2.58}{2(.001)} \right]^2 = 1,664,100$$

This example shows that if you want lots of accuracy, you can get it provided you are willing to take a large sample. If you made 1 toss per second, it would take about 19 days and 7 hours to make the required number of tosses.

Summary

In many research problems it is important to know the value of some parameter. We have considered techniques for estimating three parameters; the population mean, population standard deviation, and population proportion. These are estimated by selecting a random sample and then using the sample mean, sample standard deviation, and sample proportion as estimates, respectively. In symbols, $\bar{X}$ estimates μ; s estimates σ; and $\hat{p}$ estimates p.

The accuracy of an estimate is described by a confidence interval, which consists of giving the estimate of the unknown parameter and a bound on the error, together with the degree of confidence you have that the error will not exceed this bound. Sometimes, before a sample is taken, it is desirable to obtain an estimate with specified accuracy. This may be achieved by taking a sample that is large enough.

Important terms

Confidence coefficient

Confidence intervals

Law of large numbers

Parameters

Population

Population data

Random sample

Sample data

Width of a confidence interval

Additional problems

5-8 In the firing of ceramics, pyrometric cones are used to determine when the temperature in the kiln reaches the desired level. At this temperature the cone will melt. A random sample of 100 cones are chosen from a large lot of cones which are supposed to melt at 2300°F. The average melting temperature of the sample is found to be 2290°F with a standard deviation of 8°. Find the 99% confidence interval for the mean melting point of the lot.

5-9 To determine the mean amount of time people spend watching TV each day, a preliminary study finds that the standard deviation of this time is 4 hour. Assuming this to be the correct value of the standard deviation, how large a sample must be taken in order that the 95% confidence interval for the mean will be no wider than 1 hour?

5-10 A random sample of 64 lightbulbs made by a certain electrical firm is tested to determine the length of time before the bulbs burn out. It is found that the average lifespan is 44 hours with a standard deviation of 3 hours. Find the 95% confidence interval for the mean life of the bulbs.

5-11 On a certain test it is known that the mean score for men is 80. A sample of 50 women are given the test, and their mean is 77 with a standard deviation of 5. Is this strong evidence for the conclusion that men score higher?

5-12 Suppose that for a radar system the error in pinpointing the longitude of the position of an airplane several miles away has standard deviation 1/2 mile.

   a. If there are several radar sights focusing on an aircraft, how could you estimate the longitude?

   b. How many observations are needed so that the width of the 95% confidence interval for this estimate is less than .1 mile?

5-13 A dairy claims to sell butter in packages weighing 1 pound. In a random sample of 36 packages the average is found to be .95 pound with a standard deviation of .1 pound. Are these results consistent with the dairy's claim?

5-14 A sociologist wants to know the mean score on a social involve-
ment test. Suppose the test is administered to 30 people and the
following scores are observed:

3 2 5 6 2 8 6 5 9 6

8 7 3 2 5 6 6 7 5 6

8 9 8 1 1 8 7 5 5 7

a. What should be the estimate of the mean score?

b. Find the 90 and 95% confidence intervals for the mean.

5-15 The following weights were recorded for 36 newborn babies:

5.2	6.3	6.1	7.6	7.2	6.5
6.8	10.2	7.3	5.1	7.1	6.4
6.5	7.2	9.4	9.1	8.9	9.6
7.5	7.2	6.8	6.9	7.2	8.0
8.2	7.1	7.9	6.5	5.2	9.6
7.6	4.9	5.1	6.5	7.5	7.6

a. Find the 90 and 99% confidence intervals for the mean.

b. Find the 90% confidence interval for the probability p that a
baby chosen at random will weigh at least 6 pounds.

5-16 When double-parking, a person received traffic tickets 15 out of
120 times.   What is the 95% confidence interval for the
probability of receiving a ticket when double-parked?

5-17 A resident of a large apartment building observed that in 90 out of
200 times the elevator was on the ground floor when he entered
the building.   What is the 90% confidence interval for the
probability of finding the elevator on the ground floor?

5-18 Two brands of aspirin, brands A and B, are compared to determine which one is more effective.   In a study of 50 aspirin
consumers, it is found that 53% prefer brand A.

a. Is it safe to conclude that brand A is better?

b. How large a sample should be taken so that the sample
proportion has probability .95 of being within .02 of the
population proportion?

5-19 In estimating the proportion of defective items in a large lot of
transistors, how many transistors must be sampled so that the
width of the 95% confidence interval for the proportion of defectives in the lot is .02?

*5-20 A certain life insurance company can make money on a policy only if the holder dies after age 55. Suppose that in a sample of 200 past policies the mean lifespan is 60.2 with a standard deviation of 11 years.

a. Give an estimate of p, the proportion of profitable policies.

b. Give a 95% confidence interval for p based on the estimate computed in part (a).

*5-21 Assume that during the summer the number of mosquitoes living in a certain area is normally distributed with standard deviation 1,000. Suppose that a malaria epidemic breaks out if more than 10,000 mosquitoes live in the area.

a. If a malaria epidemic breaks out in 18 out of 100 randomly chosen years, what is the 90% confidence interval for the probability of a malaria epidemic?

b. How could you estimate the mean number of mosquitoes living in the area during the summer? Is it possible to give a confidence interval for the mean?

*5-22 Over the past 10 years there have been 2,500 earth tremors on the San Andreas fault. Of these, 18% registered at least 5 on the Richter scale and 60% registered less than 4. Assume these readings follow a normal distribution.

a. Estimate the mean and standard deviation of the distribution of the Richter-scale readings for tremors on the San Andreas fault.

b. Find a 95% confidence interval for the mean reading.

c. Find a 95% confidence interval for the proportion of readings over 6.

True-false test

Circle the correct letter.

T F   1. Sample estimates without a margin of error and a given degree of confidence give no information as to their reliability.

T F   2. As the degree of confidence increases so does the width of the corresponding confidence interval.

T F   3. If we use a 95% confidence interval to estimate a certain population mean μ, and if we use this procedure over and over again, our interval will include μ approximately 95% of the time.

T F   4. The larger we make the sample size, the more accurate our estimate becomes.

T F   5. For a fixed interval width, we can obtain a confidence coefficient as high as we like, providing we take a large enough sample size.

T F   6. When better methods of obtaining point estimates are devised, confidence intervals will become obsolete.

T F   7. If you select a point at random from a 95% confidence interval, then it will be the parameter you are trying to estimate with probability .95.

T F   8. When tossing a certain coin, a 100% confidence interval for the probability of heads is the interval from 0 to 1.

T F   9. The central limit theorem enables us to compute confidence intervals regardless of the underlying distribution of the population data.

T F   10. If the confidence interval is too wide, then we may suspect that our sample is not randomly selected.

Introduction to hypothesis testing 6

In Chap. 5 we studied a basic problem of statistical inference, that of estimating an unknown parameter. Using a confidence interval, we were able to estimate the location of an unknown mean or proportion to within a certain small interval and with a prescribed degree of reliability or confidence. Now, we shall consider a different method of attacking the same sort of problem, the method of testing a statistical hypothesis.

Your role is still that of a statistical detective. Through the results of a random sample, you are presented with incomplete information about a population under study and you wish to make an inference concerning the true value of an unknown parameter. This time, however, instead of trying to estimate the parameter, you want to test the validity of a claim or hypothesis made about the parameter. This is not to imply that you will be doing something radically different. The procedures we shall study now are closely related to the methods presented in Chap. 5. Consider, then, the following example.

Example 6-1. A pharmaceutical company claims to have developed a pill which has some effect in preventing blistering due to poison ivy. A consumers' group wishes to test this claim and takes a random sample of 100 poison ivy sufferers in order to test the pill. After a reasonable period the sample members are examined to determine whether they have developed significant blisters. Suppose it is known that approximately 50% of poison ivy sufferers develop significant blisters when no treatment is given. How should the consumers' group use the evidence provided by the results of the sample to make a decision?

SOLUTION: We may restate the problem in a more precise way by letting p denote the probability that a poison ivy sufferer who takes the pill will develop blisters. The quantity p is unknown. If the pill is ineffective, then $p = .5$; on the other hand, if the pill is effective, then $p < .5$. In these terms we are interested in reaching a decision as to whether $p = .5$ or $p < .5$. These two possibilities represent *hypotheses* about the pill's effectiveness, which we label H_0, standing for the *null hypothesis*, and H_a, standing for the *alternative hypothesis*. In these terms we wish to choose between the null hypothesis H_0: $p = .5$ and the alternative hypothesis H_a: $p < .5$.

As is customary, we let the null hypothesis be the hypothesis which asserts that "there is no difference," or, more generally, that "the treatment is ineffective or no better than the old one." Accordingly, the alternative hypothesis is the hypothesis which asserts that "there is a difference," or that "the treatment is effective."

The decision as to which is the correct hypothesis to choose must be based on the number of sample members who develop blisters since this number measures the effectiveness of the pill. We denote this number by S. If the pill is ineffective, we expect the proportion of sample members who develop blisters to be about .5; in this case, we expect S to be about $.5(100) = 50$. However, if the pill is effective, we expect the proportion of sample members who develop blisters to be less than .5; accordingly, S will probably be less than 50. Thus, a "small" value of S would be strong evidence in favor of the pill's effectiveness, and a "large" value of S would indicate that the pill is ineffective. These considerations imply that our procedure for choosing between H_0 and H_a should have the following general form:

Reject H_0 if S is small.

Accept H_0 if S is large.

In order to determine which values of S are small and which are large, we must ask ourselves how much we want to limit the chance of rejecting H_0 when, in fact, H_0 is true, an error we wish to avoid. Consider one possible decision procedure:

Reject H_0 if $S \leq 40$.

Accept H_0 if $S > 40$.

In other words, using this decision procedure, we would say that the pill is effective and $p < .5$ if we observe that 40 people or fewer develop blisters.

If the pill is ineffective and H_0 is therefore true, then S is a binomial variable with success probability $p = .5$ and $n = 100$. Using the above decision procedure, we say the pill is effective whenever we observe $S \leq 40$. Now, assuming that H_0 is true, we use the normal approximation to the binomial distribution to compute

$$P(S \leq 40) = P\left(Z \leq \frac{40 - np}{\sqrt{npq}}\right)$$

$$= P\left(Z \leq \frac{40 - 50}{5}\right)$$

$$= P(Z \leq -2)$$

$$= .0228$$

In other words, when we use this decision procedure, the probability of rejecting the null hypothesis when it is really true is .0228.

We shall base our decision procedures for tests of hypotheses on exactly this type of error probability, which we shall call the *significance level* of the test.

Definition. The significance level of a statistical test for a particular decision procedure is the probability of rejecting H_0 when H_0 is really true. We shall use the Greek letter α (alpha) to denote significance level.

Although determining the significance level of a particular test is a subjective matter, based on the strength of your desire to avoid mistakenly rejecting H_0, there are a few significance levels which have become almost universally used and which we shall use throughout this book. These significance levels are $\alpha = .01$, $\alpha = .05$, and $\alpha = .1$. That these significance levels are the popular ones is a matter of custom and convenience.

Example 6-1 revisited. We return now to Example 6-1 and devise a decision procedure for which $\alpha = .05$; in other words, we wish to determine the value x for which $P(S \leq x) = .05$. If we let our decision procedure be to reject H_0 and say the pill is effective whenever we observe $S \leq x$, we will ensure that the probability of mistakenly rejecting H_0 is only .05. Rather than find this value directly, we shall convert our observations to the Z scale. Looking at the normal table in the Appendix, we see that $P(Z \leq -1.65) = .05$; therefore, the following decision procedure will have a significance level of $\alpha = .05$:

$$Reject\ H_0\ if \frac{S - np}{\sqrt{npq}} \leq -1.65.$$

$$Accept\ H_0\ if \frac{S - np}{\sqrt{npq}} > -1.65. \quad (\text{See Fig. 6-1.})$$

We have $n = 100$; and if H_0 is true, then $p = .5$ and our decision procedure is thus

$$Reject\ H_0\ if \frac{S - 50}{5} \leq -1.65.$$

$$Accept\ H_0\ if \frac{S - 50}{5} > -1.65.$$

Thus if we observe that 41 members of the sample develop blisters, we would compute $Z = (41 - 50)/5 = -1.8$. Since $-1.8 < -1.65$, we would reject H_0, stating that such an observation is significant evidence that the pill is effective at a significance level of $\alpha = .05$. It turns out that 41 is the highest value for which we would make such a conclusion, for if we observed $S = 42$, we would compute $Z = (42 - 50)/5 = -1.6$. Since $-1.6 > -1.65$, we would conclude that an observation of 42 (or more) in the sample who develop blisters is not enough evidence to conclude that the pill is effective at a significance level of $\alpha = .05$. The chance is just too great that this result could be due to the random fluctuation of S rather than to the effectiveness of the pill.

Significance probabilities

Previously we computed that $P(S \leq 40) = .0228$. Also we observed that the value on the Z scale corresponding to $S = 40$ is $Z = -2$ and that since $-2 < -1.65$, such an observation would lead us to reject H_0. Another way to see this is to notice that $.0228 < .05$. In fact, since we are testing at a significance level of $\alpha = .05$, any value x for which $P(S \leq x) < .05$ would fall in the rejection region. Probabilities such as these provide useful information and have a name of their own, *significance probabilities*.

> **Definition.** The significance probability of observing a result x of the test statistic is the probability of observing an outcome as extreme as x [in the present case, $P(S \leq x)$]. The significance probability of a particular observation tells how close the observation is to the rejection region. If the test is conducted at level α, H_0 is rejected if the associated significance probability is $\leq \alpha$.

Continuing with Example 6-1, suppose we observe that 44 members of the sample develop blisters. We know already that such an observation leads us to accept $H_0(Z = -1.2 > -1.65)$. The significance probability of this observation is $P(S \leq 44) = .1151$, which, of course, is larger than .05.

Notice that our test does not detect exactly how effective the pill is, but only whether or not it is effective at all. Along the same lines, a confidence interval for p could be determined using the methods of the

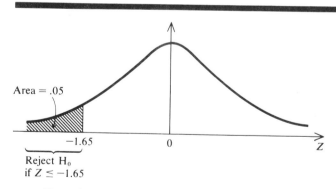

Figure 6-1 $P(Z \leq -1.65) = .05$.

last chapter. Also, we have not mentioned the possible degree of blistering of the poison ivy sufferer. This may be another relevant factor in the study. It is up to you, the experimenter, to decide which questions and hypotheses are the relevant ones for a particular problem and to devise your analyses accordingly.

We shall summarize the results of Example 6-1:

H_0:   Pill is ineffective ($p = .5$).

H_a: Pill is effective ($p < .5$).

Significance level $\alpha = .05$.

Sample size $n = 100$.

We observe S, the number of people in the sample who develop blisters.

Under H_0, S is binomially distributed with $\mu = 100 \times .5 = 50$ and $\sigma = \sqrt{100 \times .5 \times .5} = 5$.

Decision procedure: We reject H_0 and decide that the pill is effective if we observe $(S - 50)/5 \leq -1.65$.

Example 6-2. A claim is made that the proportion of men in Metro City who prefer long sideburns is .5. Pierre Poupée, trendsetting men's hair stylist, disagrees with this claim, thinking that the true proportion is higher. As a test, 400 randomly selected men are asked whether or not they prefer the long sideburns, and 230 say they do. Is this enough evidence at a significance level of $\alpha = .05$ to support Pierre's assertion? At a significance level of $\alpha = .01$?

SOLUTION: We follow the same procedure as in Example 6-1. First, we label the hypotheses to be tested:

H_0: $p = .5$

H_a: $p > .5$

We shall use $S =$ the number of men in the sample who prefer long sideburns as the test statistic. Since large values of S contradict H_0 and support H_a, our decision procedure should be to reject H_0 if S is sufficiently large. Recall that in Example 6-1 we rejected H_0 if S was sufficiently small, which illustrates the fact that the rejection area depends on the alternative hypothesis. Since we are testing at significance level $\alpha = .05$, and since S is a binomial variable with large sample size, again we shall use the central limit theorem and, instead of using S directly, (from the normal tables) compute $Z = (S - np)/\sqrt{np(1 - p)}$, rejecting H_0 if $Z \geq 1.65$. Under H_0,

$$np = 400 \times .5 = 200$$
$$\sqrt{np(1 - p)} = \sqrt{400 \times .5 \times .5}$$
$$= 10$$

and so our rejection region shall be any value of S for which $(S - 200)/10 \geq 1.65$. Since the sample yielded $S = 230$, we compute $Z = (230 - 200)/10 = 3$; and since $3 \geq 1.65$, we reject H_0 in favor of Pierre Poupée's assertion.

To determine the rejection region at a significance level of $\alpha = .01$, again we go to the normal tables and discover that $P(Z \geq 2.33) = .01$ (see Fig. 6-2). Thus, H_0 should be rejected at level $\alpha = .01$ if $Z \geq 2.33$. Since we have already computed $Z = 3$, we also reject H_0 at a significance level of $\alpha = .01$. In fact, we can compute the significance probability of our result: $P(S \geq 230) = P(Z \geq 3) = .0013$. Thus we would reject H_0 for any value of $\alpha \geq .0013$.

Example 6-3. A winemaker has developed a new wine-making process and wishes to determine if wine made with the new method tastes better than the winery's usual wine of this type. He takes a sample of 90 people and pours three glasses of wine for each person,

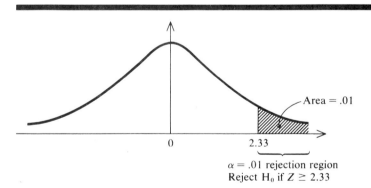

$\alpha = .01$ rejection region
Reject H_0 if $Z \geq 2.33$

Figure 6-2   Standard normal curve: $P(Z \geq 2.33) = .01$.

two of the glasses containing the usual wine, the third containing wine made by the new method. The winemaker asks each person to taste the three wines (they don't know which is which) and to specify one glass of wine which tastes *best*. Even if a person cannot detect any difference in the wines, he must specify one which tastes best. If there is no improvement in the wine by using the new method, the selection of the best glass will be done at random. The winemaker wishes to test at a significance level of $\alpha = .05$ whether or not there is a difference between the new and old methods. Suppose 35 of the 90 tasters select the new wine as the kind which tastes best.   What should the winemaker conclude?

SOLUTION: The winemaker has set up the two hypotheses as follows:

$$H_0: \quad p = \frac{1}{3}$$

$$H_a: \quad p > \frac{1}{3}$$

where $p = P$(taster selects new wine). The possibility that the new wine tastes *worse* than the old is to be considered part of H_0 since the winemaker wishes to reject H_0 only if the new method seems better, in which case he might adopt the new procedure. In this way a test at a significance level of $\alpha = .05$ will ensure the winemaker that the probability of adopting the new method when it is no better than the old one is only .05. The other type of error, in this case the error of not converting to the new method when it really is better, is not explicitly guarded against in the design of the test although the tests we discuss are effective in controlling this type of error. (In fact, you should notice that in all the examples we have presented, no mention has been made of the error of accepting H_0 when it is really false.   Determining the significance level of a particular test does not control this type of error, it controls only the error of rejecting H_0 when it is really true.)

The winemaker wishes to determine the appropriate rejection region. He knows that if there is no difference in the wines, then the tasters would essentially be choosing at random when they decide which glass tastes best. Since only 1 of the 3 glasses contains the new wine, the probability of selecting the new wine would be $p = 1/3$. Thus, if there is no difference and if we let $S =$ the number of times the tasters "successfully" select wine made by the new method, then S is binomially distributed with sample size $n = 90$ and success probability $p = 1/3$. In this case,

$$np = 90 \times 1/3 = 30$$

$$\sqrt{np(1 - p)} = \sqrt{90 \times 1/3 \times 2/3} = 4.47$$

If the new method is really better, the number of successes S should be larger than what is expected if H_0 is true, namely, 30. Again using the normal approximation, the winemaker should reject H_0 and decide in favor of the new method if $(S - 30)/4.47 \geq 1.65$.

Since the winemaker observed $S = 35$, we compute $Z = (35 - 30)/4.47 = 1.12$. Since $1.12 < 1.65$, the winemaker would not reject H_0. Notice, however, that 1.12 is reasonably close to 1.65. In fact, computing the associated significance probability shows $P(S \geq 35) = P(Z \geq 1.12) = .1314$. Although it does not meet the standards for rejection, this probability is reasonably low. Since this is an important decision, the winemaker may decide that the information he has obtained is inconclusive and that he should gather more information before making a final decision. (Remember that a statistical test should not be the sole factor in determining an important decision, but rather only one indication of what the proper course of action should be.)

Note that every test we have discussed so far has involved a test statistic S, which is binomially distributed. Tests such as these are called *binomial tests*. Also, every sample size has been large enough to employ the normal approximation to the binomial distribution, enabling us to transform S to Z. If the sample size for a binomial test is too small to use the normal approximation, we would use the binomial tables to compute the rejection region. In the situations we have been discussing, the sample size has been large enough to employ the normal approximation; in Chap. 9 we shall discuss similar situations involving small sample sizes.

Types of errors; determining the null hypothesis

Sometimes it is not clear which hypothesis should be the null hypothesis. For instance, in Example 6-3, why couldn't the winemaker have let H_0 be that the new wine is better and H_a be that there is no difference in the wines? We mentioned previously that H_0 is always "there is no difference," but we didn't say why. To decide which hypothesis should be the null hypothesis, remember how tests of a given significance level are constructed. The rejection region is determined by limiting to a certain level the probability of making the error of rejecting the null hypothesis when it is really true. This kind of error is sometimes called a *type I error*. Another kind of error that is possible, namely, accepting the null hypothesis when it is really false, is called a *type II error*.

Definition. A type I error is the rejection of H_0 when H_0 is true. A type II error is the acceptance of H_0 when H_0 is false.

Of course it would be ideal if we could make the probability of both types of error as low as we wanted. Unfortunately, this is not possible. In fact, for a predetermined sample size, when we reduce the probability of making a type I error, we simultaneously increase the probability of making a type II error. In order to make both error probabilities arbitrarily low we would have to increase the sample size. For this reason we must choose which type of error we want to reduce; the convention has been to always control the probability of making a type I error. This is what we do when we choose a significance level for a given test. In designing a test, we should decide which kind of mistake we want to avoid most and set up our hypotheses appropriately. In Example 6-3, the mistake the winemaker wants to avoid most is to start using the new method, a costly procedure, when this method is really not an improvement. Therefore he lets H_0 be that there is no difference. Testing at level $\alpha = .05$ will ensure that there is only a .05 probability of using the new method when there is no improvement.

Unfortunately, making determinations like these can be quite difficult. Fortunately, the rule we have given, namely, to let H_0 be "there is no difference," is a useful one and invariably sets up the test in the proper way. We shall always use this rule and let H_0 always be "there is no difference."

A great deal of study can be done at a more advanced level to see which tests do the best at minimizing type II errors for a given significance level; such tests are sometimes called *best tests*. We shall not explore this topic, however. We promise to present only good and widely used tests for you to study, thus enabling you to reap the benefits of modern statistical theory.

Example 6.4. A survey is done to determine whether children prefer violent cartoons to nonviolent ones. Two cartoons, each of the same general type but with cartoon A containing more violence than cartoon B, are shown to 400 children. To eliminate possible experimental bias, 200 children are shown cartoon A first while 200 children are shown cartoon B first. The cartoons are shown separately to each child. After seeing the cartoons, each child is asked which cartoon he prefers, and S, the number of children who prefer cartoon A (the more violent cartoon), is recorded. Testing at a significance level of $\alpha = .05$, what should the conclusion be if $S = 210$?

SOLUTION: We label the hypotheses to be tested as follows:

H_0: $p = .5$

H_a: $p > .5$

where $p = P$(child prefers cartoon A). Assuming that H_0 is true, S, the number of children in the sample who favor violent cartoons, is binomially distributed with

$$np = 400 \times .5 = 200$$

$$\sqrt{np(1 - p)} = \sqrt{400 \times .5 \times .5} = 10$$

The rejection region, then, is any result for which $Z = (S - 200)/10 \geq 1.65$. Since we observed $S = 210$, we compute $Z = (210 - 200)/10 = 1$ and conclude that there is insufficient evidence that children prefer violent cartoons. The associated significance probability is $P(S \geq 210) = P(Z \geq 1) = .1587$, which is larger than .05.

Summary

We have introduced a classic method of statistical decision making, testing a statistical hypothesis. We have set up a framework by which we can use information from a random sample to help in deciding whether or not a certain claim or hypothesis is true. This structure has been used when the information from the sample is in the form of a test statistic which is binomially distributed. We have seen that we can control the probability of making the error of falsely rejecting H_0, this error probability being called the *significance level* of the test. We have used the normal approximation to the binomial distribution as an essential tool for determining the rejection region of the test, this method being effective whenever the sample size is large. Finally, we have emphasized that the use of these methods does not dictate with certainty which hypothesis is true but provides a mathematical model in which we can effectively utilize sample information.

Important terms

Alternative hypothesis

Null hypothesis

Rejection region

Significance level

Significance probability

Statistical test

Test statistic

Type I error

Type II error

Problems

For all problems, state H_0 and H_a and whether you would accept or reject H_0 after observing the data.

6-1 You suspect that a certain gambler's coin is biased in favor of heads, but the gambler claims it is fair. The coin is tossed 45 times, and you observe 9 heads. Test at a significance level of $\alpha = .05$. (*Hint:* Let H_0: $p = 1/2$; H_a: $p > 1/2$.)

6-2 An automobile manufacturer asserts that the seat belts on his cars are 90% effective. A consumers' group doubts this, claiming that the seat belts are not that effective. Out of 50 cars that are tested, the seat belts work on 37 of them. Test at a significance level of $\alpha = .05$, letting H_0: $p = .9$; H_a: $p < .9$.

6-3 A beermaker thinks a costly new process will improve the taste of his beer. He decides to test the new process on a group of 301 people by offering them each 4 glasses of beer to taste, 1 of which is made with the new process (they don't know which one), and then asking them to select the glass of beer which tastes best. (Even if they all taste the same, each person must select 1 glass.) The experiment was performed, and 1 person became too drunk to select a best glass. Of the remaining 300, 90 selected the beer made with the new process. At a significance level $\alpha = .05$ is this evidence for the beermaker to use the new process? At a significance level of $\alpha = .01$? Would the beermaker be likely to use a high or low significance level in this situation?

6-4 A claim is made that 20% of the people in a certain rural community are malnourished, but a concerned citizens' group thinks the percentage is higher. A random sample of size 100 is taken, and 30 people are found to be malnourished. At a significance level of $\alpha = .05$ is this evidence to reject or accept the claim?

6-5 A new vaccine is claimed to be effective 90% of the time. As a test, a researcher gives the vaccine to 900 laboratory animals who have been exposed to the associated disease; 72 animals contract the disease. At a significance level of $\alpha = .01$ test whether or not the claim is true.

6-6 Of those who take a certain review course in preparation for the law bar exam, 60% usually pass the exam. The course is altered with a new and supposedly better teaching method being used. If 65 out of 100 people taught by the new method pass the exam, would you conclude at a significance level of $\alpha = .05$ that the new method is better?

6-7 With the former judge, 20% of convicted first-offenders were sentenced to jail terms. During the first 6 months with the new judge, 227 of 1,000 convicted first-offenders were sentenced to jail terms. At a significance level of $\alpha = .05$, is this evidence that the new judge gives harsher sentences than the old judge?

6-8 In an area free of radioactive fallout, the rate of babies born with birth defects is 1% per year. In a 1-year study in an area of heavy radioactive fallout due to atomic testing, it was found that of 10,000 births 150 babies had birth defects. Is it safe to conclude that the rate of birth defects is increased significantly by atomic tests? Use a level of significance of $\alpha = .05$.

6-9 With the use of a certain powerful pesticide only 15% of a certain crop was killed by pests. After this pesticide was outlawed for being harmful to the ecology, a new, less toxic pesticide was used. During the first year the new pesticide was used, 120 out of a sample of 1,500 plants were killed. Is this evidence that the new pesticide is less effective than the old one at a .05 level of significance?

6-10 In the general populace it is known from the Nielsen ratings that

30% of TV viewers prefer movies with a high level of violence. In a sample of 100 people who have authoritarian personalities (as measured by the F scale used by psychologists to measure authoritarianism), 50 people indicated a preference for movies with a high level of violence. Is this evidence that people with authoritarian personalities tend to have a greater preference for violent movies than the general populace at a .01 level of significance?

True-false test

Circle the correct letter.

T F   1. If a hypothesis is rejected at the .05 level of significance, then it will also be rejected at the .01 level of significance.

T F   2. A coin is tossed a large number of times, and $S = $ total number of heads recorded. If we are testing H_0: $P(\text{head}) = 1/2$ versus H_a: $P(\text{head}) > 1/2$, then we would reject H_0 for large values of S.

T F   3. If a wrong decision is reached after analyzing the data, we have necessarily committed a type I error.

F F   4. If a type II error has occurred, then the wrong test must have been applied.

T F   5. Results of a statistical test should be reported only if H_0 is rejected.

T F   6. A type I error occurs when H_0 is falsely rejected.

T F   7. Tests of hypotheses aid us in making decisions on the basis of limited information obtained from a random sample.

T F   8. Experienced statisticians tend to use lower significance levels for their tests.

T F   9. A test is significant at the .05 level of significance if the probability of rejecting H_0 when it is actually true is .05.

T F   10. If a test is significant at the .01 level of significance then the probability of making a type I or type II error is less than .01.

One-sample
tests
of a
population mean

7

In Chap. 6 we developed a method for testing hypotheses involving population proportions. In the present chapter we shall discuss an analogous procedure for testing population means. It may be necessary to determine whether the average score on a certain test differs in a year when a new teaching method is used from the average score in past years. A bridge-construction firm intending to use a certain kind of steel supporting rods might want to know whether the average strength of the rods exceeds a certain minimal level. The Federal Drug Administration tests certain agricultural products to determine if the average amount of insecticides per item is too high. In all such situations a test of a single population mean is made. Some hypothesized value is specified for the mean, and then a decision is made as to whether the mean of the observed sample data differs significantly from the hypothesized value. This test is called a *one-sample test* because only one sample is taken.

It turns out that the test statistic used to evaluate the hypothesized value has approximately a normal distribution when the sample size is large. However, when the sample size is small, this approximation will not necessarily hold and thus a different procedure is required. Also, certain assumptions are required in the case of small samples which are not necessary for large samples. Accordingly, we shall divide the discussion into two sections: large- and small-sample tests.

Large-sample test of a population mean

We give an example to introduce the main ideas.

Example 7-1. In a certain city, welfare applicants complain that it takes too long to process their applications. The city welfare department responds with a new "simplified" procedure designed to reduce the processing time. To check whether the new procedure actually reduces the time, a random sample of 100 welfare applicants is selected and the time between the day they apply and the day they receive their first check is recorded. The sample average is $\bar{X} = 55$ days, and the sample standard deviation is $s = 30$ days. Under the old procedure, the average time for processing applications is known to be 58 days, as computed from complete records of applications submitted during that period. We wish to determine whether the sample data gives strong evidence that the average processing time has been reduced under the new procedure. (Assume, for simplicity, that the new procedure is not any worse than the old one but possibly the same or better.)

> SOLUTION: This problem is examined in the same spirit as in Chap. 6. As before, we set up a null hypothesis and an alternative hypothesis which reflect the possibilities in which we are interested. The quantity to be examined is the average time it takes to process an application using the new procedure, which we denote by μ. This is an unknown population mean. Its value could be determined exactly only if we knew how long it would take to process each person who ever applied for welfare—past, present, and future—using the new procedure. Of course, this is impossible, and so therefore we must use statistical methods to discuss μ. The hypotheses to be tested are
>
> H_o: $\mu = 58$
>
> H_a: $\mu < 58$
>
> The choice of the value of μ expressed in the null hypothesis is

based on the rationale of "nothing is new," the conservative perspective used to specify the null hypothesis. The alternative hypothesis H_a: $\mu < 58$ reflects our assumption that the new procedure is no worse than the old one.

Since the sample mean $\bar{X}$ is an estimate of μ, the quantity $\bar{X} - 58$ estimates the difference between the actual value of the population mean μ and the hypothesized value of 58. But some difference between $\bar{X}$ and 58 is expected owing to the random nature of our data even if, in fact, $\mu = 58$. This random fluctuation is accounted for by taking the test statistic to be

$$Z = \frac{\bar{X} - 58}{s/\sqrt{n}}$$

where the sample standard deviation $s = 30$ and the sample size $n = 100$.

If H_o is true and $\mu = 58$, then $\bar{X}$ will be near 58 and Z will be near 0. If, however, H_o is false and $\mu < 58$, then $\bar{X}$ will probably be less than 58 and Z will be less than 0. If the observed value of Z is too far below 0, it would indicate that $\mu < 58$ and lead us to reject H_o. Exactly how far below 0 is "too far" is determined by the level of significance and the fact that the sampling distribution of Z is approximately standard normal for large sample sizes when H_0 is true. From the normal tables we find that $P(Z \leq -1.65) = .05$. Hence, H_0 is rejected at the .05 level of significance if $Z \leq -1.65$. From the data,

$$Z = \frac{\bar{X} - 58}{s/\sqrt{n}} = \frac{55 - 58}{30/\sqrt{100}} = -1$$

Since $Z = -1 > -1.65$, H_0 is accepted. There is not sufficient evidence to indicate that the new procedure really reduces the processing time at a .05 level of significance.

Example 7-1 illustrates a class of tests called *tests of means*. These tests must be used whenever it is necessary to determine if an unknown population mean μ equals some specified value μ_0. Three pairs of hypotheses are possible: One pair is H_0: $\mu = \mu_0$ versus H_a: $\mu < \mu_0$, which must be used if it can be assumed that μ is no greater than μ_0. A similar setup is expressed by H_0: $\mu = \mu_0$ versus H_a: $\mu > \mu_0$, where it is assumed that μ is no less than μ_0. The tests used to evaluate these pairs of hypotheses are called *one-sided tests* because the alternative specifies that μ may be only on one side or in one direction of μ_0. A third pair of hypotheses must be used when no such assumptions can be made; these are H_0: $\mu = \mu_0$ versus H_a: $\mu \neq \mu_0$. The test used to decide between these hypotheses is called a *two-sided test*.

To test any of these pairs of hypotheses a random sample is drawn and the sample mean $\bar{X}$ and sample standard deviation s are computed. As Example 7-1, the test statistic is always

$$Z = \frac{\bar{X} - \mu_0}{s/\sqrt{n}}$$

where n = sample size. According to the central limit theorem for sample means, Z will have an approximately standard normal distribution for large n when H_0 is true. The approximation is accurate enough for most purposes whenever n is at least 30, permitting the use of the normal tables to find the rejection region for any particular level of significance. The procedures used when the level of significance is $\alpha = .05$ or $\alpha = .01$ are given in Fig. 7-1.

	Rejection region	
Hypothesis	$\alpha = .05$	$\alpha = .01$
H_0: $\mu = \mu_0$ H_a: $\mu < \mu_0$	Reject H_0 if $Z \leq -1.65$	Reject H_0 if $Z \leq -2.33$
H_0: $\mu = \mu_0$ H_a: $\mu > \mu_0$	Reject H_0 if $Z \geq 1.65$	Reject H_0 if $Z \geq 2.33$
H_0: $\mu = \mu_0$ H_a: $\mu \neq \mu_0$	Reject H_0 if $Z \geq 1.96$ or $Z \leq -1.96$	Reject H_0 if $Z \geq 2.58$ or $Z \leq -2.58$

Figure 7-1

Example 7-2. A consumers' group is suspicious about the weight of a certain brand of cereal for which the boxes are labeled as containing 10 ounces. A random sample of 50 boxes yields an average of 9.6 ounces with a standard deviation of 1.45 ounces. Test the hypothesis that the population mean weight is 10 ounces at a level of significance of $\alpha = .05$.

SOLUTION: Let μ = population mean weight of a box. We wish to test the hypothesis

H_0: $\mu = 10$

H_a: $\mu < 10$

Notice that we have chosen H_a: $\mu < 10$ as the alternative hypothesis because it is reasonable to assume that the manufacturer would not put in more than 10 ounces per box. Thus, we have a one-sided test and can use the procedure given in Fig. 7-1. Substituting the observed values of $\bar{X} = 9.6$, $s = 1.4$, and $n = 50$ and the value of μ under H_0: $\mu = 10$ into the formula for Z, we obtain

$$Z = \frac{\bar{X} - \mu_0}{s/\sqrt{n}} = \frac{9.6 - 10}{1.4/\sqrt{50}} = -2.0$$

Since the observed value of $Z = -2.0$ is less than -1.65, H_0 is rejected. Thus, at a .05 level of significance we may conclude that the average weight per box is less than the 10 ounces claimed by the manufacturer. Notice that if a .01 level of significance were chosen, then H_0 would be accepted since $Z = -2.0 > -2.33$.

Example 7-2 illustrates that a decision to reject or accept H_0 depends upon the level of significance chosen by the researcher.

Example 7-3. A biology professor in Wisconsin specializes in mosquito identification. From many years of research at Lake Michigan he has found that the average wing length of a mosquito is .5 centimeters. On his sabbatical he plans to go to Lake Tahoe, Nevada, to continue his study. He wishes to determine if mosquitoes there have wing lengths that are different than those at Lake Michigan. He intends to measure the wing lengths of 1,000 mosquitoes. What procedure should he use in order to reach a conclusion if he is willing to allow 1 chance in 100 of concluding there is a difference in wing length when, in fact, there is not?

SOLUTION: Denote by μ the average wing length of mosquitoes at Lake Tahoe. Then the hypotheses to be tested are

H_0: $\mu = .5$ centimeter

H_a: $\mu \neq .5$ centimeter

Notice that choosing the null hypothesis to be H_0: $\mu = .5$ centimeter represents the assertion that "there is no difference" between the two means. The alternative hypothesis of H_a: $\mu \neq .5$ asserts that "there is a difference." The test is two-sided since the professor has no a priori reason to suspect that the average wing length of a mosquito at Lake Tahoe is larger or smaller than the average wing length of a mosquito at Lake Michigan. Since he is willing to allow 1 chance in 100 of concluding H_0 is false when, in fact, it is true, a .01 level of significance is chosen. Thus, to reach a conclusion after his study, the professor should compute the statistic

$$Z = \frac{\bar{X} - .5}{s/\sqrt{1,000}}$$

where $\bar{X}$ = sample mean and s = sample standard deviation; he should reject H_0 if $Z \geq 2.58$ or $Z \leq -2.58$.

Problems

7-1 A farmer claims that the average weight of his pumpkins is 4.1 pounds, but a consumers' group suspects that the average weight is less. A sample of 50 pumpkins is taken, with the results $\bar{X} = 3.9$ and $s = 1.2$. Do these data confirm the consumers' suspicion at the .05 level of significance?

7-2 The average score obtained by the general populace on the well-known F scale, which measures degree of authoritarianism, is 65. In a sample of 100 people given the F test who were raised in a family where the father was a career army officer, it is found that $\bar{X} = 80$ and $s = 10$. Interpret these results.

7-3 A manufacturer of shoes claims his soles are at least 1/2 inch thick. In a sample of 50 shoes chosen at random the average thickness is found to be .47 inch with a standard deviation of .15 inch. Is this strong evidence against the manufacturer's claim at the .05 level of significance?

7-4 The police chief of a certain city claims that the mean age of rapists is 24.6 years. A women's liberation group takes a random sample of 100 cases from police files, obtaining a mean age of 22.1 years with a standard deviation of 6 years. Are these data consistent with the police chief's claim at the .05 level of significance? (Use a two-tailed test.)

Small samples: the one-sample t test

As we have noted, the test presented in the last section is a large-sample test. In practice this means that the sample size is at least 30, in

which case the normal approximation is accurate enough for most purposes. Since this approximation is not reliable for small samples, a
slightly different method, called the *one-sample t test,* is used. Unlike
the corresponding large-sample test, the *t* test requires that the population data be normally distributed.

The procedure is the same as in large-sample tests. We are interested in testing the null hypothesis H_0: $\mu = \mu_0$ against one of the alternatives H_a: $\mu < \mu_0$, H_a: $\mu > \mu_0$, or H_a: $\mu \neq \mu_0$. From the
sample data the value of the test statistic

$$t = \frac{\bar{X} - \mu_0}{s/\sqrt{n}}$$

is computed. The form of the test rule is the same as in large-sample
tests. For example, if we wish to test H_0: $\mu = \mu_0$ versus
H_a: $\mu > \mu_0$, then H_0 is rejected for large values of t. To find the exact
rejection region for a particular level of significance we need to know
the sampling distribution of t. Assuming the population data are normally distributed or nearly so, the test statistic t will have a sampling
distribution called *Student's t distribution,* under H_0. The *t distribution,* as it is often abbreviated, is similar to the standard normal distribution in that it is symmetric about zero and bell shaped; however, it is
more spread out than the normal curve. Actually, there are many t distributions, each characterized by a number called its *degrees of
freedom*. This number is determined from the sample size. If n observations are collected, there will be $n-1$ degrees of freedom. The larger
n is, the closer the t distribution with n degrees of freedom is to being a
standard normal distribution (Fig. 7-2).

Notice that the test statistic t is the same as Z which we used in

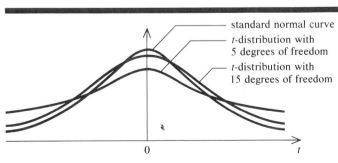

standard normal curve

t-distribution with
5 degrees of freedom

t-distribution with
15 degrees of freedom

0

t

Figure 7-2

the large-sample case. We have used a different label because the distribution of t is different from that of Z.

To compute probabilities for the t distribution, tables are provided in the Appendix. Since we shall need the t distribution only for testing hypotheses, the tables give only the values of t needed to determine the rejection region for the common values of α: .1, .05, and .01. For example, suppose a sample of size $n = 10$ is drawn from a population with normally distributed data. To find the value of t which will be exceeded with probability .05, look in the the row corresponding to the degrees of freedom $n - 1 = 10 - 1 = 9$ and the column headed by $\alpha = .05$, obtaining the value 1.83. This means $P(t > 1.83) = .05$ (Fig. 7-3).

Before using the t test you should carefully consider whether the assumption of normality is really valid. There are other tests which may be used when this assumption is dubious; we shall study such tests in Chaps. 9 and 10. As a rule of thumb, test scores and errors in physical measurements are usually normally distributed.

Example 7-4. An automobile tire manufacturer advertises that its tires are good for 15,000 miles on the average. In a sample of 20 tires, it was found that the average lifetime was 14,500 miles with a standard deviation of 1,800 miles. Are these data consistent with the manufacturer's claim at a .05 level of significance?

SOLUTION: Let μ = average lifetime of all tires manufactured by the company. We are interested in testing

H_0: $\mu = 15,000$

H_a: $\mu < 15,000$

Notice that the alternative hypothesis is chosen as H_a: $\mu < 15,000$ because it may be safely assumed that the company would not advertise a value of μ less than the true value. This leads to a one-sided test, and H_0 will be rejected if t is too far below 0. Assuming the lifetime of tires is normally distributed, we can use the t distribution to determine the rejection region. From the data given $\bar{X} = 14,500$, $s = 1,800$, and $n = 20$; therefore,

$$t = \frac{\bar{X} - \mu_0}{s/\sqrt{n}} = \frac{14,500 - 15,000}{1,800/\sqrt{20}} = -1.24$$

Since $n = 20$, there are 19 degrees of freedom. From the t tables (in the row for 19 degrees of freedom), we obtain

$P(t < -1.73) = .05$ (Fig. 7-4). Thus, the observed value of $t = -1.24 > -1.73$, and H_0 is accepted. The data are consistent with the manufacturer's claim at a .05 level of significance.

Example 7-5. The telephone company tests its directory-assistance operators at the end of a 6-month probationary period to see if they are meeting the standards required for the job. The company wishes to retain only those operators who spend on the average 1 minute or less answering a caller's request for information. Suppose that an operator's average service time is 1.25 minutes with a standard deviation of .2 minutes for 10 randomly selected calls that are monitored without the operator's knowledge. Should the operator be fired? (Assume that the telephone company allows only 1 chance in 100 of firing an operator who is really meeting the standards.)

SOLUTION: LET μ denote the average time it takes the operator to answer a call. We wish to test H_0: $\mu = 1$ minute versus H_a: $\mu > 1$ minute. Assuming that the times to answer a call are nor-

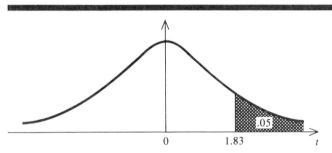

Figure 7-3 t distribution with 9 degrees of freedom.

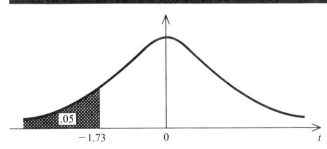

Figure 7-4 t distribution with 19 degrees of freedom.

mally distributed, the small-sample test for a population mean may be applied. From the data given we compute

$$t = \frac{\bar{X} - 1}{s/\sqrt{n}} = \frac{1.25 - 1}{.2/\sqrt{10}} = \frac{.25}{.063} = 3.97$$

The sampling distribution of t is Student's t distribution with $n - 1 = 10 - 1 = 9$ degrees of freedom. The assumption that the telephone company allows only 1 chance in 100 of falsely firing an employee means we should use a .01 level of significance. From the t tables we find that H_0 is rejected if $t \geq 2.82$. Since the observed value of $t = 3.97 > 2.82$, H_0 is rejected. This is strong evidence that the operator has an average service time per call exceeding 1 minute and therefore should be fired.

Summary

Statistical tests used to determine if a population mean μ equals a specified value μ_0 are called one-sample tests of means. These apply to hypotheses of the following form:

H_0: $\mu = \mu_0$ H_0: $\mu = \mu_0$ H_0: $\mu = \mu_0$

H_a: $\mu < \mu_0$ H_a: $\mu > \mu_0$ H_a: $\mu \neq \mu_0$

Which pair of hypotheses to test in a given situation depends on the assumption one is willing to make about the possible values of μ. The test statistic used to discriminate between H_0 and H_a when $n \geq 30$ is

$$Z = \frac{\bar{X} - \mu_0}{s/\sqrt{n}}$$

where $\bar{X} = $ sample mean, $s = $ sample standard deviation, and $n = $ sample size. For the first pair of hypotheses, H_0 is rejected for values of Z that are too far below 0. Similarly, for the second pair, H_0 is rejected when Z is too large. For the third pair of hypotheses, H_0 is rejected when the absolute value Z is too big, in other words, if Z is *either* too large or too small.

If the sample size n is large ($n \geq 30$), then Z will approximate a standard normal distribution. This fact allows the use of the normal tables to arrive at an exact description of the rejection region for any given level of significance. For small values of n, it must be assumed that the population data are normally distributed. Then the test statistic t will have a t distribution with $n - 1$ degrees of freedom. Tables of the t distribution in the Appendix give the value of t necessary to determine the rejection region for common levels of significance and small sample sizes.

Important terms

Degrees of freedom

Large-sample tests

One-sample t test

One-sided tests

One-tailed tests

Small-sample tests

t distribution

t test

Tests of a population mean

Two-sided tests

Two-tailed tests

Additional problems

7-5   A certain laboratory wishes to test the hypothesis that the average tread lifetime of a certain brand of tire is 30,000 miles against the alternative hypothesis that it is less than 30,000 miles. Suppose a sample of 10 tires yields $\bar{X} = 28,500$ and $s = 3,000$. Assuming that tread lifetimes are normally distributed, test the hypotheses at the .05 level of significance.

7-6   A certain type of rope is required to have a mean breaking strength of 1,200 pounds. A sample of 100 pieces of rope are tested, with the results $\bar{X} = 1,150$ and $s = 230$. Is it reasonable to conclude that the mean breaking strength is significantly lower than 1,200 pounds at the .01 level of significance?

7-7   A restaurant owner who is trying to sell his restaurant claims that the average gross income per day is $850. A prospective buyer examines the record books for the past 150 days, finding the average daily gross to be $800 with a standard deviation of $275.

Is this strong evidence against the owner's claim at a .05 level of significance?

7-8 A vending-machine distributor knows that his machines will sell an average of 612 drinks per week. He tests 15 new vending machines with neon flashing lights to see if more customers will be attracted. He finds $\bar{X} = 620$ and $s = 50$. Testing at the .05 level of significance, should he conclude that the flashing lights improve sales?

7-9 A department store wants to test the effectiveness of its special sales. Its average profits per day on nonsale days are $1,508. Over a period of 20 sale days the average daily profits are found to be $1,620, with a standard deviation of $s = \$175$. Do these data indicate that significantly more profits are obtained on sale days, or are these results what might be expected due to random fluctuation? Test at a .01 level of significance.

7-10 It is known that the average family spends 20.8 hours per week with the television set turned on. A study involving 200 families in which the head of the household is a psychologist is conducted, and it is found that for these families the television is turned on for an average of 23.1 hours per week with a standard deviation of 4. State reasonable hypotheses and test at a level of significance $\alpha = .05$.

True-false test

Circle the correct letter.

T F   1. The t test can be applied with no assumptions about the distribution of the underlying population data.

T F   2. When applying a two-tailed test, H_0 will be rejected for extreme values of the test statistic in either the positive or negative direction.

T F   3. The larger the number of degrees of freedom in the t distribution, the closer it is to the standard normal distribution.

T F   4. If you are testing H_0: $\mu = 6$ versus H_a: $\mu > 6$, and you observe a value of $z = 5$, you would necessarily reject H_0.

T F   5. If H_0 is rejected by a one-sided test, then it will be rejected by a two-sided test with the same level of significance.

T F   6. For large n, the statistic t has approximately a standard normal distribution.

T F   7. A two-sided test can be converted into a one-sided test at the same level of significance by eliminating the rejection region in one direction.

T F   8. The number of degrees of freedom for the t statistic are independent of the sample size.

T F   9. One-sided tests are often used to further reinforce the bias of a researcher.

T F   10. The larger s is, the more extreme $\bar{X}$ must be in order to reject H_0.

Comparisons: two-sample tests of means **8**

Whenever you read the newspapers, watch TV, or listen to the radio, there are continual references to results of comparative experiments. The studies may compare various groups of people, brands of merchandise, or the state of something before and after a treatment has been applied. There are several statistical tests which can be used to analyze the results of these experiments. The choice of which test to apply depends on several factors, including the number of things compared, the number of populations sampled, the sample size, and the distribution of the population data. The problem of comparing two population means when two independent samples are drawn is treated in this chapter. The discussion is divided into the cases of large and small samples. In Chap. 9 several other possibilities are examined.

Large-sample comparison of two population means

Here is an example of the type of problem we shall be studying.

Example 8-1. A newspaper reported the results of a research project designed to determine if frequent attendance at rock concerts is associated with hearing loss. At a high school located near a popular concert hall, two samples were chosen from the twelfth-grade class at the time of their annual hearing test. Group 1 included 60 students who had attended more than 5 concerts, and group 2 contained 50 students who had attended fewer than 2 concerts. Hearing tests were administered to each group with the following results, expressed in terms of the minimum decibel level which could be recognized. From group 1 the sample mean and standard deviation were $\bar{X}_1 = 25.8$ and $s_1 = 10.3$; from group 2 they were found to be $\bar{X}_2 = 18.1$ and $s_2 = 8.1$. The newspaper article concluded that the results imply that those students who attend rock concerts frequently have worse hearing than those who do not attend frequently. Was this conclusion justified?

> SOLUTION: First we construct a statistical model that is appropriate for comparing the hearing-test results of the two groups. This is done, as we have done before, by specifying two hypotheses representing the possible conclusions that might be drawn. Then it is necessary to specify a procedure or decision rule for deciding which hypothesis is best supported by the data.
>
> Let μ_1 and μ_2 denote the average minimum levels of sound recognizable by those who attend at least five concerts and by those who attend fewer than two concerts, respectively. Notice that low values of these means would indicate good hearing, as they would imply a person can detect very soft sounds. We wish to decide between the two hypotheses
>
> H_0: $\mu_1 = \mu_2$
>
> H_a: $\mu_1 > \mu_2$
>
> The alternative hypothesis H_a: $\mu_1 > \mu_2$ has been chosen since it is reasonable to assume that a person's hearing can be only hurt or unaffected by exposure to loud noise, not improved.
>
> Perhaps a reasonable test procedure should be based on the size of the difference between the sample means $\bar{X}_1 - \bar{X}_2$. However, some difference is expected due to random sampling, and this is taken into account by using as the test statistic the quantity
>
> $$Z = \frac{\bar{X}_1 - \bar{X}_2}{\sqrt{s_1^2/n_1 + s_2^2/n_2}}$$
>
> where $n_1 =$ size of sample from group 1 and $n_2 =$ size of sample from group 2: $n_1 = 60$ and $n_2 = 50$. A large positive difference

between $\bar{X}_1$ and $\bar{X}_2$ is reflected in a large value of Z. Thus, intuitively the form of the test procedure should be to reject H_0 if Z is too large. It turns out that since n_1 and n_2 are each greater than 30, the central limit theorem applies and the sampling variable Z has approximately a standard normal distribution. This fact allows us to determine the rejection region from the normal tables once a level of significance has been chosen.

Let us use a .05 level of significance for our test. From the normal tables we find that a standard normal variable exceeds 1.65 with probability .05. Hence, H_0 should be rejected only if the observed value of Z exceeds 1.65. Making the appropriate computation, we find

$$Z = \frac{\bar{X}_1 - \bar{X}_2}{\sqrt{s_1^2/n_1 + s_2^2/n_2}} = \frac{25.8 - 18.1}{\sqrt{\dfrac{(10.3)^2}{60} + \dfrac{(8.1)^2}{50}}}$$

$$= \frac{7.7}{\sqrt{3.08}} = 4.39$$

leading to a rejection of H_0. That is, at a .05 level of significance, students who attend rock concerts frequently have significantly worse hearing than those who do not attend frequently.

This test procedure of Example 8-1 is easily summarized in general terms. Let μ_1 and μ_2 denote the means of two populations. If it can be assumed that the mean of population 1 is greater than or equal to the mean of population 2, then the appropriate hypotheses are

H_0: $\mu_1 = \mu_2$

H_a: $\mu_1 > \mu_2$

On the other hand, if no such assumption can be made, then the hypotheses are

H_0: $\mu_1 = \mu_2$

H_a: $\mu_1 \neq \mu_2$

In either case, to test the hypotheses, samples of sizes n_1 and n_2 are drawn from populations 1 and 2, respectively, and the sample means $\bar{X}_1$ and $\bar{X}_2$ and sample standard deviations s_1 and s_2 are found. From these, the test statistic is computed:

$$Z = \frac{\bar{X}_1 - \bar{X}_2}{\sqrt{s_1^2/n_1 + s_2^2/n_2}}$$

When both n_1 and n_2 are large (at least 30), Z will have an approximately standard normal distribution. The test procedure is summarized in Fig. 8-1 for levels of significance of $\alpha = .05$ and $\alpha = .01$.

Hypotheses	$\alpha = .05$	$\alpha = .01$
$H_0:$ $\mu_1 = \mu_2$	Reject H_0 if $Z \geq 1.65$	Reject H_0 if $Z \geq 2.33$
$H_a:$ $\mu_1 > \mu_2$		
$H_0:$ $\mu_1 = \mu_2$	Reject H_0 if $Z \geq 1.96$	Reject H_0 if $Z \geq 2.58$
$H_a:$ $\mu_1 \neq \mu_2$	or if $Z \leq -1.96$	or if $Z \leq -2.58$

Figure 8-1

The terms *one-sided* and *two-sided* apply to these two-sample test procedures just as they did in the one-sample tests discussed in Chap. 7. For tests involving the alternative hypothesis $H_a:$ $\mu_1 > \mu_2$, the tests are said to be *one-sided* since H_0 is rejected only if Z is too large. For tests involving the alternative hypothesis $H_a:$ $\mu_1 \neq \mu_2$, the tests are said to be *two-sided* since H_0 is rejected if Z is too large or too small (negatively).

Example 8-2. A doctor wishes to determine which of two diets is more effective in reducing weight. A sample of 100 obese adults who are interested in losing weight is randomly divided into groups of 50. The average weight loss for those undergoing diet 1 is 10 pounds with a standard deviation of 5 pounds. For diet 2 the average loss is 11 pounds with a standard deviation of 6 pounds. Do these results indicate a significant difference between the two diets at a .05 level of significance?

SOLUTION: Let μ_1 and μ_2 denote the mean weight loss due to diets 1 and 2, respectively. A priori, the doctor has no reason to suspect that one diet is better than the other, and so the hypotheses to be tested are

$H_0:$ $\mu_1 = \mu_2$

$H_a:$ $\mu_1 \neq \mu_2$

From the data $\bar{X}_1 = 10$, $s_1 = 5$, $\bar{X}_2 = 11$, and $s_2 = 6$. The test statistic is

$$Z = \frac{\bar{X}_1 - \bar{X}_2}{\sqrt{s_1^2/n_1 + s_2^2/n_2}} = \frac{10 - 11}{\sqrt{25/50 + 36/50}} = \frac{-1}{\sqrt{1.22}} = -.91$$

From Fig. 8-1, the test procedure is to reject H_0 if $Z \geq 1.96$ or $Z \leq -1.96$ at the .05 level of significance. Since the observed value of Z falls between -1.96 and $+1.96$, H_0 is accepted and the doctor concludes that there is no significant difference between the two diets.

Example 8-3. A government agency investigating the claims of advertisers decides to compare two brands of deodorants. Each brand is advertised as being the most effective in eliminating body odor. A sample of 50 people is chosen and asked to apply deodorant A, and another sample of 50 is given deodorant B. Each person is then asked to stand in front of a sensitive aroma-detecting machine used in many TV commericals, and a reading is made. The average reading for deodorant A is found to be 5.2 with a standard deviation of 2, and the average for deodorant B is 4.7 with a standard deviation of 2.3. Do these results indicate any significant difference in body-odor-reducing ability between the two deodorants at a .05 level of significance?

SOLUTION: Let μ_1 and μ_2 denote the mean body-odor level for people using deodorants A and B, respectively. We wish to test

H_0: $\mu_1 = \mu_2$

H_a: $\mu_1 \neq \mu_2$

Using the given data, we compute

$$Z = \frac{\bar{X}_1 - \bar{X}_2}{\sqrt{s_1^2/n_1 + s_2^2/n_2}} = \frac{5.2 - 4.7}{\sqrt{(2)^2/50 + (2.3)^2/50}} = \frac{.5}{.431} = 1.16$$

From Fig. 8-1 we see that the appropriate test procedure is to reject H_0 if $Z \geq 1.96$ or $Z \leq -1.96$. Since the observed value of $Z = 1.16$ lies between -1.96 and 1.96, H_0 is accepted. The agency may conclude that the evidence is not sufficient to claim that one deodorant is better or worse than the other at a .05 level of significance.

Problems

8-1 Two methods of teaching French, methods A and B, are compared by administering the same test to 50 students who have been taught by each method. Method-A test results show $\bar{X}_1 = 65$ and $s_1 = 8$; and method-B results show $\bar{X}_2 = 70$ and $s_2 = 10$. Do these results indicate that one method is more successful than the other? Perform your test at the .05 level of significance.

8-2   A study is conducted to compare the amount of protein in two brands of hot dogs. A sample of 100 hot dogs is randomly selected from each brand and analyzed for their protein content. Brand A is found to have 18% protein with a standard deviation of 5%, and brand B is found to have 15% protein with a standard deviation of 3%. Do these results indicate any difference in the amount of protein in the two brands at a .01 level of significance?

8-3   A customer is trying to determine which of two hamburger stands has quicker service, McDougals or Dial-a-Burger. The average waiting time at McDougals over a period of 45 randomly selected days is 5.3 minutes with a standard deviation of 2.1 minutes. The average waiting time at Dial-a-Burger over a 40-day period is found to be 10.1 minutes with a standard deviation of 3.7 minutes. Can the customer conclude that one of the hamburger stands gives faster service than the other at a .01 level of significance?

Small samples: the two-sample t test

The decision rules given in Fig. 8.1 are appropriate only when both the sample sizes n_1 and n_2 are large. In practice, this means that n_1 and n_2 must each be at least 30. When one or both of the sample sizes are small, a slightly different procedure is required and certain assumptions must be met. In this case, the data from each population must be normally distributed and the two population standard deviations must be equal. Note that the sample standard deviations do not have to be equal. The sample statistics are random and are likely to differ even when the population standard deviations are equal.

Suppose samples are drawn independently from two populations and at least one of the samples is small. Assume that both sets of population data are normally distributed and have the same standard deviation. We shall test the null hypothesis H_0: $\mu_1 = \mu_2$ against one of the alternatives H_a: $\mu_1 > \mu_2$ or H_a: $\mu_1 \neq \mu_2$. The test statistic is

$$t = \frac{\bar{X}_1 - \bar{X}_2}{s\sqrt{1/n_1 + 1/n_2}}$$

where $\bar{X}_1$ and $\bar{X}_2$ denote the means of the samples drawn from each population and n_1 and n_2 are the sample sizes. The quantity s is the estimate of the common population standard deviation and is defined by

$$s = \sqrt{\frac{(n_1 - 1)s_1^2 + (n_2 - 1)s_2^2}{n_1 + n_2 - 2}}$$

Just as in the large-sample case, if the hypotheses are H_0: $\mu_1 = \mu_2$ and H_a: $\mu_1 > \mu_2$, then H_0 is rejected for large values of t. If the hypotheses are H_0: $\mu_1 = \mu_2$ and H_a: $\mu_1 \neq \mu_2$, then H_0 is rejected if t is too large or too small (negatively). When the required conditions on the population data are satisfied, the test statistic t will have a Student's t distribution if H_0 is true. This is the same t distribution used in the one-sample t test in Chap. 7. The same tables can be used, but now we compute the degrees of freedom by $n_1 + n_2 - 2$. The tests of H_0: $\mu_1 = \mu_2$ that we have just described are called *two-sample t tests*.

Example 8-4. A certain business uses two makes of automobiles, and it wishes to determine if the maintenance costs during the first 10,000 miles are different for the two makes. After 25 cars of make 1 and 20 cars of make 2 have been driven 10,000 miles, it is found that the average maintenance cost for cars of make 1 is $265 with a standard deviation of $20, and the average for cars of make 2 is $250 with a standard deviation of $15. Do these data indicate a real difference between the average maintenance costs of makes 1 and 2 at a .01 level of significance?

> SOLUTION: Let μ_1 and μ_2 denote the mean maintenance costs during the first 10,000 miles for all cars of makes 1 and 2, respectively. We wish to test the hypotheses
>
> H_0: $\mu_1 = \mu_2$
>
> H_a: $\mu_1 \neq \mu_2$
>
> (Notice that the alternative hypothesis is H_a: $\mu_1 \neq \mu_2$ because there is no a priori reason to believe one make is cheaper to maintain than the other.) From the given data $\bar{X}_1 = \$265$, $\bar{X}_2 = \$250$, $s_1 = \$20$, $s_2 = \$15$, $n_1 = 25$, and $n_2 = 20$. Thus, the estimate of the common population standard deviation is
>
> $$s = \sqrt{\frac{(n_1 - 1)s_1^2 + (n_2 - 1)s_2^2}{n_1 + n_2 - 2}}$$
>
> $$= \sqrt{\frac{24(20)^2 + 19(15)^2}{25 + 20 - 2}}$$
>
> $$= 17.96$$

Hence, the test statistic

$$t = \frac{\bar{X}_1 - \bar{X}_2}{s\sqrt{1/n_1 + 1/n_2}}$$

$$= \frac{265 - 250}{17.96\sqrt{1/25 + 1/20}}$$

$$= \frac{15}{5.39}$$

$$= 2.78$$

From the t tables, using $n_1 + n_2 - 2 = 25 + 20 - 2 = 43$ degrees of freedom, we see that H_0 should be rejected if $t \geq 2.70$ or $t \leq -2.70$. Since the observed value of $t = 2.78 > 2.70$, H_0 is rejected. There is strong evidence that the average maintenance costs for the two makes of automobiles are different. Looking back at the data, we see that the sample mean $\bar{X}_1$ for make 1 exceeds the sample mean $\bar{X}_2$ for make 2. Thus, the test implies that make 1 costs significantly more than make 2 to maintain at the .01 level of significance.

Example 8-5. An unscrupulous butcher wished to determine if his beef sales could be increased by adding red dye to his older pieces of meat, making them appear fresh. Over a 20-week period (immediately preceding his apprehension by a consumers' group) he would alternately dye the old meat during one week and not dye it during the next week. He found that the average sales per week during the 10 weeks when he added dye were $485 with a standard deviation of $40, and the average sales for the other 10 weeks were $470 with a standard deviation of $35. Would the butcher be justified in concluding that the addition of dye to old meat increases sales at a .05 level of significance?

SOLUTION. Let μ_1 and μ_2 denote the mean weekly sales when dye is added and when it isn't, respectively. We wish to test H_0: $\mu_1 = \mu_2$ versus H_a: $\mu_1 > \mu_2$. The alternative hypothesis is chosen to reflect the assumption that adding dye to the meat can only increase sales, not decrease them. From the sample data $\bar{X}_1 = \$485$, $\bar{X}_2 = \$470$, $s_1 = \$40$, $s_2 = \$35$, $n_1 = 10$, and $n_2 = 10$. Hence, the common population standard deviation is estimated by

$$s = \sqrt{\frac{(n_1 - 1)s_1^2 + (n_2 - 1)s_2^2}{n_1 + n_2 - 2}}$$

$$= \sqrt{\frac{(9)(40)^2 + (9)(35)^2}{10 + 10 - 2}}$$

$$= 37.58$$

The test statistic

$$t = \frac{\bar{X}_1 - \bar{X}_2}{s\sqrt{1/n_1 + 1/n_2}}$$

$$= \frac{485 - 470}{37.58\sqrt{1/10 + 1/10}}$$

$$= .89$$

The sampling distribution of 5 is the Student's t distribution with $n_1 + n_2 - 2 = 18$ degrees of freedom. From the t tables we find that H_0 should be rejected if the observed value of t is at least 1.73. Since $t = .89 < 1.73$, the butcher should conclude that adding red dye to old meat does not increase sales.

Summary

Experiments performed to detect a difference between several items, such as products, treatments, or methods, are called comparative experiments. A variety of statistical tests can be applied to these experiments. Which test to use depends on the number of items compared, how the sample or samples are drawn, the sample size, and the underlying distribution of the population data. In this chapter we have examined procedures for comparing the means of two populations, denoted by μ_1 and μ_2 when independent samples are drawn from each population. The possible alternatives to H_0: $\mu_1 = \mu_2$ are H_a: $\mu_1 \neq \mu_2$, H_a: $\mu_1 < \mu_2$, or H_a: $\mu_1 > \mu_2$.

Two cases were considered: large and small samples. In both cases a sample is drawn from each population and the sample means $\bar{X}_1$ and $\bar{X}_2$ and sample standard deviations s_1 and s_2 are computed. In the case where both samples are large, the test statistic is defined by

$$Z = \frac{\bar{X}_1 - \bar{X}_2}{\sqrt{s_1{}^2/n_1 + s_2{}^2/n_2}}$$

and will have an approximately standard normal distribution when H_0 is true. In the case where one or both samples are small, the test statistic is defined differently and certain assumptions regarding the population data are required. Then it is necessary to have equal standard deviations for each population, and both populations must have normally distributed data. Slight departure from these assumptions will not affect things too much. The test statistic is defined by

$$t = \frac{\bar{X}_1 - \bar{X}_2}{s\sqrt{1/n_1 + 1/n_2}}$$

where s is the estimate of the common standard deviation and is defined by the formula

$$s = \sqrt{\frac{(n_1 - 1)s_1^2 + (n_2 - 1)s_2^2}{n_1 + n_2 - 2}}$$

The sampling distribution of t is Student's t distribution with $n_1 + n_2 - 2$ degrees of freedom.

The appropriate test procedures for each of the three possible hypotheses were summarized. In the case of small samples the test is called the two-sample t test.

Important terms

Comparative experiments

Independent samples

One-sided tests

Student's t distribution

Two-sample t test

Two-sided tests

Additional problems

8-4 In comparing the gas mileage for two types of cars, makes X and Y, the following mileages were observed:

Make X

28	29	27.5	28.2	30.5
27.3	29.4	28	30.1	29.1

Make Y

27.3	28.2	29.1	27.8	29.1
30.2	27.3	29.7	28.0	27.5

Assuming that gas mileages are normally distributed, perform the appropriate test to determine if there is a significant difference between the two makes at a .05 level of significance.

8-5 A farmer wished to determine which of two stimulants A or B is more effective in increasing the number of eggs laid by his chickens. He added stimulant A to the diets of 50 randomly selected chickens and added stimulant B to the diets of another 50 randomly selected chickens. He observed that for stimulant A the average number of eggs laid per chicken was 15.3 with a standard deviation of 5, and the average for B was 17.2 with a standard deviation of 4.3. Do these results indicate a significant difference in the effect of the two drugs at a .05 level of significance?

8-6 A commuter is trying to decide between two different routes to work, freeway 1 and freeway 2. On each work day over a 6-month period he tosses a coin to determine which route to take, taking freeway 1 if the coin lands heads and freeway 2 if it lands tails. On the 72 days the coin landed heads his average travel time was 47.2 minutes with a standard deviation of 8.4 minutes. On the 58 days the coin landed tails his average travel time was 43.1 minutes with a standard deviation of 5.8 minutes. Is one route significantly faster than the other, testing at a .05 level of significance?

8-7 A clever waiter wishes to determine which of two methods of behavior, being overly complimentary or being highly aloof, yields higher tips. His tips average $8.75 a day with a standard deviation of $4.20 over 35 randomly selected days when he acts overly complimentary. On 42 days when he acts highly aloof, his tips average $10.62 with a standard deviation of $5.83. Based on these data is there sufficient evidence to conclude that one method of behavior yields higher tips than the other? Use a .01 level of significance.

8-8 Two beer-bottling machines are compared to determine if there is a difference in the average time between breakdowns. The average time between 10 randomly selected breakdowns and the breakdown immediately following is recorded for machines 1 and 2 with $\bar{X}_1 = 9.3$ hours, $\bar{X}_2 = 10.8$ hours, $s_1 = 2$ hours, and

$s_2 = 3.7$ hours. Assuming that the time between breakdowns is normally distributed with common standard deviations for both machines, test H_0: $\mu_1 = \mu_2$ versus H_a: $\mu_1 \neq \mu_2$ at a .05 level of significance.

8-9 A comparison is made between two airlines to determine if the arrival times of their regular flights from Los Angeles to Cleveland are off schedule by the same amount of time. For 100 randomly selected flights, airline A was observed to be off schedule by an average time of 17 minutes with a standard deviation of 11.3 minutes. For 150 randomly selected flights, airline B was found to be off schedule by an average time of 23 minutes with a standard deviation of 13.8 minutes. Do these data indicate any real difference in the off-schedule times? Use a .01 level of significance.

8-10 Describe a situation in which a one-tailed test would be appropriate and a situation in which a two-tailed test would be appropriate.

True-false test

Circle the correct letter.

T F 1. The two-sample test of population means applies to any two samples no matter how they are obtained.

T F 2. The two-sample t test only applies when the two sets of population data are normally distributed.

T F 3. In the two-sample t test the number of degrees of freedom for the test statistic increases as the sample sizes increase.

T F 4. The two-sample test of population means can be used only when equal sample sizes are drawn.

T F 5. If, upon applying the two-sample t test, a value of $t = 0$ is observed, then H_0 would be accepted no matter which level of significance is used.

T F 6. A two-sample test is twice as accurate as a one-sample test because twice as much data is used.

T F 7. If it is known that μ_1 is at least as big as μ_2, then the appropriate t test will be one-sided.

T F   8. The standard normal tables can be used to determine the rejection region for tests comparing two population means no matter what the sample sizes are.

T F   9. If t is negative, then there is strong evidence that H_0 is false.

T F   10. A one-sided t test is usually appropriate when comparing a new treatment with an old one because we are interested only in determining whether the new treatment is better.

Nonparametric tests for comparing two treatments **9**

The tests considered in the previous chapters require special conditions. For example, when the sample is large, the central limit theorem is applied to the test statistic for tests involving population means; when the sample is small, it is necessary to assume the underlying population data are normally distributed. In addition, in the tests we have considered it is implicitly assumed that the sample data are measurement data, which are needed to compute means and standard deviations. In many experiments, however, these assumptions are not satisfied. Different tests are needed when a small sample is drawn and no normality assumptions can be made, or when we have or may wish to use ordinal data instead of measurement data. In this chapter we shall introduce some effective and easy-to-use tests which can be applied to ordinal data in small-sample situations.

As in Chap. 8, we shall examine tests appropriate for comparative experiments involving two items. Now, however, we shall study *nonparametric tests,* so denoted because no assumptions about the distribution of the population data are required. In these tests we first reduce measurement data (if that is the form of our data) to ordinal data by replacing measurements with ranks. If we have ordinal data to begin with, the tests can still be applied. As mentioned in the discussion of types of data in Chap. 2, transforming measurements to rank does entail some loss of information. However, in this process the data are simplified enough to compute probability distributions so that tests may be conducted without making any assumptions about the underlying nature of the data. You will not have to determine these probabilities; the relevant ones are tabulated in the Appendix, and we shall use them in applying the tests we shall study. The first and easiest test to apply is the *sign test*.

The sign test

The sign test gives an easy method for checking whether or not the centers of distribution are the same in a situation with *paired data*. Paired data occur when we obtain observations from two different populations or from one population at two different times and when we can pair up each observation in the first sample with a unique observation in the second sample. Paired data typically occur when we are testing the effects of a treatment or process and have "before-and-after" observations for each sample element. For example, we may wish to test the effects of a certain diet, a hair-growth hormone, or a method of teaching and therefore must obtain before-and-after data. The main advantage of paired data is that by obtaining related pairs of observations, the variation between samples of our measurements is reduced, making it easier to detect changes or differences in the effects of dissimilar treatments.

Sometimes we may not have before-and-after observations but may be able to pair data originating from two independent samples. For example, if we were comparing two different teaching methods and had two groups of students, one group being taught with each method, we might wish to make an analysis based on test scores from each group. Before making this analysis, it may be possible to pair the students, one from each group, according to items we consider relevant to our study, such as IQ scores, ages, or economic backgrounds. (In the examples we shall study, the data will already be paired.) Consider now, the following example.

Example 9-1. A researcher wishes to determine whether or not a certain 6-week crash diet is effective over a long period of time. A sample of 15 dieters is selected. Each person's weight is recorded before starting the diet and 1 year after it is concluded. Based on the results shown in Fig. 9-1, should you conclude that the diet is effective over a 1-year period, or should you conclude instead that the possible short-term effects of the diet disappear as the dieters return to their normal eating habits.

Weights of 15 dieters before and 1 year after a 6-week "crash-diet" program

Before	1 year after	
160	163	−
180	182	−
170	175	−
170	120	+
145	150	−
162	160	+
195	190	+
140	137	+
128	126	+
200	190	+
210	230	−
155	120	+
135	134	+
140	90	+
182	183	−

Figure 9-1

SOLUTION: Statistically speaking, we wish to test the following hypotheses:

H_0:   The crash diet has no long-term benefits (i.e., the average weight difference is zero).

H_a: The diet is effective (before > after).

(Notice that the two-sample t test may not be appropriate in this situation because the data do not arise from two independent samples.)

We proceed in the following way. We look at the before-

and-after weight difference $B - A$ for each person, recording a $+$ if $B - A > 0$, recording a $-$ if $B - A < 0$. We record nothing if $B - A = 0$, in effect, eliminating this person from the study. In this way we obtain a list of $+$'s and $-$'s, as shown in the right-hand column of Fig. 9-1. If H_0 is really true, namely, if the crash diet has no long-term effects, we would expect to observe roughly the same number of $+$'s as $-$'s; statistically speaking, then, $P(+) = P(-) = 1/2$. This would make S, the total number of $+$'s in our data, a binomial variable with $n = 15, p = 1/2$. If H_a is true, that is, if the diet is effective, we would expect to see more larger "before" weights than "after" weights; in other words, we would expect to see a lot of $+$'s ($+$ is recorded only when $B > A$). Thus, we would tend to reject H_0 if S is sufficiently large.

We shall, in fact, use S as the test statistic. We have converted our paired data into a binomial situation and thus rewrite our hypotheses as

H_0: $P(+) = P(-) = 1/2$

H_a: $P(+) > 1/2$

To decide when to reject H_0, say, at significance level $\alpha = .05$, we need only find the number c such that under H_0, $P(S \geq c) = .05$. We would reject H_0 if we observe c or more $+$'s. This is similar to the procedure followed in Chap. 6, except that in Chap. 6 we had a large sample and could use the normal approximation for S. Now our sample is not large enough to do this; instead, we must obtain our information from the binomial tables in the Appendix. Looking, then, at the binomial tables for $p = 1/2, n = 15$, we obtain the following probabilities:

$P(S \geq 13) = .004$

$P(S \geq 12) = .018$

$P(S \geq 11) = .059$

If H_0 is true and the diet is ineffective, the probability of obtaining 13 or more $+$'s is only .004; the probability of obtaining 12 or more $+$'s is only .018; and the probability of obtaining 11 or more $+$'s is .059.

This last probability slightly exceeds the significance level we wish to use, namely, .05. There is no c for which $P(S \geq c) = .05$; the probabilities jump from .018 to .059. If we wish to relax our type I error probability slightly and use a test with a significance level of $\alpha = .059$, our rejection region should be to reject H_0 if we observe 11 or more +'s; then P(type I error) $= .059$. On the other hand, if we wish to use a lower significance level of $\alpha = .018$, our rejection region should be to reject H_0 if we observe 12 or more +'s; then, P(type I error) $= .018$. (This decision is a subjective one; it is up to you to decide at what significance level to conduct your test.)

Returning to the data from our experiment, we see $S = 9$; therefore we would accept H_0 for either of the above tests. There is insufficient evidence at level $\alpha = .05$ (.059) to indicate that the diet has any long-term effects.

The test we have just developed is called the *sign test*. It converts paired data into plus or minus signs depending upon whether the differences in pairs are positive or negative. The sign test is a useful test for a number of reasons: It is easy to apply, makes no assumptions about the underlying distribution of the data, and makes use of the fact that the data are paired. However, converting every difference to a plus or minus sign, no matter how large it is, causes us to lose some information. (We shall soon introduce a refinement of the sign test, the Wilcoxon test, which uses more information from the data.)

The steps used in applying the sign test are summarized as follows.

STEPS IN APPLYING THE SIGN TEST

1   State the hypotheses to be tested, determining whether the test is to be one- or two-tailed. The hypotheses will always be

H_0: $P(+) = P(-) = 1/2$

H_a: $P(+) > 1/2$ [or $P(+) < 1/2$]   (one-tailed)

or

$P(+) \neq 1/2$ (two-tailed)

2   Determine the significance level and rejection region. Under H_0, $P(+) = P(-) = 1/2$, and so the number of +'s is binomial with success probability $p = 1/2$ and n trials.

In the one-tailed test using significance level α, find the c such that under H_0, $P(S \geq c) = \alpha$; or find the c such that $P(S \leq c) = \alpha$, depending upon the direction of the test. Then H_0 will be rejected if you observe $S \geq c$ (or $S \leq c$). In the two-tailed test, the procedure is analogous. Usually there will be no c which exactly satisfies the previous equations. You must then choose the c closest to the desired significance level α (or perhaps the c closest to but not bigger than the desired level), being careful to indicate in your analysis exactly what significance level you finally chose.

3   Gather the data, take the necessary differences, recording +'s and −'s (excluding zero differences), compute S, and determine the result of the test.

Example 9-2. A producer of X-rated movies believes that the closing of a movie in one city because it is declared obscene will have a positive effect on its box office sales in other cities owing to the publicity it will receive. To test this hypothesis, the producer obtains the records for 14 movies which have been declared obscene and compares the nationwide weekly sales averaged over the 4 weeks immediately preceding their first closings with the 4 weeks following the closings, as shown in Fig. 9-2. If the sign test is used at a significance level of $\alpha = .05$, what should the movie producer conclude?

Average weekly sales of X-rated movies (in thousands of dollars)

Movie	4 weeks before first closing	4 weeks after first closing	Sign of difference
"The 10-Week Caress"	12	24	−
"Inside a Love Commune"	12.5	8	+
"The Lab Assistant"	18	20.5	−
"Massage Madness"	8.4	7	+
"The Flesh Merchants"	13	25.3	−
"The Vicious Circle"	10	28	−
"Finders Keepers"	8	24	−
"The Groupies"	1	8.5	−
"The Orgy"	3	11	−
"The French Friar"	25	45	−
"The Organ Recital"	21.1	19	+
"Suburban Love Slaves"	15.3	14.1	+
"The City Model"	11.5	23	−
"Triple Play"	19	18	+

Figure 9-2

SOLUTION: We follow the instructions for applying the sign test. The hypotheses to be tested are as follows.

H_0:   There is no difference in sales before and after the first obscenity closing $[P(+) = P(-) = 1/2]$.

H_a: An obscenity closing has a positive effect on sales $[P(+) < 1/2]$.

Notice that the "direction" of the rejection region is different than its direction in Example 9-1. If the obscenity closing has a positive effect, we would expect the "after" sales to be bigger than the "before" sales; that is, there would be more $-$'s than $+$'s and $P(+) < 1/2$. In other words, we would reject H_0 if S is sufficiently small.

We let $\alpha = .05$. Since we shall reject H_0 when S is sufficiently small, we wish to find the number c such that under H_0, $P(S \leq c) = .05$. Looking at the binomial tables in the Appendix, we see that under H_0, $P(S \leq 3) = .029$ and $P(S \leq 4) = .09$. We choose the lower significance level $\alpha = .029$, deeming .09 to be too great a chance of making a type I error. (Of course, this is a subjective decision.) Thus, we will reject H_0 if we observe $S \leq 3$.

From Fig. 9-2 we observe that $S = 5$ and is therefore not in the rejection region. Thus, based on the sign test, we do not reject H_0. In other words, there is insufficient evidence to conclude that the closing of an X-rated movie in one city will boost its sales in another city.

Note that other factors may affect the results of our study, for example, the sizes of the cities in which the movies are closed. In a more detailed analysis we would try to take into account as many such factors as possible.

LARGE SAMPLES

In the case of large samples, the normal approximation to the binomial can be used for the sign test in much the same way as it was used in Chap. 6. Of course, after the conversion to $+$'s and $-$'s, the sign test itself is a binomial test. Under H_0, S is binomial with n trials and success probability $p = 1/2$, and so $np = n(1/2)$, $\sqrt{np(1-p)} = \sqrt{n(1/2)(1/2)} = (1/2)\sqrt{n}$. We would then use the procedures of Chap. 6.

Problems

9-1 An encyclopedia company wishes to determine whether its new salesmanship course is effective in increasing the average weekly sales of its door-to-door salesmen. Of those who have taken the course, 10 persons are randomly selected.. Their gross weekly sales average over 8 weeks before and 8 weeks after taking the course are recorded below. Testing at a significance level of $\alpha = .05$, use the sign test to determine if the course is effective.

Gross weekly sales (in dollars)

8 weeks before taking course	8 weeks after taking course
480	620
310	290
175	225
560	570
230	295
360	320
415	418
421	456
281	481
75	78

9-2 A study is done to determine whether or not vitamin C reduces the incidence of colds. A random sample of 15 persons is selected. These people are given large daily doses (2 grams per day) of vitamin C over a 1-year period. The number of colds they contract is recorded and is compared with the number of colds they contracted in the previous year, as tabulated below. Using the sign test at a significance level of $\alpha = .05$, determine whether or not vitamin C is effective.

Incidence of colds in a 1-year period

Cold sufferer	Without vitamin C	With vitamin C
A	6	1
B	5	1
C	2	0
D	1	0
E	9	4

(*continued*)

Cold sufferer	Without vitamin C	With vitamin C
F	0	0
G	4	3
H	4	5
I	3	1
J	5	2
K	4	2
L	7	3
M	2	1
N	6	1
O	3	0

9-3   A study is done to determine whether there is a significant loss of hearing incurred by workers who must spend most of their time near noisy machinery. In a large factory 12 workers are randomly selected and their hearing is tested before and after their initial 2-year period on the job. With the data below, use the sign test at a significance level of $\alpha = .05$ to decide whether there is significant hearing loss.

Sound detection (measured in decibels on the ISO scale)*

Worker	Before job experience	After 2 years on job
A	18.2	26.1
B	23.6	28.2
C	19.0	28.1
D	12.7	22.3
E	38.1	37.2
F	42.6	40.0
G	11.9	20.6
H	13.2	22.5
I	10.3	24.7
J	25.1	24.8
K	21.5	21.4
L	17.6	38.1

*The numbers indicate the decibel level of the softest sound recognized, and so a high number indicates hearing loss.

9-4   A study is done by the parapsychology institute to determine whether spending lengthy periods in an orgone box increases

ESP abilities. A sample of 12 volunteers is selected, and these people spend 12 hours a day in an orgone box over a 2-month period. Before and after this period, the volunteers are asked to guess which of the three colored balls (red, yellow, or blue) the experimenter has secretly selected; this experiment is performed 25 times. The number of correct guesses, both before and after, is then recorded. Using the sign test at a level of $\alpha = .1$, determine whether sitting in the orgone box increases ESP abilities.

Number of correct guesses in 25 trials

Before orgone-box treatment	After orgone-box treatment
8	6
9	10
5	7
6	8
11	6
13	4
9	8
4	7
8	6
7	10
10	9
3	7

The Wilcoxon Test

The *Wilcoxon matched-pairs signed-ranks test* is another nonparametric test which makes use of paired data; however, this test uses more information from the data than the sign test. The setting for the Wilcoxon test is exactly the same as for the sign test, but the Wilcoxon test employs the Wilcoxon statistic W, which is computed as follows.

First, the absolute differences of the data pairs are computed and ranked according to size: The lowest absolute difference receives rank 1; the second lowest, rank 2; etcetera. Then, a plus or minus sign is affixed to each rank according to whether the differences are positive or negative (as in the sign test). We now have a list of *signed ranks*, rather than a list of only +'s and −'s (as in the sign test), and we obtain W by summing these signed ranks. The idea behind W is similar to the ratio-

nale of the sign test; in fact, the Wilcoxon test can be considered a refinement of the sign test. In the sign test we considered only the
number of positive paired differences; now we consider the ranks of
these differences as well, W supplying this added information which
could not be supplied by S in the sign test.

   Under the null hypothesis that the two populations have the same
distributions, we would expect the ranks of the plus and minus differences to be evenly distributed. In this case, if we added the signed
ranks, we would expect the plus and minus ranks to roughly cancel
each other, the result being that W would be close to 0. If H_0 is false,
we would expect W to be large (either positively or negatively) in the
two-tailed case or to be in a particular direction in the one-tailed case.
(We shall consider only the one-tailed case, the two-tailed case would
be treated in an analogous way.) When the sample size n is small, the
exact distribution of W under H_0 is known; this result is used in obtaining the Wilcoxon table in the Appendix. When n is large, it turns out
(not surprisingly) that W can be approximated by a normal curve.

   We return now to Example 9-2 and apply the Wilcoxon test. In
Fig. 9-3 we repeat the data of Fig. 9-2, but instead of listing +'s and −'s,
we list the absolute differences (with signs in parentheses) and then the

Average sales of X-rated Movies (in thousands of dollars)

Movie	4 weeks before first closing	4 weeks after first closing	(Sign) $|B - A|$	Signed rank of $|B - A|$
"The 10-Week Caress"	12	24	(−)12	(−)10
"Inside a Love Commune"	12.5	8	(+) 4.5	(+) 6
"The Lab Assistant"	18	20.5	(−) 2.5	(−) 5
"Massage Madness"	8.4	7	(+) 1.4	(+) 3
"The Flesh Merchants"	13	25.3	(−)12.3	(−)11
"The Vicious Circle"	10	28	(−)18	(−)13
"Finders Keepers"	8	24	(−)16	(−)12
"The Groupies"	1	8.5	(−) 7.5	(−) 7
"The Orgy"	3	11	(−) 8	(−) 8
"The French Friar"	25	45	(−)20	(−)14
"The Organ Recital"	21.1	19	(+) 2.1	(+) 4
"Suburban Love Slaves	15.3	14.1	(+) 1.2	(+) 2
"The City Model"	11.5	23	(−)11.5	(−) 9
"Triple Play"	19	18	(+) 1	(+) 1

$$W = -73$$

Figure 9-3

ranks of these differences; the lowest absolute difference is given rank 1, the second lowest rank 2, etcetera. At the bottom of the rank column is W, the sum of the signed ranks. Note that $W = -73$.

Recall the hypotheses we are testing:

H_0: There is no difference in sales before and after the first obscenity closing $[P(+) = P(-) = 1/2]$.

H_a: An obscenity closing has a positive effect on sales $[P(+) < 1/2]$.

Since H_a asserts that the sales are increased by an obscenity closing, we would expect that under H_a high ranks would correspond to negative differences. This would make the Wilcoxon statistic W extremely low (negative), and accordingly we would reject H_0 if $W \leq$ a specified number. We shall use the same significance level as before: $\alpha = .05$. Turning to the Wilcoxon table in the Appendix, we find the significance probability of the observed result to be $P(W \leq -73) = .01$, which places the observed value of W in the .05 rejection region. Therefore, we reject H_0, concluding that there is a difference in sales before and after an obscenity closing.

You should now be thinking, "Hey, wait a minute! How can we be rejecting H_0 now when we just accepted it using the sign test only a few pages ago?" Actually, such situations are not infrequent in statistical analyses. Different tests make different use of the data at hand and may give conflicting results. In the present example, we observed that the Wilcoxon test utilizes the ranks of the differences rather than just their signs. In the data presented, the number of plus differences was not different enough from the number of minus differences to warrant rejecting H_0 with the sign test. However, the minus differences were mostly quite large while the plus differences were relatively small, giving the minus differences a large number of high ranks and making W extremely low. Thus, W fell in the .05 rejection region.

From the information at hand we cannot determine which test is "correct." Although the Wilcoxon test makes better use of the data and one would be inclined to rely on it over the sign test, it is possible that H_0 is really true and the sign test yields the correct decision. If you obtain conflicting results, as we just did, you may wish to reserve making a final decision until more data have been gathered.

Example 9-3. In a certain city a study is done to determine whether or not the use of police helicopters reduces burglary rates. The study includes 8 cities, and in Fig. 9-4, the data are presented

Average no. of burglaries (per thousand)

2 years before using helicopters	2 years after using helicopters	(Sign) $\lvert B - A \rvert$	Signed rank of $\lvert B - A \rvert$
3.1	2.8	(+) .3	(+)3
7.9	8.1	(−) .2	(−)2
5.1	2.7	(+)2.4	(+)8
11.2	10.0	(+)1.2	(+)5
9.6	11.0	(−)1.4	(−)6
14.2	12.7	(+)1.5	(+)7
12.8	12.7	(+) .1	(+)1
8.7	9.1	(−) .4	(−)4
			$W = 12$

Figure 9-4

which gives the average number of burglaries per 1,000 people during the 2 years immediately preceding and immediately following the use of police helicopters. Using the Wilcoxon test at a significance level of $\alpha = .05$, what should be concluded?

SOLUTION: This is a one-sided test, and the hypotheses are

H_0:   The helicopters have no effect.

H_a: The helicopters are effective in reducing burglary rates.

We have stated H_a this way because it is reasonable to assume that the helicopters don't increase burglary rates.

In Fig. 9-4 we have given the computations necessary to compute W, obtaining $W = 12$. Since H_a asserts that the burglary rates are reduced, if H_a is true, we would expect large positive differences and a large (positive) value of W. Thus we shall reject H_0 if W is sufficiently large. Looking at the Wilcoxon table, we obtain $P(W \geq 20) = .098$. Since $P(W > 12) > P(W \geq 20) = .098$ and since $.098 > .05$, we conclude that W does not lie in the $\alpha = .05$ rejection region. Based on this result, we conclude that there is insufficient evidence to justify the claim that the use of police helicopters reduces burglary rates.

Of course, in making a decision of this importance, we would wish to obtain more data than was presented and analyze the data from many different viewpoints. You should view this example as an application of the Wilcoxon test, not as a thorough analysis of the police-helicopter problem. As an exercise, see if you can list some other aspects of this problem which should be investigated. What further data would you want to obtain if you were in charge of conducting this study?

LARGE SAMPLE SIZES

If the sample size n is too large for the Wilcoxon table, we can use the normal approximation. If H_0 is true, then the statistic W has a mean $\mu = 0$ and a standard deviation of $\sigma = \sqrt{n(n+1)(2n+1)/6}$, and $Z = (W - \mu)/\sigma$ has approximately a standard normal distribution. Therefore the normal tables can be used to carry out the appropriate test.

TIES

If we obtain zero differences (*tied pairs*), we include them in the data and rank them along with the other scores. This is different from the sign test where tied pairs are eliminated.

MIDRANKS

Another type of tie is in obtaining two or more identical differences. For instance, the pairs $(6,3)$, $(5,2)$, and $(2, -1)$ each have a difference of 3. In such situations we do what is called *taking midranks*, namely, we take the average of the ranks consumed by the tied values and assign this average to each value, the other ranks being unaffected. Then W can be computed in the ordinary way. For example, if we had differences of -1, 1.1, 3, 3, 3, and 8, we would assign ranks as shown in Fig. 9-5. If there are many ties, we might try to obtain more data or refine our measuring techniques.

Assigning midranks

Difference	Rank
−1	1
1.1	2
3	4
3	4 → Average of ranks 3 to 5
3	4
8	6

Figure 9-5

Problems

9-5   Apply the Wilcoxon test to the data in Prob. 9-1, testing at a significance level of $\alpha = .05$. Are the results conflicting? Explain.

9-6   Apply the Wilcoxon test to the data in Prob. 9-3, testing at a significance level of $\alpha = .05$. Are the results conflicting? Explain.

9-7   Make up paired data for which the sign test accepts H_0 and the Wilcoxon test rejects H_0 (at a significance level of $\alpha = .05$). Explain why this conflict may occur and what you would do when encountering such a situation.

The Mann-Whitney test

We return now to the case of unpaired data. As in the last chapter, we wish to compare two populations on the basis of independent samples drawn from each. Now, however, we shall consider a nonparametric test called the *Mann-Whitney test*. This test requires neither a large sample nor the assumption that the population data must be normally distributed. Although we shall discuss this test for small sample situations, it can be used for large samples as well.

As with the Wilcoxon and sign tests, the Mann-Whitney test can be used with data having measurements on only an ordinal scale and with no parametric assumptions being made. Like the Wilcoxon test, the Mann-Whitney test is a rank test. As usual, we wish to test a hypothesis about the distributions of the two populations under study, for example, whether they have the same distributions or whether one is generally larger than the other. Various statistics are appropriate for this test, all of which lead to the same conclusion for any particular set of data. We shall use one statistic that is easy to compute and denote it by R.

We compute R in the following way. First, we combine all the data from both samples and rank them (assigning rank 1 to the lowest overall observation, 2 to the second lowest, etcetera). We then arbitrarily label the populations 1 and 2 (with sample sizes, say, n_1, n_2) and let R be the sum of ranks of population 1. If H_0 is true and the populations have the same distributions, we would expect that the values of population 1 would be randomly sprinkled among the scores in the combined list of data from both populations. This means that relative to the respective sample sizes, the value of R would be about the same no matter which population is labeled 1 and which is labeled 2. Since we know the sum of all ranks is $N = 1 + 2 + \cdots + (n_1 + n_2)$, R should be close to $[n_1/(n_1 + n_2)]N$, or the proportion of N assigned to population 1, based on the sample sizes.

Carrying out this line of reasoning in much greater detail actually enables us to compute exact probabilities for R, for small sample sizes, assuming that H_0 is true. These probabilities are used to determine the Mann-Whitney table in the Appendix, which lists two-tailed rejection regions for selected sample sizes. We shall illustrate the two-tailed situation, in which H_0 is rejected if R is either too large or too small. (The one-tailed test is treated analogously.) Notice that the sample sizes n_1 and n_2 may be different when using this test, a situation which cannot occur when using paired data.

For large sample sizes, a normal approximation is available for R. However, in large-sample situations, the methods of Chap. 8 can be used whenever the data is measure data.

TIES

We may sometimes encounter tied observations. As in the Wilcoxon test, we assign them midranks, the average of the ranks they would take. For instance if the third, fourth, and fifth largest observations were the same, they would each receive rank $(3 + 4 + 5)/3 = 4$, with the next higher observation receiving rank 6. If there are a large number of ties, we may wish to gather more data or use a different test.

Example 9-4. An airplane manufacturer wishes to compare production times of two different assembly schemes for jumbojets. For this purpose, scheme A is used in the production of seven jets and scheme B is used in the production of eight jets; the data are recorded in Fig. 9-6. Using the Mann-Whitney test at a significance level of $\alpha = .05$, what should the manufacturer conclude?

SOLUTION: The hypotheses to be tested are as follows:

H_0: There is no significant difference between the two schemes (the production times are distributed the same for both schemes).

H_a: One scheme is faster than the other.

This is a two-tailed test. The manufacturer has no knowledge of which scheme, if any, is better and wishes to detect a difference in either direction.

Looking at Fig. 9-6, note that we have labeled scheme A

*Production times (in months) and ranks
for two Jumbo-jet assembly schemes*

Scheme A (Pop. 1)		Scheme B (Pop. 2)	
Production time	Rank	Production time	Rank
4.2	2	4.3	3
4.4	4	4.5	5
5.0	10	5.2	12
4.8	8	5.3	13
4.9	9	5.8	15
4.6	6	4.7	7
5.1	11	5.4	14
		3.7	1

R = sum of scheme A ranks = 50

Figure 9-6

production times as population 1. Since we wish to detect a difference in times in either direction, we shall reject H_0 if R is either too low (indicating scheme A is faster) or too high (indicating scheme B is faster). We note from Fig. 9-6 that $R = 50$. Turning to the Mann-Whitney tables, in the Appendix, we determine that at a significance level of $\alpha = .05\,(.054)$, H_0 is to be rejected if either $R \leq 39$ or $R \geq 73$. Since we have observed $R = 50$, which is not in this rejection region, we accept H_0, concluding that there is no significant difference in the production times between the two schemes.

Additional problems

9-8 A new type of document shredder has been devised. In order to test whether or not the new shredder is more efficient than the conventional one, each shredder is fed 100-pound batches of politically sensitive documents and the shredding times are recorded below. Using the Mann-Whitney test at a significance level of $\alpha = .05$, would you accept or reject the hypothesis that one shredder is better than the other?

Shredding times (in minutes) for 100-pound batches of documents

Old shredder	New shredder
10.2	13.2
9.8	8.7
10.1	12.1
8.6	11.9
11.5	16.2
12.7	18.1
13.2	10.5
7.1	11.4

9-9   A teacher wishes to compare two methods of teaching how to type, method A and method B. Each method is used on a group of nine students, and the typing speeds are recorded (in words per minute) at the end of the course. The data are given below. Use the Mann-Whitney test at a significance level of $\alpha = .05$ to detect whether or not there is a difference between the methods.

Typing speeds

Method A	Method B
47	62
51	65
50	59
53	40
39	71
48	53
52	58
50	60

Summary

We have discussed three tests for comparing two sets of data when we have information obtained from small samples. The tests we have studied are called nonparametric tests because we need make no assumptions about the underlying distribution or parameters of the populations under study. The first two sets we considered, the sign test and the Wilcoxon test, were used in situations where paired data are available. Paired data occur commonly in situations where we wish to assess the effect of a new treatment or influence, and as a result we have "before-and-after" measurements. In other situations data can sometimes be paired by associating common factors with elements of two samples. Pairing tends to lower the sample variance, and so it is often a useful technique.

The third test we introduced, the Mann-Whitney test, was applied to situations in which we had data from two independent samples, the same setup as in Chap. 8. All the tests we studied can be applied in the large-sample situation using the central limit theorem and appropriate normal approximations; however, our emphasis has been on the small-sample case.

Important terms

Absolute differences

Independent samples

Mann-Whitney statistic R

Mann-Whitney test

Midranks

Nonparametric tests

Paired data

Ranks

Sign test

Signs

Wilcoxon statistic W

Wilcoxon test

True-false test

Circle the correct letter.

T   F   1. The sign test and Wilcoxon test are applied in the same situations.

T F 2. The Wilcoxon test and Mann-Whitney test are applied in the same situations.

T F 3. If the sign test rejects H_0, so will the Wilcoxon test since the Wilcoxon test uses more information from the sample data.

T F 4. The t test can be applied whenever the Mann-Whitney test can be applied.

T F 5. The tests in this chapter are useful in cases where no assumptions about the distribution of the underlying population can be made.

T F 6. One drawback to nonparametric tests is that significance levels cannot always be determined.

T F 7. Transforming data to ranks loses some information but enables exact distributions (under H_0) to be determined.

T F 8. Although the sign test involves loss of much data information, its simplicity of application makes it a useful tool.

T F 9. The tests in this chapter cannot be used for large-sample situations.

T F 10. Ranking the data makes the corresponding test less sensitive to a few "wild" or extreme values.

Chi-square tests: goodness of fit and test of independence **10**

In this chapter we shall examine two useful nonparametric tests: The *chi-square goodness-of-fit test* is often used to determine if a certain probability model is consistent with observed sample data. The *chi-square test of independence* is used to check whether or not two variables are independent. The methods employed in both tests are very similar. Both tests use *nominal*, or *count, data*, observations which are classified into certain categories, or groups. The goodness-of-fit test is presented first because it is simpler.

Chi-square goodness-of-fit test

In all the problems of hypothesis testing considered so far, we have outlined procedures for carrying out tests of particular parameters like μ and ρ. Sometimes we are interested, not in a parameter of a population, but in whether the population has a specified probability distribution. This problem often arises in exploratory research, particularly when it is necessary to establish the assumptions required for the application of some parametric test. For example, it may be important to know if scores on a certain test are normally distributed or if accidents on a certain highway tend to occur uniformly through the year. In such situations the *chi-square goodness-of-fit test* is used, as in the following example.

Example 10-1. Suppose a fire department of a large city suspects that the percentage of fires in different parts of the city is changing. From records of the past several years it is known that 40% of the fires in a year occur on the north side of town, 20% occur on the south side, 30% occur on the east side, and 10% occur on the west side. During the year just passed, however, the following frequencies of fires occurred:

No. of fires in parts of city

North	South	East	West
110	50	80	60

Figure 10-1

The fire department wants to know if these observed frequencies are consistent with the old ones or if they are so different that they could not be attributed to chance fluctuations alone.

SOLUTION: We must choose between the following hypotheses:

H_0:   The percentage of fires in different parts of the city is not changing.

H_a: The percentage of fires in different parts of the city is changing.

We may compute the expected frequency of fires in different parts of the city, assuming H_0 is true. Since there were 300 fires all year, we expect about $.4(300) = 120$ in the north side of the city, $.2(300) = 60$ in the south side, $.3(300) = 90$ in the east side, and $.1(300) = 30$ in the west side. The amount of discrepancy between the observed frequencies and the expected frequencies is measured by the sample statistic

$$\chi^2 = \Sigma \frac{(O - E)^2}{E}$$

where χ is the greek letter *chi*. The statistic χ^2 is called the *chi-square statistic*. In this formula, O denotes the observed frequency, E denotes the expected frequency, and Σ indicates we should sum the quantities $(O - E)^2/E$, one value for each of the four sides of the city.

If H_0 is true, we expect $(O - E)^2$ to be small for each side and hence χ^2 will be small. Thus, we reject H_0 if χ^2 is too large. The computation of the χ^2 statistic is given in Fig. 10-2.

| | No. of fires in part of city | | | |
	North	South	East	West
O = *Observed frequency*	110	50	80	60
E = *Expected frequency*	120	60	90	30
$\dfrac{(O - E)^2}{E}$	.83	1.67	1.11	30

Figure 10-2

For example, the quantity $(O - E)^2/E$ entered in the column headed by "North" is computed according to

$$\frac{(110 - 120)^2}{120} = \frac{(10)^2}{120} = \frac{100}{120} = .83$$

For "South" it is

$$\frac{(50 - 60)^2}{60} = \frac{(10)^2}{60} = \frac{100}{60} = 1.67$$

and so on. Adding the values of $(O - E)^2/E$ on the bottom row, we obtain

$$\chi^2 = .83 + 1.67 + 1.11 + 30 = 33.61$$

It turns out that under the null hypothesis, χ^2 has, for large n, approximately a *chi-square distribution*. Each chi-square distribution is characterized by a number called the *degrees of freedom* (df). As we shall explain in a moment, in our example, the distribution of χ^2 has 3 degrees of freedom (3 df). From the tables of chi-square distributions for various degrees of freedom given in the Appendix, we find that a χ^2 variable with 3 df will exceed 11.3 with probability .01 if H_0 is true. Thus, at a .01 level of significance, we reject H_0 if $\chi^2 \geq 11.3$ and accept H_0 if

$\chi^2 < 11.3$. Since the observed value of χ^2 is $33.61 > 11.3$, H_0 is rejected. This means there is less than 1 chance in 100 of observing frequencies as extreme as ours if H_0 were true. Thus, the fire department can conclude that the percentage of fires in different parts of the city is changing.

In general, suppose an experiment is repeated n times and on each trial there are k possible categories numbered $1, 2, \ldots, k$. Let p_1, $p_2, \ldots, p_k$ denote the suspected values of the probabilities of the k possible categories. These are the values of the probabilities of the categories specified by the null hypothesis. We wish to determine if these values are consistent with the sample data. In Example 10-1, $n = 300$, $k = 4$, and $p_1 = .4$, $p_2 = .2$, $p_3 = .3$, and $p_4 = .1$. The two hypotheses to be tested are:

H_0:   The probability of outcome $1 = p_1$.
The probability of outcome $2 = p_2$.
$$\vdots$$
The probability of outcome $k = p_k$.
H_a: At least one of these equalities does not hold.

A random sample is chosen, and the observed frequencies of each of the k categories are counted; these are denoted by $O_1, O_2, O_3, \ldots$, O_k. Then the expected frequencies $E_1, E_2, \ldots, E_k$ are computed from the formulas $E_1 = np_1$, $E_2 = np_2$, $\ldots$, and $E_k = np_k$. Finally, the quantity $(O - E)^2/E$ is found for each of the k categories. The result of this procedure is summarized in Fig. 10-3. The measure of discrepancy between the observed data and what would be expected if H_0 were true is

$$\chi^2 = \Sigma \frac{(O - E)^2}{E}$$

$$= \frac{(O_1 - E_1)^2}{E_1} + \frac{(O_2 - E_2)^2}{E_2} + \cdots + \frac{(O_k - E_k)^2}{E_k}$$

	Possible categories			
	1	*2*	*3*	*k*
$O = Observed\ frequency$	O_1	O_2	O_3 $\cdots$	O_k
$E = Expected\ frequency$	E_1	E_2	E_3	E_k
$\dfrac{(O - E)^2}{E}$	$\dfrac{(O_1 - E_1)^2}{E_1}$	$\dfrac{(O_2 - E_2)^2}{E_2}$	$\dfrac{(O_3 - E_3)^2}{E_3}$	$\dfrac{(O_k - E_k)^2}{E_k}$

Figure 10-3

This sample statistic will have approximately a chi-square distribution when the sample size is large if H_0 is true. As we have mentioned earlier, associated with each chi-square distribution is a number called the *degrees of freedom*. Whenever a chi-square goodness-of-fit test is carried out according to the model we have just described, with k possible categories, there are $k - 1$ degrees of freedom. The word "freedom" refers to the maximum number of category frequencies which may vary independently of each other. In this case, the number of degrees of freedom is $k - 1$ because the frequencies must add up to the total sample size; hence any particular category frequency may be determined by subtracting the sum of the other frequencies from the total sample size.

Using the chi-square distribution to approximate the sampling distribution of χ^2 will be accurate enough for most purposes when each of the expected frequencies exceeds 5. When this is not the case, it is advisable to redefine the categories so as to combine those with low expected frequencies. There is no definite way to do this, and a certain amount of subjectivity is necessarily involved. However, keep in mind that the goal is to obtain categories which have expected frequencies of at least 5.

The chi-square distribution has the general shape of the curve in Fig. 10-4. The exact shape depends on the number of degrees of freedom. The number χ_α^2 is the value of χ^2 that will be exceeded with probability α. Since large values of χ^2 indicate large discrepancies between observed and expected data, we

Reject H_0 if $\chi^2 \geq \chi_\alpha^2$.

Accept H_0 if $\chi^2 < \chi_\alpha^2$.

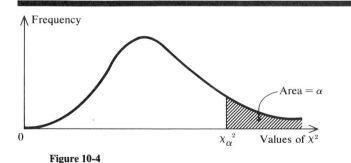

Figure 10-4

at a level of significance α. The value of χ^2 for any particular α and degree of freedom may be found in the tables of the chi-square distribution in the Appendix.

Example 10-2. A statistics professor suspects his students are so attuned to the concept of randomness that a table of random numbers might be constructed by merely having each student jot down in rapid succession the digits which spontaneously enter his mind. Upon collecting 100 numbers in this manner, the professor observed the frequencies given in Fig. 10-5. Are these numbers really random, and can the professor conclude that his students are attuned to the concept of randomness?

	Digit									
	0	*1*	*2*	*3*	*4*	*5*	*6*	*7*	*8*	*9*
O = Observed frequency	6	6	9	15	8	12	14	10	14	6

Figure 10-5

SOLUTION: We wish to test the null hypothesis that the numbers are randomly selected. If we let

$p_0 = P$(particular digit is a 0)

$p_1 = P$(particular digit is a 1)

and so on; then the hypotheses are

H_0: $p_0 = p_1 = \cdots = p_9 = .1$.

H_a: At least one of the probabilities does not equal .1.

From Fig. 10-5, we obtain Fig. 10-6:

	Digit									
	0	*1*	*2*	*3*	*4*	*5*	*6*	*7*	*8*	*9*
O = Observed frequency	6	6	9	15	8	12	14	10	14	6
E = Expected frequency	10	10	10	10	10	10	10	10	10	10
$\dfrac{(O-E)^2}{E}$	1.6	1.6	.1	2.5	.4	.4	1.6	0	1.6	1.6

Figure 10-6

We compute

$$\chi^2 = \frac{\Sigma(O - E)^2}{E} = 1.6 + 1.6 + .1 + 2.5 + .4 + .4 + 1.6 + 0 + 1.6$$
$$+ 1.6 = 11.4$$

Since k = no. of possible categories on any trial = 10, the statistic χ^2 has 9 degrees of freedom. Taking as our level of significance $\alpha = .05$, from the tables in the Appendix we find $\chi_\alpha^2 = 16.92$. Thus, $\chi^2 = 11.4 < 16.92$, and we accept H_0 at the .05 level of significance. The observed data are consistent with the hypothesis that the numbers were actually chosen at random, and the professor can conclude that his students are attuned to the concept of randomness.

Incidentally, the experiment described in Example 10-2 is fun to try in class. It's quick and easy to make the required computations. Every time we tried it, H_0 was accepted at the .05 level of significance.

Example 10-3. Some people say having children is a binomial experiment like tossing a coin. This means that the sex of a child on any birth is independent of the sex on the other births, and that the probability of a male or female being born is 1/2 on each birth. We check this hypothesis by choosing 50 families, each having 3 children, with the breakdown as to the number of boys as shown in Fig. 10-7. Are these results consistent with the hypothesis that having children is a binomial experiment?

	No. of boys			
	0	*1*	*2*	*3*
No. of families	9	22	14	5

Figure 10-7

SOLUTION: Consider the possible sex distributions of three children: *MMM, MMF, MFM, FMM, FFM, FMF, MFF, FFF.* If having children is a binomial experiment, these 8 outcomes will be equally likely, each having probability 1/8. Then

$$p_0 = P(0 \text{ boys}) = P(FFF) = 1/8$$

$$p_1 = P(1 \text{ boy}) = P(MFF, FMF, FFM) = 3/8$$

and similarly $p_2 = 3/8$, $p_3 = 1/8$. Hence, we wish to test

H_0: $p_0 = 1/8$ $p_1 = 3/8$ $p_2 = 3/8$ $p_3 = 1/8$

H_a: At least one of these probabilities is false.

Thus, under H_0, the expected frequencies are $np_0 = np_3 = 50(1/8) = 6.25$, $np_1 = np_2 = 50(3/8) = 18.75$. The observed and expected frequencies and the values of $(O - E)^2/E$ are given in Fig. 10.8.

| | No. of boys | | | |
	0	1	2	3
$O = $ Observed frequency	9	22	14	5
$E = $ Expected frequency	6.25	18.75	18.75	6.25
$\dfrac{(O - E)^2}{E}$	1.21	.56	1.20	.25

Figure 10-8

From Fig. 10-8 we obtain

$$\chi^2 = \Sigma \frac{(O - E)^2}{E} = 1.21 + .56 + 1.20 + .25 = 3.22$$

Since there are four categories, $k = 4$. Hence the number of degrees of freedom is $4 - 1 = 3$. At a level of significance of $\alpha = .05$ with 3 degrees of freedom, we find from the Appendix tables $\chi_{.05}^2 = 7.82$. Therefore we accept H_0 since the observed value of $\chi^2 = 3.22 < 7.82$. In other words, the evidence is not sufficient to reject the hypothesis that having children is a binomial experiment.

Example 10-4. The pioneer research in genetics by Gregor Mendel provides a classic example of the use of the chi-square goodness-of-fit test. Mendel postulated a genetic theory which predicts that the outcomes of a certain cross will result in three kinds of offspring, which we shall call XX, Xx, and xx, having probabilities 1/4, 1/2, and 1/4, respectively. Testing his theory on pea plants, he reported the following counts of offspring: 35XX, 67Xx, and 30xx. Test Mendel's theory at a .05 level of significance.

SOLUTION: Associating with the three categories XX, Xx, xx the numbers 1, 2, and 3, respectively, we obtain $p_1 = 1/4$, $p_2 = 1/2$, and $p_3 = 1/4$, the hypothesized probabilities for the categories. Thus, we want to test

H_0: $p_1 = 1/4$ $p_2 = 1/2$ $p_3 = 1/4$
H_a: At least one equality is false.

The observed and expected frequencies and the values of $(O - E)^2/E$ are given in Fig. 10.9.

	Categories			
	XX	*Xx*	*xx*	*Total*
$O = $ *Observed frequency*	35	67	30	132
$E = $ *Expected frequency*	33	66	33	
$\dfrac{(O - E)^2}{E}$	.121	.015	.273	

Figure 10-9

The observed value of the test statistic is

$$\chi^2 = \Sigma \frac{(O - E)^2}{E} = .121 + .015 + .273 = .409 \approx .41$$

Since $k = 3$, there are $k - 1 = 2$ degrees of freedom. For the chi-square distribution with 2 degrees of freedom we find from the tables in the Appendix that $\chi_{.05}^2 = 5.99$. Since the observed value of $\chi^2 = .41 < 5.99$, we accept H_0. Thus, at a .05 level of significance, there is no evidence to indicate Mendel's theory is false.

Problems

10-1 A park ranger believes that a certain lake contains 20% bass, 50% perch, and 30% catfish. Of 50 fish that are netted, the following data are recorded: 12 bass, 22 perch, and 16 catfish. Are these results consistent with the ranger's belief? Test at the .05 level of significance.

10-2 A sociologist wishes to determine whether people's attitudes toward legalized abortion are changing. A survey taken 3 years earlier showed 60% were opposed to legalized abortion, 25%

were indifferent, and 15% were in favor. The sociologist takes a new sample of size 200, and the following data are obtained: 105 are opposed, 40 are indifferent, and 55 are in favor. Use the chi-square goodness-of-fit test at a significance level of $\alpha = .05$ to determine if attitudes towards legalized abortion are changing.

10-3 A study is conducted to determine if the ethnic composition of people in managerial positions is changing. Several large companies are randomly selected, and the breakdown by ethnic group of people in managerial positions is recorded. From past records it is found that managerial positions were held by 1% blacks, 5% Orientals, 2% chicanos, 92% whites. In the present study, for the 1,020 managerial positions, the following data are obtained: 80 blacks, 60 Orientals, 25 chicanos, and 855 whites. Do these data indicate a change in ethnic group composition? Test at the .05 level of significance.

Chi-square test of independence

As we have seen, the chi-square goodness-of-fit test is used to check whether sample data are consistent with a hypothesized distribution of a single variable. A slight modification of this test handles the problem of whether two variables are related. Perhaps the Internal Revenue Service wishes to know if geographic area and the amount of cheating on income tax returns are related in order to allocate their auditors more efficiently. Medical researchers may be interested in determining whether people who consume large quantities of meat tend to have heart trouble. A controversial issue in modern times is whether children who see lots of violence on television tend to have more aggressive behavior than other children. These problems can be analyzed by the *chi-square test of independence*. We shall illustrate this test by the following example.

Example 10-5. Critics of student political demonstrations have argued that these demonstrations hinder the learning process because they detract from time spent in class and doing homework. Suppose a sample of 200 students is chosen from a certain university which is known for its numerous political demonstrations. Each student is evaluated according to his extracurricular political activity, and his grade-point average (GPA) is recorded. The results are given in Fig. 10-10. Are the data consistent with the hypothesis that political activity and grades are independent?

GPA	Low	*Political activity* Moderate	High	Totals
3.0–4.0	35	15	10	60
2.0–3.0	45	10	5	60
Less than 2.0	50	25	5	80
Totals	130	50	20	200

Figure 10-10

SOLUTION: We wish to test the following hypotheses:

H_0: Political activity and grades are independent.

H_a: Political activity and grades are not independent.

This problem is attacked in the same spirit as other hypotheses-testing problems we have encountered. First, we decide what characteristics the sample data are likely to have if H_0 is true, and then we devise a test statistic which reflects these characteristics. If the observed value of the test statistic is too improbable under the assumption that H_0 is true, then we reject H_0.

Suppose, for a moment, that H_0 is true, that students' political activity and grades are independent. We are interested in the kind of pattern that is likely to arise in this case. For example, how many students can be expected to have a GPA of 3.0 to 4.0 and low political activity? Reason it this way: The proportion of students with low political activity is $130/200 = .65$. Among these, the proportion who have a GPA of 3.0 to 4.0 should be the same as the proportion who have a GPA of 3.0 to 4.0 among the whole population, which is $60/200 = .3$. Hence you could expect the proportion who fall in both categories to be $(.65)(.3) = .195$. In other words, if a student's political activity and grades are independent, then you could expect that among a group of 200 students, about $(.195)(200) = 39$ will have a GPA of 3.0 to 4.0 and low political activity.

Another way to arrive at the expected frequency should occur to those readers who recall the discussion of independence in Chap. 3. Imagine a large population of students containing 30% with a GPA of 3.0 to 4.0 and 65% with low political activity. The assertion made in the null hypothesis H_0 means that knowing a student's political activity tells you nothing about his GPA; in other words, the two events (a student has low political activity) and (a student has a GPA of 3.0 to 4.0) are independent. This implies P(a student has low political activity and a GPA of 3.0 to 4.0) $= P$(a student has low political activity) $\times P$(a student has a GPA of 3.0 to 4.0) $= (.65)(.3) = .195$. If a large number of students, say, 200, are drawn from the population, it is likely that about $(.195)(200) = 39$ will have both low political activity and a grade-point average of 3.0 to 4.0.

By the same reasoning we can compute the expected frequency for each box in Fig. 10-10 under the assumption that H_0 is true. These expected frequencies are given in Fig. 10-11.

Expected frequencies of political activities

GPA	Low	Moderate	High
3.0–4.0	$\left(\frac{60}{200}\right)\left(\frac{130}{200}\right)(200) = 39$	$\left(\frac{60}{200}\right)\left(\frac{50}{200}\right)(200) = 15$	$\left(\frac{60}{200}\right)\left(\frac{20}{200}\right)(200) = 6$
2.0–3.0	$\left(\frac{60}{200}\right)\left(\frac{130}{200}\right)(200) = 39$	$\left(\frac{60}{200}\right)\left(\frac{50}{200}\right)(200) = 15$	$\left(\frac{60}{200}\right)\left(\frac{20}{200}\right)(200) = 6$
Less than 2.0	$\left(\frac{80}{200}\right)\left(\frac{130}{200}\right)(200) = 52$	$\left(\frac{80}{200}\right)\left(\frac{50}{200}\right)(200) = 20$	$\left(\frac{80}{200}\right)\left(\frac{20}{200}\right)(200) = 8$

Figure 10-11

The test statistic should reflect the differences between the observed sample data and what would be expected if H_0 were true. Thus, the test statistic is chosen to measure the discrepancy between the entries in Figs. 10-10 and 10-11. This discrepancy is measured by the quantity χ^2, defined as before by

$$\chi^2 = \Sigma \frac{(O - E)^2}{E}$$

where O denotes the observed frequency, E denotes the expected frequencies, and Σ means we are adding all the numbers $(O - E)^2/E$, one for each box in the table. If the various values of $(O - E)^2$ are large, then χ^2 will be large. As before, we shall reject H_0 when χ^2 is too large.

The sample statistic χ^2 has a chi-square distribution for large n if H_0 is true. As we shall explain in a moment, there are 4 degrees of freedom. (For this use of the χ^2 statistic, degrees of freedom are computed in a different manner than before.) Suppose we take $\alpha = .05$. Then, with 4 degrees of freedom, from the tables in the Appendix, $\chi_{.05}{}^2 = 9.49$. The test rule is to reject H_0 if $\chi^2 \geq 9.49$ and accept H_0 if $\chi^2 < 9.49$. The observed value of χ^2

computed from the sample data is

$$\chi^2 = \Sigma \frac{(O - E)^2}{E}$$

$$= \frac{(35 - 39)^2}{39} + \frac{(45 - 39)^2}{39} + \frac{(50 - 52)^2}{52}$$

$$+ \frac{(15 - 15)^2}{15} + \frac{(10 - 15)^2}{15} + \frac{(25 - 20)^2}{20}$$

$$+ \frac{(10 - 6)^2}{6} + \frac{(5 - 6)^2}{6} + \frac{(5 - 8)^2}{8}$$

$$= .41 + .92 + .08 + 0 + 1.67 + 1.25 + 2.67 + .17 + 1.13$$

$$= 8.30$$

Thus, at the .05 level of significance, we accept H_0. We conclude that the data are consistent with the hypothesis that political activity and grades are independent.

CONTINGENCY TABLES

In general terms, the chi-square test of independence is used to determine whether or not two variables are independent. Let us call the two variables X and Y. The hypotheses are:

H_0: X and Y are independent.

H_a: X and Y are dependent.

Suppose there are r categories of X and c categories of Y. A sample is drawn from the population, and each member is classified according to its category for each variable. The observed data are summarized in a table, such as in Fig. 10-12, called a *two-way contingency table*. This table gives the frequency of sample members in each of the possible pairs of categories, one for each variable. The positions in the table where frequencies appear are called *cells*.

If H_0 is true and the variables are independent, we expect a certain pattern in the table: the proportion of sample members in different categories of Y among those in a specific category of X should not depend upon the particular X category, and vice versa. This pattern implies that the expected frequency E of a particular cell is given by the formula

$$E = \frac{n_X}{n} \frac{n_Y}{n} n = \frac{n_X n_Y}{n}$$

Two-way contingency table

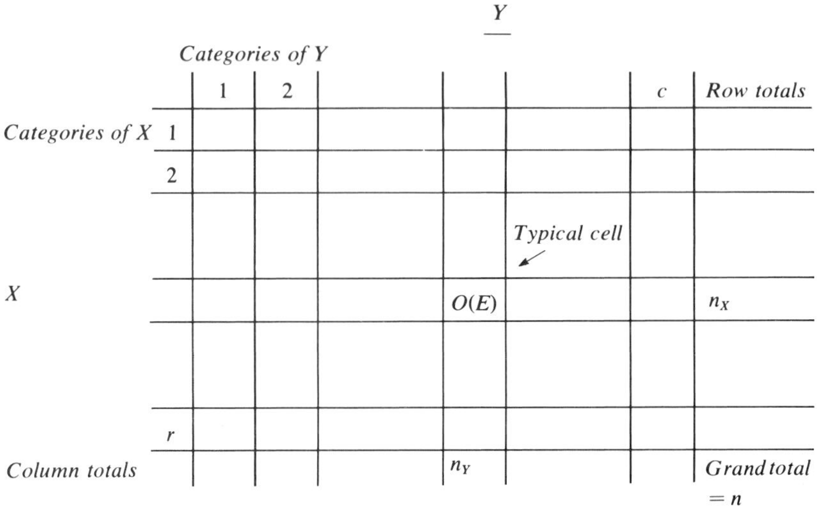

$$Y$$

Categories of Y

	1	2			*c*	Row totals
Categories of X 1						
2						
			Typical cell ↙			
X			O(E)			n_X
r						
Column totals			n_Y			Grand total $= n$

$O =$ Observed frequency

$E =$ Expected frequency $= \dfrac{n_X n_Y}{n}$

Figure 10-12

where n_X is the row total for the row containing the cell and n_Y is the corresponding column total. It is convenient to have one table containing both the observed and the expected frequencies. The expected frequencies may be placed in parentheses in their appropriate cells so as to distinguish these numbers from the observed frequencies.

The degree to which the observed frequencies differ from the expected frequencies in Fig. 10-12 is measured by the χ^2 statistic $\chi^2 = \Sigma(O - E)^2/E$ just as in the chi-square goodness-of-fit test. In words, χ^2 is computed by first calculating the value of $(O - E)^2/E$ for each cell in the table and then adding these numbers. The form of the test rule is to reject H_0 if χ^2 is large and accept H_0 if χ^2 is small.

As usual, the exact rejection region for a particular test is determined by the sampling distribution of the test statistic χ^2 and the chosen level of significance. If the sample size n is sufficiently large, then χ^2 will have approximately a chi-square distribution with $(r - 1)(c - 1)$ degrees of freedom when H_0 is true. Suppose α denotes the level of significance. Let χ_α^2 denote the value of χ^2 which is exceeded with probability α, as determined from the table of the chi-square distribution in the Appendix for $(r - 1)(c - 1)$ degrees of freedom. In these terms, H_0 is rejected if the observed value of χ^2 is $\chi^2 \geq \chi_\alpha^2$; otherwise, H_0 is accepted.

As we have mentioned, the chi-square distribution is only an approximation to the true sampling distribution of the test statistic χ^2. This approximation will be accurate enough for most purposes if the expected frequency in each cell is at least 5. If a contingency table has been set up for a particular set of sample data and the expected frequencies are fewer than 5, then some of these categories should be combined.

Example 10-6. A potter was having trouble obtaining a beautiful red color he desired in a certain glaze. Sometimes the color was fine; other times it was not. He conjectured that possibly the difference was caused by the position of the piece in the kiln when it was fired, and he collected the data in Fig. 10-13. Do these data indicate that position in the kiln affects the quality of the glaze?

SOLUTION: Using the formulas in Fig. 10-12, we obtain the expected frequencies shown in parentheses in Fig. 10-14. We then compute

Quality of glaze	*Observed frequencies, position in kiln*			
	Bottom	*Middle*	*Top*	*Totals*
Ugly	10	15	20	45
Satisfactory	12	20	8	40
Beautiful	3	5	12	20
Totals	25	40	40	105

Figure 10-13

Quality of glaze	*Observed and expected frequencies, position in kiln*			
	Bottom	*Middle*	*Top*	*Totals*
Ugly	10 (10.7)	15 (17.1)	20 (17.1)	45
Satisfactory	12 (9.5)	20 (15.2)	8 (15.2)	40
Beautiful	3 (4.8)	5 (7.6)	12 (7.6)	20
Totals	25	40	40	105

Figure 10-14

$$\chi^2 = \Sigma \frac{(O-E)^2}{E}$$

$$= \frac{(10-10.7)^2}{10.7} + \frac{(12-9.5)^2}{9.5} + \frac{(3-4.8)^2}{4.8}$$

$$+ \frac{(15-17.1)^2}{17.1} + \frac{(20-15.2)^2}{15.2} + \frac{(5-7.6)^2}{7.6}$$

$$+ \frac{(20-17.1)^2}{17.1} + \frac{(8-15.2)^2}{15.2} + \frac{(12-7.6)^2}{7.6}$$

$$= .05 + .66 + .68 + .26 + 1.52 + .89 + .49 + 3.41 + 2.55$$

$$= 10.51$$

Since $r = 3$ and $k = 3$, there are $(r-1)(k-1) = (2)(2) = 4$ degrees of freedom. From the tables of the χ^2 distribution in the Appendix, with 4 degrees of freedom, we find $\chi_{.05}{}^2 = 9.49$. Thus, at a .05 level of significance we reject H_0 since $\chi^2 = 10.51 > 9.49$. This is strong evidence that the position in the kiln affects the quality of the glaze.

Example 10-7. A random sample of 190 adults is chosen from a large population, and each sample member is classified according to the number of pounds of meat consumed per week and heart condition. These data are shown in the observed frequency table in Fig. 10-15. Do these data constitute strong evidence of a relation between meat consumption and heart condition?

Meat consumption, pounds per week	Observed frequencies, heart condition			
	Good	Fair	Bad	Totals
0–1	30	15	4	49
1–2	38	6	11	55
2–3	26	5	10	41
More than 3	16	12	17	45
Totals	110	38	42	190

Figure 10-15

SOLUTION: Computing the expected frequencies and entering them in parentheses in the observed frequency table gives Fig. 10-16. From this we compute

Meat consumption, pounds per week	Observed and expected frequencies, heart condition			
	Good	*Fair*	*Bad*	*Totals*
0–1	30 (28.4)	15 (9.8)	4 (10.8)	49
1–2	38 (31.8)	6 (11.0)	11 (12.2)	55
2–3	26 (23.7)	5 (8.2)	10 (9.1)	41
More than 3	16 (26.1)	12 (9.0)	17 (9.9)	45
Totals	110	38	42	190

Figure 10-16

$$\chi^2 = \Sigma \frac{(O - E)^2}{E}$$

$$= \frac{(30 - 28.4)^2}{28.4} + \frac{(38 - 31.8)^2}{31.8} + \frac{(26 - 23.7)^2}{23.7} + \frac{(16 - 26.1)^2}{26.1}$$

$$+ \frac{(15 - 9.8)^2}{9.8} + \frac{(6 - 11.0)^2}{11.0} + \frac{(5 - 8.2)^2}{8.2} + \frac{(12 - 9.0)^2}{9.0}$$

$$+ \frac{(4 - 10.8)^2}{10.8} + \frac{(11 - 12.2)^2}{12.2} + \frac{(10 - 9.1)^2}{9.1} + \frac{(17 - 9.9)^2}{9.9}$$

$$= 22.3$$

Since there are 4 categories of meat consumption and 3 categories of heart conditions, $r = 4$ and $c = 3$. Hence the number of degrees of freedom is $(r - 1)(c - 1) = 3(2) = 6$. From the tables in the Appendix, at a .01 level of significance for 6 degrees of freedom, we find $\chi_{.01}^2 = 16.8$. Since the observed value of χ^2 is $\chi^2 = 22.3 > 16.8$, we reject H_0 even at a significance level of $\alpha = .01$. We conclude that there is strong evidence of a relation between meat consumption and heart condition.

It is tempting to conclude that a cause-and-effect relationship exists between two variables when a chi-square test of their independence has resulted in a rejection of the null hypothesis. However, no such conclusion is justified. A rejection of the null hypothesis indicates only that there may be some relationship between the variables. This means that knowing the level of one variable gives some information about the level of the other variable and vice versa. This point is illustrated by the well-known example of smoking and lung cancer. Many experiments comparing these two variables have yielded significant results, leading to the rejection of the null hypothesis that smoking and lung cancer are independent. Although there is a variety of evidence which identifies smoking as a causal factor in the development of cancer, such a conclusion is not implied by the results of a chi-square test alone. Looking only at the chi-square test, it would be just as plausible to say that lung cancer causes cigarette smoking or that some variable associated with smoking causes lung cancer. To emphasize this point, consider an experiment which tests the independence of having lung cancer and having yellow fingers. If a chi-square test of independence were applied to the data of such an experiment, it is reasonable to believe that the null hypothesis would be rejected, since people who smoke tend to have yellow fingers due to tobacco stains.

Summary

The chi-square goodness-of-fit test is used to determine whether a set of population data has a specified distribution. It is assumed that the population may be divided into groups, or categories. When a member of the population is selected at random, the outcome is described by giving the category to which the chosen member belongs. We denote the categories by $1, 2, 3, \ldots, k$, and let $p_1, p_2, \ldots, p_k$ be probabilities of the k categories specified in the null hypothesis.

To determine if these numbers are actually the true probabilities of the k categories, a sample of size n is drawn and the results recorded. Let $O_1, O_2, \ldots, O_k$ denote the observed frequency in each category. If the null hypothesis is true, we could expect np_1 occurrences of category 1, np_2 occurrences of category 2, etcetera. These expected frequencies are denoted by $E_1, E_2, \ldots, E_k$. The discrepancy between the distribution specified in the null hypothesis and the observed frequencies is measured by $\chi^2 = \Sigma(O - E)^2/E$. At a level of significance α, H_0 is rejected if $\chi^2 \geq \chi_\alpha^2$, where χ_α^2 is the value of the chi-square distribution with $k - 1$ degrees of freedom, which is exceeded with probability α. This test requires that each expected frequency exceed 5.

The chi-square test of independence is used to determine whether two variables are independent. The null hypothesis asserts they are independent, and the alternative hypothesis states they are not. Sample data are collected and recorded in a two-way contingency table, which gives the observed frequencies for each pair of categories, one for each variable. These are entered in the cells of the table. The test statistic is $\chi^2 = \Sigma(O - E)^2/E$, where O is the observed frequency and E is the expected frequency as computed from the formula

$$E = \frac{(\text{Row total})(\text{column total})}{\text{Total sample size}}$$

according to the particular row and column, including the cell being considered. Σ indicates that each of the quantities $(O - E)^2/E$ are added, one for each cell. At a level of significance α, the null hypothesis is rejected if $\chi^2 \geq \chi_\alpha^2$, where χ_α^2 is defined as in the goodness-of-fit test.

Important terms

Cells

Chi-square distribution

Chi-square goodness-of-fit test

Chi-square test of independence

Two-way contingency table

Additional problems

10-4 A well-known study claims that children raised by nursemaids are less neat than other children. A sample of children was selected from each category, and each child was classified as neat or sloppy by a psychologist. Using the chi-square test on the following data, decide if the claim of the study is justified. (Use a .01 level of significance.) Do you think the method of classifying the children is a good one?

	Sloppy	*Neat*
Raised by nursemaid	80	20
Not raised by nursemaid	220	180

10-5 A study is done to determine whether the degree of lighting on city streets is related to the number of muggings. Comparable streets with various degrees of lighting are randomly selected from several cities. Over a 6-month period the following data was observed. Test whether the type of street lighting and number of muggings are independent at the .05 level of significance.

	Type of lighting		
No. of muggings	*No lighting*	*Poor lighting*	*Adequate lighting*
Fewer than 10	10	20	30
10–20	20	30	30
20–30	50	25	25
More than 30	20	10	10

10-6 A sociologist was awarded a grant to study regional differences in longevity. Public records from areas in three regions—urban, rural, and suburban—were randomly selected, and the following data were obtained. Do these data indicate that longevity and region are related in some way? (Conduct your test at a significance level of $\alpha = .01$.)

	Region		
Lifespan, years	*Urban*	*Rural*	*Suburban*
Fewer than 50	60	25	35
50–60	48	53	49
60–70	50	100	70
Over 70	28	50	32

10-7 A medical researcher wishes to determine if there is a relationship between a person's age and recovery time from a sprained ankle. A random sample taken from hospital records in a large city produced the following data. Perform the chi-square test of independence, using a .05 level of significance, to determine if a relationship exists.

	Recovery time (in weeks)			
Age, years	*Fewer than 2*	*2 to 4*	*4 to 6*	*More than 6*
Under 20	180	100	70	50
20–30	20	50	20	30
30–40	20	25	35	20
Over 40	7	13	15	45

True-false test

Circle the correct letter.

T F   1. The number of degrees of freedom in the chi-square goodness-of-fit test depends on the sample size.

T F   2. To test whether a die is evenly balanced the chi-square goodness-of-fit test should be used.

T F   3. The chi-square tests can be applied only to data which are normally distributed.

T F   4. If H_0 is rejected on the chi-square test of independence, then you can safely conclude that there is a cause-and-effect relationship.

T F   5. If you obtain a negative value for the chi-square statistic, then you have made a mistake in your calculations.

T F   6. The lower the level of significance for a chi-square test, the higher the value of χ^2 needed to reject H_0.

T F   7. The chi-square test is a situation where nominal data are used.

T F   8. For a fixed level of significance, the greater the degree of freedom, the greater the value of χ^2 necessary to reject H_0.

T F   9. For a 3×9 contingency table, the number of degrees of freedom would be 27.

T F   10. If the expected frequency in some category is too small to use the chi-square test, it is often possible to reduce the number of categories.

Nonparametric tests for multiple comparisons

11

In Chaps. 8 and 9 we studied comparative experiments involving two sets of data. Recall that the Wilcoxon test and sign test are used when related samples based on paired data are drawn, and the *t* test and Mann-Whitney test are used with independent samples. Now we shall consider tests appropriate for comparing observations from more than two samples, the *Friedman test* for comparing related samples and the *Kruskal-Wallis test* for comparing independent samples. These are highly effective and easy-to-apply *nonparametric tests*. The Kruskal-Wallis test is an alternative to the classical analysis of variance treated in Chap. 12, a parametric test which requires that the population data be normally distributed.

The Friedman test for related samples

Example 11-1. A study is conducted to determine whether or not an individual's pulse rate is affected differently by three types of cigarettes: mild, strong, and menthol. A sample of 10 smokers participate in this experiment. At various times of the day over a 1-week interval they smoke one type of cigarette and their pulse rates are measured. Then, each person's average pulse rate for the week is recorded. Each week a different type of cigarette is smoked, and this procedure is repeated for 3 weeks. The results of this study are recorded in Fig. 11-1. Based on these data, can we conclude that there is some difference in the effects of the three types of cigarettes on an individual's pulse rate?

| | Average pulse rates | | |
Smoker	Strong	Mild	Menthol
A	110	105	107
B	95	92	108
C	90	80	85
D	135	138	142
E	85	86	83
F	70	68	69
G	87	84	83
H	94	97	96
I	98	97	101
J	76	75	77

Figure 11-1

SOLUTION: We wish to make our analysis without any assumptions about the underlying distribution of the sample data. Notice that our data are analogous to the paired data we studied in Chap. 9. The data cannot be considered as arising from independent samples since the same people smoked each of the three types of cigarettes. If we had only two types of cigarettes to compare, the Wilcoxon test would be applicable. We shall proceed to develop a statistic which can handle the comparison of three or more treatments. As in Chap. 9, we shall use ranks instead of the raw data.

We begin by ranking separately the three pulse rates observed for each person in the study, 1 being assigned to the type causing lowest pulse rate, 2 to the second lowest and 3 to the highest. This information, along with the rank sum R for each column, is recorded in Fig. 11-2.

Smoker	Ranks of pulse rates		
	Strong	*Mild*	*Menthol*
A	3	1	2
B	2	1	3
C	3	1	2
D	1	2	3
E	2	3	1
F	3	1	2
G	3	2	1
H	1	3	2
I	2	1	3
J	2	1	3
	$R_1 = 22$	$R_2 = 16$	$R_3 = 22$

Figure 11-2

We define H_0: there is no difference in the effects of the three types of cigarettes on an individual's pulse rate. If there is no difference, on the average, we would expect the ranks to be evenly distributed; that is, the number of 1s in each column should be about the same, the number of 2s in each column should be about the same, and likewise for the 3s. In this case we would expect the rank sum for each column to be about the same, namely, 1/3 of the total of all ranks. The total of all ranks must be 60 because there are 10 rows and the sum of each row is $1 + 2 + 3 = 6$. Thus, assuming no difference between types of cigarette, each column total should be about $(1/3)(60) = 20$. If we then look at the squared differences between the observed rank sum for each column and 20 and sum these differences, we would obtain a measure of how close the observed rank sums are to the expected sum under H_0; a large value would indicate that the ranks are not evenly distributed among the columns. Thus, if the sum of these squared differences is small, we would tend to accept H_0 (that there is no difference in the effects of the three types of cigarettes on an individual's pulse rate), and if the sum is large, we would reject H_0.

The procedures for looking at the differences between observed and expected sums is similar to the methods of Chap. 10. In fact, if we modify our procedures a bit, we shall obtain a statistic which has a chi-square distribution; then we may use the chi-square table to carry out our test. The modification entails multiplying the sum of squared differences by the number $12/n(k)(k + 1)$ where k is the number of treatments (in this case 3) and n is the number of participants in the study (in this case 10). The resulting statistic is called the *Friedman statistic*, denoted by Q, which has approximately a chi-square distribution with $k - 1$ degrees of freedom under H_0. In the present example we compute Q as follows:

$$Q = \frac{12}{[10(3)(3+1)]} \ [(R_1 - 20)^2 + (R_2 - 20)^2 + (R_3 - 20)^2]$$

$$= \frac{1}{10} \ [(22 - 20)^2 + (16 - 20)^2 + (22 - 20)^2]$$

$$= \frac{1}{10} \ (4 + 16 + 4)$$

$$= 2.4$$

Turning to the chi-square tables in the Appendix, we see that for $3 - 1 = 2$ degrees of freedom, $X_{.05}^2 = 5.99$. Since $2 < 5.99$, we accept H_0. The evidence is not sufficient to believe that there is a difference in the effects of the three types of cigarettes on an individual's pulse rate.

We now define Q and state the general procedure for applying the Friedman test. Suppose that we have k treatments to compare, where k is a number bigger than 2, and suppose we have n participants in the study. The participants are the recipients of the treatments. In Example 11-1 we had $k = 3$ types of cigarettes and $n = 10$ people in the experiment. We obtain our data on the responses of the treatments and list it as in Fig. 11-1. Then, considering each participant separately, the k treatments are ranked according to the magnitude of their responses, 1 denoting the smallest, 2 the second smallest, etcetera, as we did in Fig. 11-2. We sum the ranks for each column, calling these sums $R_1, R_2, \ldots, R_k$. We shall test the hypotheses

H_0: There is no difference in the effects of the treatments.

H_a: There is a difference in the effects of the treatments.

We use the Friedman statistic

$$Q = \frac{12}{nk(k + 1)} \ (R_1^2 + R_2^2 + \cdots + R_k^2) - 3n(k + 1)$$

Notice that the above formula for Q appears different from the formula we used in Example 11-1. Actually, the formula is the same. We developed it earlier in an intuitive way, and now it has been rewritten to make computations easier. The statistic Q has (approximately) a chi-square distribution with $k - 1$ degrees of freedom under H_0. To apply the Friedman test at a significance level α, from the Appendix tables we obtain χ_α^2 from the chi-square table for $k - 1$ degrees of freedom; then we compute Q, rejecting H_0 if $Q \geq \chi_\alpha^2$ and accepting H_0 if $Q < \chi_\alpha^2$.

The Friedman test as we have defined it is a *two-tailed*, or *non-directional*, test. It is used to detect the existence of some difference in the effects of certain treatments. It does not detect which treatment is different nor the direction of that difference. Although this information is often obvious by inspection of the data, to obtain more precise information concerning which treatments are different it is necessary to use a method of *multiple comparisons*. (A discussion of this method can be found in more advanced texts.)

Example 11-2. A vegetable gardener wishes to compare the effects of four different fertilizers on the crops he is growing to determine if there is any overall difference. He digs four experimental gardens in comparable locations and randomly selects one fertilizer for each garden. He records the total yield for each crop and then ranks the crop yields for each garden; these data are displayed in Figs. 11-3 and 11-4. Using the Friedman test at a significance level of $\alpha = .05$, what should the gardener conclude?

Crop yield (in bushels) according to fertilizer type

Crop	Organic	Garden (fertilizer type) Mulch	Brand X	Brand Y
Corn	9.3	4.1	5.7	7.0
Tomatoes	10.4	5.3	10.6	6.9
String beans	2.1	.5	1.0	1.4
Lettuce	8.7	3.3	5.8	4.8
Zucchini	6.8	6.1	1.9	7.7

Figure 11-3

Ranks of crop yield according to fertilizer (1 means lowest yield; 4 means highest yield)

Crop	Organic	Garden (fertilizer type) Mulch	Brand X	Brand Y
Corn	4	1	2	3
Tomatoes	3	1	4	2
String beans	4	1	2	3
Lettuce	4	1	3	2
Zucchini	3	2	1	4
	$R_1 = 18$	$R_2 = 6$	$R_3 = 12$	$R_4 = 14$

Figure 11-4

SOLUTION: At a significance level of $\alpha = .05$, we shall test the hypotheses

H_0:   There is no overall difference in crop yields according to fertilizer.

H_a: There is a difference in crop yields according to fertilizer.

We compute the Friedman statistic Q, noting that $k = 4$ fertilizers and $n = 5$ crops:

$$Q = \frac{12}{nk(k+1)} (R_1^2 + R_2^2 + R_3^2 + R_4^2) - 3n(k+1)$$

$$= \frac{12}{5(4)(5)} (18^2 + 6^2 + 12^2 + 14^2) - 3(5)(5)$$

$$= 9$$

Looking at the chi-square tables in the Appendix, with $4 - 1 = 3$ degrees of freedom, we see $\chi_{.05}^2 = 7.82$. Since $Q = 9 > 7.82$, we reject H_0 at level $\alpha = .05$. Thus the gardener should conclude that there is a difference in crop yields according to fertilizer.

Another look at the data indicates that the organic fertilizer seems to have produced the best overall yield and the mulch produced definitely the lowest. If the gardener gardens for a hobby, this analysis may be enough. If gardening is his livelihood, he may wish to proceed to do a multiple comparison or to consider other relevant factors such as costs and long-term effects of each fertilizer upon the soil.

TIES

In applying the Friedman test, ties in treatment response might occur, either between more than one participant or between different treatments for the same participant. Since the statistic Q involves ranking the treatment effects for each participant separately, we would encounter difficulty only in the case of ties in different treatments for the same participant. In such a situation, we would continue to follow the procedure of assigning midranks as developed in Chap. 10. For example, if we are comparing the effects of four treatments and for one participant had the responses 6.8, 8.1, 6.8, and 9.7, we would assign the ranks 1.5, 3, 1.5, and 4, respectively because the lowest two responses are ties at 6.8 and $(1 + 2)/2 = 1.5$. If there are many ties, we may wish to obtain more data or to refine our measuring techniques.

The Friedman test for ranking products

The Friedman test can be used to determine whether or not there is general agreement or a difference of opinion about k products being evaluated by n judges. The idea is that if there is no difference of opinion regarding the quality of the products being judged, then the rankings of the judges will be similar and the Friedman statistic Q will be high. Presumably this would correspond to the situation when one product is significantly better or worse than the others. On the other hand, if there is disagreement among the judges or if there is little difference among the products and the rankings are done essentially at random, Q would tend to be low. In this situation, the judges are the sample elements and the different products are the "treatments." Example 11-3 illustrates this use of the Friedman test.

Example 11-3. Ten book critics who usually have similar opinions are each asked to rank four novels as to their degree of suspense. The results are tabulated in Fig. 11-5. Do these results indicate that there is general agreement among the critics about these books?

SOLUTION: We use the Friedman test at a significance level of $\alpha = .05$. The hypotheses to be tested are:

H_0: There is little difference in suspense content among the novels (the critics do not agree).

H_a: The critics have some agreement about suspense content; most likely there is a difference.

Ratings by suspense content of four mysteries (1 means most suspenseful; 4 means least suspenseful)

Critic	"The Boy Scout Murders"	"The Secret Ballot"	"The Mystery of the White Tree"	"The Haunted Doghouse"
A	1	2	4	3
B	2	3	1	4
C	3	4	1	2
D	4	2	1	3
E	1	2	3	4
F	1	2	3	4
G	1	2	4	3
H	3	1	2	4
I	2	1	4	3
J	4	1	3	2
	$R_1 = 22$	$R_2 = 20$	$R_3 = 26$	$R_4 = 32$

Figure 11-5

Notice that the rankings are in *reverse order*: in this the most suspenseful book is to be given rank 1, the least suspenseful rank 4, and so on. This makes no difference as long as all judges rank in the same way. Noting that $n = 10$ and $k = 4$, we compute the Friedman statistic:

$$Q = \frac{12}{10(4)(5)} (22^2 + 20^2 + 26^2 + 32^2) - 3(10)(5)$$

$$= 5.04$$

Looking at the chi-square tables in the Appendix for $4 - 1 = 3$ degrees of freedom, we see that $\chi_{.05}^2 = 7.82$. Since $Q = 5.04 < 7.82$, we accept H_0. We conclude that there is little difference in suspense content among the novels (the critics do not agree).

Problems

11-1 A study is done to compare incomes among four religious groups. To reduce sample variance, average incomes from random samples in each group at various age levels are compared. For the following data use the Friedman test at a significance level of $\alpha = .05$ to test whether or not the incomes differ.

*Average annual income by age of religious groups
(in thousands of dollars)*

Age group	Catholics	Protestants	Jews	Transcendental meditators
20–30	11.0	11.6	10.9	8.1
30–40	12.1	12.9	13.1	10.2
40–50	12.3	13.0	12.9	8.0
Over 50	10.8	10.2	11.1	6.0

11-2 Ten people who suffer persistent headaches are asked to rank four brands of aspirin according to their ability to stop headache pain, and the rankings are tabulated as follows. Use the Friedman test at a significance level of $\alpha = .01$ to determine whether or not there is a difference in preference.

Preference for aspirin based on pain-killing ability

Sample member	Brand X	Brand Y	"Aspirex"	"Pain-O-pause"
A	3	2	4	1
B	2	3	1	4
C	4	3	1	2
D	2	4	1	3
E	3	1	2	4
F	4	1	3	2
G	2	3	1	4
H	3	4	2	1
I	1	2	3	4
J	3	2	1	4

11-3 Ten randomly selected students in an introductory statistics course are asked to rank five types of examples of statistical studies according to which they prefer to read (1 being the favorite, 2 the second favorite,…, 5 the least popular). The results are tabulated as follows. Using the Friedman test at a significance level of $\alpha = .05$, determine whether or not there are any differences in preferences.

Types of statistical studies

Student	Economic	Medical	Industrial	Agricultural	Sociological
J.T.	4	2	3	5	1
A.R.	5	3	4	1	2
M.M.	4	5	1	3	2
P.S.	3	5	4	2	1
H.H.	5	4	3	2	1
Q.T.	5	4	1	3	2
L.L.	3	4	5	2	1
Z.Z.	4	2	3	5	1
A.B.	1	5	2	4	3
F.Y.	5	3	2	4	1

11-4 Six randomly selected businessmen were asked to rank three massage parlors on the basis of overall quality. Using the Friedman test at a significance level of $\alpha = .05$, decide whether or not there are any differences of opinion among the businessmen.

Massage parlor preference

Businessman	Lolita's	Jane's	Pierre's
A	1	2	3
B	3	1	2
C	2	1	3
D	2	3	1
E	1	3	2
F	3	2	1

The Kruskal-Wallis test

We now turn to the problem of comparing k treatments where our data are in the form of k *independent* samples (not necessarily all of the same size), one sample for each treatment. We shall introduce the Kruskal-Wallis test with the following example.

Example 11-4. A consumers' group conducted a study comparing five brands of waterbeds. One aspect of the study involved determining the maximum weight a waterbed would support before bursting or springing a leak. Six waterbeds were used for each brand, and the maximum loads are listed in Fig. 11-6. Based on these data, should the consumers' group conclude that there is a difference in load-supporting capacity among the brands of waterbeds?

Maximum loads (in hundreds of pounds) for five brands of waterbeds

Brand A	Brand B	Brand C	Brand D	Brand E
8.2	2.1	13.2	9.8	4.1
10.4	10.9	12.9	11.2	12.7
7.0	12.6	14.1	10.4	11.2
6.8	10.4	13.0	14.1	9.7
5.1	11.8	13.7	16.8	10.1
11.9	11.7	12.8	14.7	13.5

Figure 11-6

SOLUTION: As in the case of the Mann-Whitney test, we shall first rank the data over all the brands, giving rank 1 to the lowest weight, etc. As for previous tests, we assign midranks (that is, the average of the ranks included) to tied values. The results of our rankings are shown in Fig. 11-7. Notice that a few waterbeds (especially one for brand B) seem to be defective in that they have extremely low breaking points. The effect of these extreme values is lessened a bit by taking ranks. Also, transforming to ranks enables us to use a statistic with a distribution that can be computed under H_0 without making any parametric assumptions about the data.

Ranks for maximum loads of waterbeds

Brand A	Brand B	Brand C	Brand D	Brand E
6	1	24	8	2
11	13	22	14.5	20
5	19	27.5	11	14.5
4	11	23	27.5	7
3	17	26	30	9
18	16	21	29	25
$R_1 = 47$	$R_2 = 77$	$R_3 = 143.5$	$R_4 = 120$	$R_5 = 77.5$

Figure 11-7

We now devise a statistic to test the hypotheses:

H_0:   There is no difference in load-supporting capacity among the brands of waterbeds.

H_a: There is a difference in load-supporting capacity among the brands of waterbeds.

If H_0 is true, we would expect the ranks to be rather evenly distributed among the brands; equivalently, we would expect the column sums to be about the same. Since there are 30 items being ranked, the total sum of ranks has to be $1 + 2 + \cdots + 30 = 465$ (using midranks for ties does not change this). Divided equally among the five brands, this would mean that if H_0 were true, we would expect each column sum to be roughly $465/5 = 93$. It would seem natural, as it did before, to look at the squared differences between observed and expected rank sums and to reject H_0 if the sum of these squared differences is large. This procedure is quite similar to the procedure used in the Friedman test, except now, because our samples are unrelated, we rank everything together instead of each row separately.

The Kruskal-Wallis statistic H is based on this idea. The formula has been modified to make calculations easier and to give a variable with a distribution that can be approximated, assuming H_0 is true. Noting that there are $N = 30$ observations altogether, $k = 5$ brands being compared, with $n_1 = n_2 = n_3 = n_4 = n_5 = 6$ waterbeds of each brand, we compute the Kruskal-Wallis statistic:

$$H = \frac{12}{N(N+1)}\left(\frac{R_1{}^2}{n_1} + \cdots + \frac{R_k{}^2}{n_k}\right) - 3(N+1)$$

$$= \frac{12}{30(31)}\left[\frac{(47)^2}{6} + \frac{(77)^2}{6} + \frac{(143.5)^2}{6} + \frac{(120)^2}{6} + \frac{(77.5)^2}{6}\right]$$
$$- 3(31)$$

$$= 105.7 - 93$$

$$= 12.7$$

It turns out that H has approximately a chi-square distribution with $k - 1$ degrees of freedom. In the present case, $k = 5$, and so there are 4 degrees of freedom. If we test at a significance level of $\alpha = .05$, with 4 degrees of freedom, $\chi_{.05}{}^2 = 9.49$. Since $H =$

12.7 > 9.49, we reject H_0, concluding that there is a difference in load-supporting capacity among the brands of waterbeds.

Another look at the data reveals that brand C seems to have a heavy load capacity while brand A waterbeds seem to be weaker than the others. We observe that although the results of the test may have seemed obvious merely by glancing at the data, using the test enables us to have a mathematical foundation to support our claim. It gives us a uniform standard for evaluating similar situations and enables us to make distinctions which often cannot be made with the naked eye.

We now define the Kruskal-Wallis test. Suppose that we wish to compare k different treatments or methods by testing the hypotheses

H_0:   There is no difference in the treatments.

H_a: There is a difference in the treatments.

Suppose, also, that we have collected data in the form of k independent samples, with n_1 observations from treatment 1, n_2 observations from treatment 2, . . . , and $n_k = N$ observations altogether. Then, if we rank all the data together and let R_1 = sum of ranks from treatment 1, . . . , R_k = sum of ranks from treatment k, we compute the Kruskal-Wallis statistic

$$H = \frac{12}{N(N+1)}\left(\frac{R_1^2}{n_1} + \cdots + \frac{R_k^2}{n_k}\right) - 3(N+1)$$

H has approximately a chi-square distribution with $k - 1$ degrees of freedom under H_0. Testing at significance level α, we reject H_0 if $H \geq \chi_\alpha^2$; otherwise, we accept H_0. The Kruskal-Wallis test is nondirectional and detects only whether or not there is some difference in the treatments. It is not used in the same situations as the Friedman test. For the Kruskal-Wallis test the data must come from independent samples, while in the Friedman test the data are in the form of related samples, or *blocks,* of information.

Example 11-5. Three different methods of weather prediction are to be compared: the use of astrology, western science (meteorology), and eastern science (Chinese weather prediction). Some forecasters of each type are selected, and the number of correct day-to-day predictions over a 1-year period are recorded (Fig. 11-8). Testing at a significance level of $\alpha = .05$, would you accept or reject the hypothesis that

	No. of correct weather predictions over 1-year period	
Astrologers	Western scientists (meteorology)	Eastern scientists (Chinese weather prediction)
121	188	284
110	164	109
35	201	162
160	173	212
140		224

Figure 11-8

there is no difference in accuracy among the methods of weather prediction?

SOLUTION: We note first that this is the type of experiment which may be quite difficult to set up. We must clearly define a *correct prediction* and, in fact, exactly what types of predictions are to be allowed. We must also ensure that the weathermen in the study are not aware of the predictions of the others (the samples must be independent). There are other factors which must also be closely accounted for, but since our current objective is to apply the Kruskal-Wallis test, we shall not proceed further into the design of this experiment.

To apply the Kruskal-Wallis test we must rank our data and sum the ranks, as in Fig. 11-9. We now compute H, noting that there are $n_1 = 5$ astrologers, $n_2 = 4$ western meteorologists, and

	Ranks of weather predictors	
Astrologers	Western scientists (meteorology)	Eastern scientists (Chinese weather prediction)
4	10	14
3	8	2
1	11	7
6	9	12
5		13
$R_1 = 19$	$R_2 = 38$	$R_3 = 48$

Figure 11-9

$n_3 = 5$ eastern scientists for a total of $N = 14$ participants in the experiment. We have

$$H = \frac{12}{14(15)} \left(\frac{19^2}{5} + \frac{38^2}{4} + \frac{48^2}{5} \right) - 3(15)$$

$$= 51.1 - 45$$

$$= 6.1$$

Looking at the chi-square tables in the Appendix, with $3 - 1 = 2$ degrees of freedom, we find $\chi^2_{.05} = 5.99$. Since $H = 6.1 > 5.99$, we reject H_0 and conclude that there is a difference in the accuracy among the methods of weather prediction. Returning to the data, it seems apparent that the astrologers are not as accurate as the other scientists.

Problems

11-5 A toy company wishes to compare the interest span of children for three different toys it is considering as future products. Three samples of seven children are randomly selected, one sample for each toy, and the amount of playing time for each child (until he loses interest in the toy) is recorded. Apply the Kruskal-Wallis test to decide if there is a difference in interest spans among the toys.

Interest span (in minutes)

Toy A	Toy B	Toy C
10.8	4.1	83.1
50.1	3.7	12.6
16.1	2.8	7.9
5.7	19.5	8.1
4.9	16.2	6.3
6.2	8.2	19.1
27.1	1.1	12.0

11-6 A number of randomly selected families are studied to compare the amount of conversation time per day that exists between husband and wife for four different wage levels (based on annual wage). Use the Kruskal-Wallis test at a significance level of

$\alpha = .01$ to determine if there is a difference in conversation times among the wage levels.

Amount of conversation time per day (in minutes) for 4 wage levels

Under $10,000/year	$10,000–$20,000	$20,000–$30,000	Over $30,000
30.1	35.8	51.5	6.1
40.9	31.7	19.3	47.9
63.1	22.9	26.9	54.5
20.6	5.8	1.7	31.3
18.2	48.6	41.8	50.7
27.5	40.0		

11-7 Eight people are randomly selected from each of four different job categories and given a psychological test which measures their level of hostility (a high score meaning a high level of hostility). Using the Kruskal-Wallis test at a significance level of $\alpha = .05$, determine whether or not there is a difference in hostility levels among the different job categories.

Scores on hostility test

Professors	Policemen	Butchers	Clerks
65	68	62	30
52	48	58	28
57	59	47	75
49	52	42	80
81	69	53	41
58	71	45	49
69	67	41	60
70	91	62	67

Summary

We have presented two nonparametric methods of comparing the effects of different treatments on a population under study. The applications ranged from comparing products to evaluating methods of weather prediction. We studied the Friedman test and Friedman statistic Q for experiments in which the information about each treatment was related in some way.

We studied the Kruskal-Wallis test and statistic H for experiments in which information is in the form of independent samples.

Both tests involve ranking the data—separately for each block in the Friedman test and overall for the Kruskal-Wallis test. Also, both test statistics involve studying the rank sums for each treatment and are designed so that under H_0 they will have approximately a chi-square distribution, which makes for an easy determination of rejection regions since tables of chi-square probabilities are available.

Important terms

Friedman test

Friedman test statistic Q

Kruskal-Wallis test

Kruskal-Wallis test statistic H

Nonparametric

True-false test

Circle the correct letter.

T   F   1. The Friedman test can be applied only when the data from each treatment come from independent samples.

T   F   2. The Kruskal-Wallis test tells exactly which treatments have more effects than others.

T   F   3. If eight randomly selected people are each asked to rank three products, the Friedman test might be used to determine if they are in general agreement.

T   F   4. The Kruskal-Wallis test is to the Mann-Whitney test as the Friedman test is to the Wilcoxon test.

T   F   5. In order to apply the Kruskal-Wallis test, the samples for each treatment must be of the same size.

Analysis of variance **12**

In Chap. 11 we considered two nonparametric tests used to analyze data obtained from comparative experiments in which more than two treatments are compared. Recall that the Friedman test may be applied to data obtained from related samples, and that the Kruskal-Wallis test is used when the samples are independent. In the latter case, if it can be assumed that the samples are drawn from populations having normally distributed data with equal standard deviations, then a powerful method called the *one-way analysis of variance,* or *F test,* may be used. This test detects a difference between treatments by comparing the mean responses to the various treatments. The procedure is analogous to the one used in the two-sample *t* test (studied in Chap. 8), except in the *F* test more than two treatments are considered.

The F test for equal sample sizes

The following example illustrates a situation in which the *F* test is applicable.

Example 12-1. Suppose you want to compare the IQ's of three groups of people, let us say, students, professors, and administrators at a certain university. Naturally, you select a random sample from each of the three groups and give each person selected an IQ test. Data from such an experiment could be analyzed by the Kruskal-Wallis test since the samples are independently selected. However, a more powerful test can be applied if you are willing to assume that the IQ scores in each group are normally distributed with common standard deviations. This assumption is not unreasonable since results of IQ test scores are usually normally distributed. (More advanced texts give techniques for checking this assumption.)

The *F* test makes comparisons of population means, in this case, the average IQ of students, the average IQ of professors, and the average IQ of administrators. For ease of reference, let us denote these three population means by μ_1, μ_2, and μ_3, respectively. We wish to test

H_0:   The three groups have the same average IQ.

H_a: The three groups do not have the same average IQ.

Equivalently, this may be stated as

H_0: $\mu_1 = \mu_2 = \mu_3$

H_a: Not all μ's are equal.

To choose between these hypotheses we need a test statistic and a test rule. Intuitively, it seems that the sample means $\bar{X}_1$, $\bar{X}_2$, and $\bar{X}_3$ should give us information about μ_1, μ_2, and μ_3. The test statistic should perhaps reflect the difference among sample means for each of the three groups. The test rule, then, should reject H_0 if the variation between $\bar{X}_1$, $\bar{X}_2$, and $\bar{X}_3$ is too large, but some variation between $\bar{X}_1$, $\bar{X}_2$, and $\bar{X}_3$ is expected due to the effects of random sampling. Thus, to reject H_0 we should require that the variation among the three sample means be large after first taking into account the expected random variation due to sampling.

To see this more clearly, consider the following two sets of data, which might be observed if the IQ tests were given to a sample of 10 people drawn from each of the 3 groups:

Data 1: IQ scores

Students	Professors	Administrators
124	111	117
122	112	114
127	107	117
128	111	113
126	111	117
127	112	114
121	112	118
122	108	114
127	111	118
126	105	118

$\bar{X}_1 = 125$ $\bar{X}_2 = 110$ $\bar{X}_3 = 116$

$s_1{}^2 = 6.4$ $s_2{}^2 = 6.0$ $s_3{}^2 = 4.0$

$s_1 = 2.53$ $s_2 = 2.45$ $s_3 = 2.00$

Data 2: IQ scores

Students	Professors	Administrators
103	90	138
141	93	95
140	147	93
90	116	154
158	132	138
147	89	100
124	123	149
108	134	100
92	95	104
147	81	89

$\bar{X}_1 = 125$ $\bar{X}_2 = 110$ $\bar{X}_3 = 116$

$s_1{}^2 = 625.1$ $s_2{}^2 = 536.7$ $s_3{}^2 = 650.7$

$s_1 = 25.00$ $s_2 = 23.17$ $s_3 = 25.51$

Notice that the means are the same for both sets of data. Thus, if a decision as to whether to accept or reject H_0 were based solely on the values of the three sample means, then both sets of data would lead to the same decision. But if we look at the data more closely, it is not clear that we should reach the same decision in both situations. The basic difference between the two sets of data is the amount of variation within each group. In data 1 the IQ scores range between 120 and 128 for students, 105 and 112 for professors, and 113 and 118 for administrators, making three nonoverlapping groups of scores. Moreover, the standard deviations for the scores in each group are small, implying that the sample means $\bar{X}_1$, $\bar{X}_2$, and $\bar{X}_3$ will tend to be close to the population means μ_1, μ_2, and μ_3. If the three population means are, in fact, equal, such a pattern would be highly unlikely; therefore, any reasonable test rule should reject H_0. On the other hand, it is not at all obvious which decision to make based on data 2. In this set of data the scores for the three groups overlap greatly and the sample standard deviations are large, implying a large variation between the sample means even if the population means are equal.

Let us look at the variation among $\bar{X}_1$, $\bar{X}_2$, and $\bar{X}_3$ more closely. The observed differences between these three sample means comes from two sources: any differences among μ_1, μ_2, and μ_3, and chance fluctuations occurring within each of the three groups due to random selection of the sample. Here's why: Since the sample means $\bar{X}_1$, $\bar{X}_2$, and $\bar{X}_3$ estimate the population means μ_1, μ_2, and μ_3, any difference among these population means is reflected in the sample data as variation among $\bar{X}_1$, $\bar{X}_2$, and $\bar{X}_3$. The bigger the differences among μ_1, μ_2, and μ_3, the bigger the differences among $\bar{X}_1$, $\bar{X}_2$, and $\bar{X}_3$ tend to be. As for the second source, no matter what the population means μ_1, μ_2, and μ_3 are, there will be some chance fluctuations in the values of $\bar{X}_1$, $\bar{X}_2$, and $\bar{X}_3$ due to the effects of random sampling. If different samples were chosen from the three groups, then different IQ test scores and hence different values of $\bar{X}_1$, $\bar{X}_2$, and $\bar{X}_3$ would result.

What we need is some measure of the first source of variation; this tells us exactly what we need to know. Unfortunately there is no way to do this directly. However, we can measure the combined effect of both sources of variation among $\bar{X}_1$, $\bar{X}_2$, and $\bar{X}_3$, and the effect of the second source of variation. By comparing these two measures, we can make inferences about H_0. The combined effect of both sources of variation among $\bar{X}_1$, $\bar{X}_2$, and $\bar{X}_3$ will be measured by

$$s_B{}^2 = \frac{10[(\bar{X}_1 - \bar{X})^2 + (\bar{X}_2 - \bar{X})^2 + (\bar{X}_3 - \bar{X})^2]}{2}$$

where $\bar{X}$ denotes the overall, or grand, mean of all data computed from the formula $\bar{X} = (\bar{X}_1 + \bar{X}_2 + \bar{X}_3)/3$.

The variation within each of the three samples is measured by

$$s_W^2 = \frac{s_1^2 + s_2^2 + s_3^2}{3}$$

Notice that s_1^2, s_2^2, and s_3^2 measure the variation within each of the three samples and s_W^2 is simply their average. If the population means μ_1, μ_2, and μ_3 differ significantly, then s_B^2 will tend to be much greater than s_W^2. If $\mu_1 = \mu_2 = \mu_3$, then s_B^2 will reflect only the second source of variation and hence s_B^2 will be approximately equal to s_W^2. Thus, we take as our test statistic

$$F = \frac{s_B^2}{s_W^2}$$

Our test rule is of the form

> *Reject H_0 if F is large.*
>
> *Accept H_0 if F is small.*

Returning now to our data, we compute F for each set. For data 1, $F = 104.2$; and for data 2, $F = .94$. The quantity F has what is called an *F distribution* when H_0 is true. Tables of the F distribution are included in the Appendix. In a moment we shall explain how to use these tables to obtain probabilities for the F distribution.

We may now describe the F-test procedure in more general terms. It is assumed that there are k different treatments, and responses from samples receiving the treatments are observed. We wish to detect any difference between the population mean responses. As we have stated before, the word "treatments" commonly refers to the objects being compared. These may be medical treatments, methods of instruction, groups of people, brands of a certain type of merchandise, etc.

The samples μ_1, μ_2, ..., μ_k are used to denote the mean response from the k treatments. These are population means: μ_1 represents the average response that would be observed if the entire population from which the sample is drawn were given treatment 1, and the same for μ_2, ..., μ_k. These numbers are unknown, of course. The hypotheses we are testing are

> H_0: $\mu_1 = \mu_2 = \cdots = \mu_k$
>
> H_a: Not all means are equal.

Notice that this is a nondirectional test (like a two-tailed test). Rejecting H_0 does not tell us which means are big and which are small, only that there is a difference.

To define the test statistic, suppose that n responses are measured from each of the k treatments; there are nk responses all together. (The case when there are different numbers of responses from each treatment is considered later.) Let $\bar{X}_1, \bar{X}_2, \ldots, \bar{X}_k$ denote the individual sample means observed from the k treatments, respectively, and let $\bar{X}$ be the average of all responses. That is

$$\bar{X} = \frac{\bar{X}_1 + \cdots + \bar{X}_k}{k} = \frac{\Sigma X}{kn}$$

$\bar{X}$ is usually referred to as the *grand mean*. Let $s_1, s_2, \ldots, s_k$ denote the sample standard deviations. As we explained in Example 12-1, the variation in the sample data comes from two sources. Their combined effect is reflected in the variation among $\bar{X}_1, \bar{X}_2, \ldots, \bar{X}_k$, which is measured by the quantity

$$s_B{}^2 = \frac{n[(\bar{X}_1 - \bar{X})^2 + \cdots + (\bar{X}_k - \bar{X})^2]}{k - 1}$$

The effect of chance fluctuations in sampling is reflected in the variation within each of the k samples, which is measured by

$$s_W{}^2 = \frac{s_1{}^2 + \cdots + s_k{}^2}{k}$$

If $\mu_1 = \cdots = \mu_k$ (that is, H_0 is true), then both $s_B{}^2$ and $s_W{}^2$ will be measures of the chance fluctuations due to random sampling; that is, $s_B{}^2$ and $s_W{}^2$ each estimate the square of the population standard deviation, σ^2. In this case, the ratio $s_B{}^2/s_W{}^2$ should be about 1. On the other hand, if the means $\mu_1, \ldots, \mu_k$ are not equal, $s_B{}^2$ will tend to be greater than σ^2, and the ratio $s_B{}^2/s_W{}^2$ will be greater than 1. Therefore, the test statistic is

$$F = \frac{s_B{}^2}{s_W{}^2}$$

and the test rule is of the form:

> *Reject H_0 if F is large.*
>
> *Accept H_0 if F is small.*

The quantity F is usually called an *F statistic*.

In order to specify which values of F are "large," it's necessary to know the distribution of F. Under the following assumptions, F has what is called an *F distribution* with v_1 and v_2 degrees of freedom, where $v_1 = k - 1$, and $v_2 = k(n - 1)$. Here are the assumptions needed for F to have an F distribution when H_0 is true:

1 The standard deviations of the responses from each of the k treatments must be roughly equal.

2 If n is small, the responses for each treatment must be normally distributed. (If n is large, the responses do not have to be normally distributed because of the central limit theorem.)

Using the F tables, we may find the cutoff value c needed to specify when F is "too large" for any given level of significance. In the Example 12-1, $k = 3$ and $n = 10$, and so $v_1 = k - 1 = 2$ and $v_2 = k(n - 1) = 3(9) = 27$. In this case, if we wanted to test H_0 at the .05 level of significance, the table shows $c = 3.39$. (See Fig. 12-1). Thus, if data 1 were observed, $F = 104.2 > 3.39$ and H_0 would be rejected; if data 2 were observed, $F = .94 < 3.39$ and H_0 would be accepted.

A few more examples will illustrate the use of the analysis-of-variance technique called the F test.

Example 12-2. A sociologist is interested in determining if there is a difference in income for immigrants from 5 countries during their first year in the United States. He selects a random sample of 50 immigrants from each of the 5 countries and obtains the following results:

$$\bar{X}_1 = 5,000 \quad \bar{X}_2 = 5,100 \quad \bar{X}_3 = 4,900 \quad \bar{X}_4 = 5,050 \quad \bar{X}_5 = 4,950$$

$$s_1{}^2 = 10,000 \quad s_2{}^2 = 15,000 \quad s_3{}^2 = 9,000 \quad s_4{}^2 = 12,000 \quad s_5{}^2 = 14,000$$

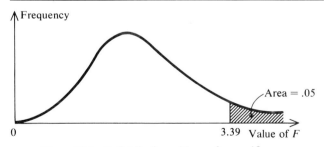

Figure 12-1 F distribution with $v_1 = 2$, $v_2 = 18$.

Using a .05 level of significance, perform an analysis of variance to determine if the mean incomes for the five countries are different.

SOLUTION: The treatments are the countries, and the responses are the incomes observed in the samples from each group of immigrants. Let

μ_1 = mean income for immigrants from country 1

μ_2 = mean income for immigrants from country 2

$$\vdots$$

μ_5 = mean income for immigrants from country 5

The hypotheses to be tested are

H_0: $\mu = \cdots = \mu_5$

H_a: The mean incomes are not all equal.

Since there are 5 countries with 50 people sampled from each country, $k = 5$ and $n = 50$. The grand mean is

$$\bar{X} = \frac{\bar{X}_1 + \cdots + \bar{X}_5}{5} = 5,000$$

and thus

$$S_B{}^2 = n \frac{(\bar{X}_1 - \bar{X})^2 + \cdots + (\bar{X}_k - \bar{X})^2}{k - 1}$$

$$= 50 \left[\frac{(5,000 - 5,000)^2 + (5,100 - 5,000)^2 + (4,900 - 5000)^2}{5 - 1} \right.$$

$$\left. + \frac{(5,050 - 5,000)^2 + (4,950 - 5,000)^2}{5 - 1} \right]$$

$$= 50 \left[\frac{25,000}{5 - 1} \right] = 312,500$$

Also,

$$S_W{}^2 = \frac{S_1{}^2 + \cdots + S_k{}^2}{k}$$

$$= \frac{10,000 + 15,000 + 9,000 + 12,000 + 14,000}{5}$$

$$= 12,000$$

Finally,

$$F = \frac{s_B^2}{s_W^2} = \frac{312,500}{12,000} = 26.04$$

Using $\nu_1 = k - 1 = 5 - 1 = 4$ and $\nu_2 = k(n - 1) = 5(49) = 245$, we may find the critical value c corresponding to the .05 level of significance from the F tables: $c = 2.37$.

Since $F = 26.04 > 2.37$, we reject H_0 and conclude that the mean incomes are not all equal. There is more variation among $\bar{X}_1, \ldots, \bar{X}_5$ than we would expect from randomness alone.

Example 12-3. A statistics professor is beginnig a new job at a certain university and he wants to determine which of four convenient-ly located banks to use. His only concern is how long it would take to do his business. He decides to use his statistical training: He sends 20 students into each of 4 banks, and each student records the length of time (in minutes) spent waiting in line in order to cash a check. Labeling the banks by 1, 2, 3, and 4, the professor obtains the following results:

$$\bar{X}_1 = 10 \qquad \bar{X}_2 = 8 \qquad \bar{X}_3 = 12 \qquad \bar{X}_4 = 18$$

and

$$s_1^2 = 60 \qquad s_2^2 = 70 \qquad s_3^2 = 55 \qquad s_4^2 = 65$$

At the .01 level of significance, should the professor reject the hypothesis that there is no difference in the mean waiting times.

SOLUTION: In this example the treatments are the banks and the responses are the observed waiting times. If we let μ_1 = mean waiting time in bank 1, μ_2 = mean waiting time in bank 2, etce-tera, then we are interested in testing the following hypotheses:

H_0: $\mu_1 = \mu_2 = \mu_3 = \mu_4$

H_a: There are differences in the mean waiting times.

Since there are four means to be compared and since 20 ob-servations are made at each bank, $k = 4$ and $n = 20$. The overall mean

$$\bar{X} = \frac{\bar{X}_1 + \cdots + \bar{X}_4}{4} = 12$$

and

$$s_B^2 = n \frac{(\bar{X}_1 - \bar{X})^2 + \cdots + (\bar{X}_k - \bar{X})^2}{k - 1}$$

$$= 20 \left[\frac{(10 - 12)^2 + (8 - 12)^2 + (12 - 12)^2 + (18 - 12)^2}{4 - 1} \right]$$

$$= \frac{20}{3} (56)$$

$$= 373.3$$

Also

$$s_W^2 = \frac{s_1^2 + \cdots + s_k^2}{k}$$

$$= \frac{60 + 70 + 55 + 65}{4}$$

$$= 62.5$$

Finally,

$$F = \frac{s_B^2}{s_W^2} = \frac{373.3}{62.5} = 5.97$$

Using $\nu_1 = k - 1 = 3$ and $\nu_2 = k(n - 1) = 76$, we may find the critical value c corresponding to the .05 level of significance from the F tables: $c = 2.76$. Since $F = 5.97 > 2.76$, we reject H_0 and conclude that there are differences in the mean waiting times.

Example 12-3 is a good illustration of the limitations of the analysis-of-variance technique. Although the F test informs the professor that the waiting times at the four banks are not the same, it doesn't specify which bank (or banks) have the shortest waiting time(s). One possible approach the professor might take at this point is to perform an F test on the data obtained by deleting the observations for the fourth bank, the one with the largest sample mean. If the results of this test on the remaining three banks are significant, he could delete the data from the third bank and perform a t test on the remaining two. If the results of the test on the three banks are not significant, he could choose one of them at random. Unfortunately, although the level of significance of any particular test in the series of tests performed has a given level of significance, the procedure as a whole has a higher level of significance.

The F test for unequal sample sizes

When the sample sizes for various treatments are unequal, a slight modification in the formulas is required. As before, assume $\mu_1, \ldots, \mu_k$ represent the (unknown) mean responses for k treatments. Let $n_1, \ldots, n_k$ denote the sizes of the samples receiving the various treatments. Also, let $\bar{X}_1, \ldots, \bar{X}_k$ be the sample means and $s_1^2, \ldots, s_k^2$ be the sample variances. The test procedure for this situation is as follows.

The hypotheses are the same as before:

H_0: $\mu_1 = \cdots = \mu_k$

H_a: The mean responses are not all equal.

Then

$$F = \text{test statistic} = \frac{s_B^2}{s_W^2}$$

where

$$s_B^2 = \frac{n_1(\bar{X}_1 - \bar{X})^2 + \cdots + n_k(\bar{X}_k - \bar{X})^2}{k - 1}$$

$$\bar{X} = \frac{n_1 \bar{X}_1 + \cdots + n_k \bar{X}_k}{n_1 + \cdots + n_k}$$

and

$$s_W^2 = \frac{(n_1 - 1)s_1^2 + \cdots + (n_k - 1)s_k^2}{n_1 + \cdots + n_k - k}$$

Under the previous assumptions, F will have an F distribution with $\nu_1 = k - 1$ and $\nu_2 = n_1 + \cdots + n_k - k$. Large values of F lead to rejection of H_0. As before, we can determine the rejection region for a given level of significance by looking at the F table.

Example 12-4. A product-testing company wishes to compare the writing lifetimes of four brands of similarly priced ballpoint pens and randomly selects 50 pens of each brand for its study. However, due to thefts by some of the testers, only 45 pens of brand 1, 48 pens of brand 2, 50 pens of brand 3, and 40 pens of brand 4 are actually tested. These are installed in a scribbling machine, and the amount of time (in hours) before they run out of ink is recorded. The data are summarized below. Is there any reason to believe that there is a difference in writing lifetimes among the four brands? (Use a .05 level of significance.)

Brand 1	Brand 2	Brand 3	Brand 4
$\bar{X}_1 = 38.5$	$\bar{X}_2 = 41.3$	$\bar{X}_3 = 37.9$	$\bar{X}_4 = 38.1$
$s_1{}^2 = 3.1$	$s_2{}^2 = 2.9$	$s_3{}^2 = 3.7$	$s_4{}^2 = 3.3$
$n_1 = 45$	$n_2 = 48$	$n_3 = 50$	$n_4 = 40$

SOLUTION: For this example, μ_1 = average writing time for pens of brand 1, μ_2 = average writing time for pens of brand 2, etcetera. We wish to test

H_0: $\mu_1 = \mu_2 = \mu_3 = \mu_4$

H_a: The means are not all equal.

Using the formulas for the F test with unequal sample sizes, we obtain

$$\bar{X} = \frac{45(38.5) + 48(41.3) + 50(37.9) + 40(38.1)}{45 + 48 + 50 + 40}$$

$$= 39.0$$

$$s_B{}^2 = \frac{45(38.5 - 39.0)^2 + 48(41.3 - 39.0)^2 + 50(37.9 - 39.0)^2}{4 - 1}$$

$$+ \frac{40(38.1 - 39.0)^2}{4 - 1}$$

$$= 119.4$$

$$s_W{}^2 = \frac{(44)(3.1) + (47)(2.9) + (49)(3.7) + (39)(3.3)}{45 + 48 + 50 + 40 - 4}$$

$$= 3.3$$

Hence,

$$F = \frac{s_B{}^2}{s_W{}^2} = \frac{119.4}{3.3} = 36.18$$

The test statistic F has an F distribution with $v_1 = 4 - 1 = 3$, and $v_2 = 45 + 48 + 50 + 40 - 4 = 179$ degrees of freedom. Using a level of $\alpha = .05$, from the F tables in the Appendix we find that the critical value is 2.60. Since the observed value of $F = 36.18 > 2.60$, H_0 is rejected at the .05 level of significance. There is strong evidence indicating that the average writing lifetimes are not all equal.

Summary

The F test, or analysis of variance, is a powerful test used to detect a difference between several population means. In this model it is assumed that there are k populations, having means $\mu_1, \mu_2, \ldots, \mu_k$. The F test is used to decide between

H_0: $\mu_1 = \cdots = \mu_k$

H_a: The means are not all equal.

The test statistic is

$$F = \frac{s_B{}^2}{s_W{}^2}$$

The quantity $s_B{}^2$ measures the variation in the data due to random sampling and to a difference among the μ's. The quantity $s_W{}^2$ measures the variation in the data due to random sampling. If H_0 is true, $s_B{}^2$ and $s_W{}^2$ should be roughly equal and F will be near 1.

Under the assumptions that the k population standard deviations are equal and that the population data are normally distributed, the test statistic F has what is called an F distribution with degrees of freedom $\nu_1 = k - 1$ and $\nu_2 = k(n - 1)$ under H_0. Tables of the F distribution are available in the Appendix for determining the rejection region in particular problems.

When the sample sizes are not all equal, the formulas must be modified slightly. Still, $F = s_B{}^2/s_W{}^2$, but the computation of $s_B{}^2$ and $s_W{}^2$ is different. Under the assumptions made in the case of equal sample sizes, F will have an F distribution with $\nu_1 = k - 1$ and $\nu_2 = n_1 + \cdots + n_k - k$ degrees of freedom.

No matter what the sample sizes are, the form of the test rule is: Reject H_0 if F is large; accept H_0 if F is small. Which values of F are "large" is determined by the given level of significance.

Important terms

F test, or analysis of variance

Response

Treatment

Problems

12-1 A scientific laboratory interested in purchasing a new computer system conducts a study to compare the average "down," or repair, time per month of three types of computers. Six computer systems of each type are studied, and the following data are recorded. Assuming normal distributions and common standard deviations for the down times (recorded in hours per month), apply the *F* test at a significance level of $\alpha = .05$.

Down times (hours per month)

System 1	System 2	System 3
$\bar{X}_1 = 15.6$	$\bar{X}_2 = 12.8$	$\bar{X}_3 = 18.1$
$s_1 = 4.1$	$s_2 = 5.9$	$s_3 = 7.5$

Do you think the assumptions of normality and equal standard deviations are reasonable in this case? What test might you apply if you are not willing to make these assumptions?

12-2 A study was made to compare the average toilet-training age in children for three different socioeconomic status (SES) levels. The following average toilet-training ages and standard deviations were recorded after studying samples of 25 children from each SES level. Assuming that these ages are normally distributed with common standard deviations, apply the *F* test at a significance level of $\alpha = .05$ to detect if there are any differences.

Toilet-training ages for different SES levels

Low	Medium	High
$\bar{X}_1 = 1.8$	$\bar{X}_2 = 2.1$	$\bar{X}_3 = 2.4$
$s_1 = .5$	$s_2 = .6$	$s_3 = .7$

12-3 Four different fertilizers are compared to see if there is a difference in average yield when each fertilizer is applied to seven randomly selected plots of corn. Assuming normality of the yields and common standard deviations, apply the *F* test at a significance level of $\alpha = .01$ to the following data and decide if there is a difference in average yields.

Corn yields (in bushels) for four fertilizers

A	B	C	D
20.6	23.6	22.8	17.2
21.7	25.7	21.6	21.5
19.8	11.1	21.0	22.8
17.1	13.2	20.3	23.1
16.2	15.5	19.4	24.1
12.0	17.1	19.8	18.0
18.5	17.1	20.7	19.6

12-4 A sociologist conducts a study to see if the "harried mother" experiences a loss in reasoning ability when she has many children. A sample is selected, including 20 women with no children, 20 mothers with 1 child, 20 mothers with 2 children, 20 mothers with 3 children, and 20 mothers with 4 children. Each woman is given a standard test to measure reasoning ability, and the following results are tabulated. Making the required assumptions, apply the F test at a significance level of $\alpha = .01$ and decide whether or not there is a difference in reasoning ability among the five groups.

Scores on reasoning-ability test

	No. of children			
0	*1*	*2*	*3*	*4*
$\bar{X}_1 = 78$	$\bar{X}_2 = 79$	$\bar{X}_3 = 72$	$\bar{X}_4 = 67$	$\bar{X}_5 = 61$
$s_1 = 12$	$s_2 = 11$	$s_3 = 11.6$	$s_4 = 9.1$	$s_5 = 7$

12-5 The McDougal's hamburger organization conducts a study to determine which is the best area of town for a hamburger stand. McDougal's hamburger stands in several cities are randomly selected from downtown shopping areas, ghetto areas, suburban shopping center areas, and freeway exit areas. The average sales per week are recorded for each hamburger stand studied. Assuming normality and common standard deviations, apply the F test at a significance level of $\alpha = .05$ to the following data.

McDougal's hamburgers: weekly gross sales (in thousands of dollars)

Downtown shopping	Downtown ghetto	Suburban shopping center	Freeway exit
5.2	8.1	7.2	4.1
5.1	4.0	6.1	4.8
4.7	5.9	5.8	3.7
6.3	5.2	5.9	6.2
6.8		6.1	5.1

Apply another suitable test to these data. Do your results for this test agree with your results for the F test?

12-6 The dean of a California state university is interested in comparing the grade-point averages of in-state students, out-of-state students, and foreign students. One phase of his study consists of comparing graduating students from each category in the current year. Each student's grade-point average over his final 2 years at the university is computed. Based on the data obtained, apply the F test at the significance level of $\alpha = .05$ to determine if there is a difference in grades.

Grade-point averages for students

In state	Out of state	Foreign
$\bar{X}_1 = 2.5$	$\bar{X}_2 = 2.8$	$\bar{X}_3 = 2.7$
$s_1^2 = .8$	$s_2^2 = .6$	$s_3^2 = .9$
$n_1 = 275$	$n_2 = 110$	$n_3 = 81$

12-7 a. In Prob. 12-6 why can the dean use a statistical test in view of the fact that he is obtaining data from all graduating students in the current year rather than taking a random sample?

b. How could the dean account for the fact that perhaps foreign students who do poorly at the university in their first 2 years must withdraw and return home while in-state students find it easier to remain at the university unless they actually flunk out, thus giving an unrepresentative group of foreign students for the test?

12-8 A consumers' group conducts a comparison of food prices in five large cities. A typical weekly food list is made up of items which a typical middle-class family of four might need in a week. Then, 10 supermarkets are randomly selected from each city, and the items on the list are purchased at the selected markets. The average amount spent is computed for each city, and the results are tabulated. Using the F test at a significance level of $\alpha = .01$, would you conclude that there is a difference in prices?

Food prices in five large cities

City 1	City 2	City 3	City 4	City 5
$\bar{X}_1 = \$42$	$\bar{X}_2 = \$45$	$\bar{X}_3 = \$39$	$\bar{X}_4 = \$44$	$\bar{X}_5 = \$47$
$s_1 = \$7$	$s_2 = \$6$	$s_3 = \$4$	$s_4 = \$5$	$s_5 = \$6$

True-false test

Circle the correct letter.

T   F   1. The F test is to the t test as the Kruskal-Wallis test is to the Mann-Whitney test.

T   F   2. In order to apply the F test, the sample size from each population must be the same.

T   F   3. In order to apply the F test, the sample standard deviations from every population sample must be the same.

T   F   4. If H_0 is rejected using the F test, then it is possible to order the population means in increasing size.

T   F   5. The F test can be applied in any situation where the Kruskal-Wallis test can be applied.

T   F   6. If some data to which the F test was applied show $\bar{X}_1 = 2.1$, $\bar{X}_2 = 2.0$, $\bar{X}_3 = 13$, $n_1 = n_2 = n_3 = 50$, then H_0 is necessarily rejected.

T   F   7. If every data value is multiplied by 2, then the F statistic is multiplied by 2.

T   F   8. Some data consist of a list of number of cars owned by 50 randomly selected families in each of six regions. In order to detect whether there is a regional difference in average number of cars per family, the F test should be applied.

T   F   9. If $\bar{X}_1 = 6.3$, $s_1 = 5.8$, $\bar{X}_2 = 6.3$, $s_2 = 12$, $\bar{X}_3 = 6.3$, $s_3 = 17$, it's possible that H_0 will be rejected when using the F test.

T   F   10. If the sample standard deviations differ greatly, then there is reason to believe that the F test may not be applicable.

The least-squares line and correlation coefficient

13

An important statistical problem encountered by the scientific experimenter is trying to determine if there is a relationship, or *correlation,* between two (or more) variables. A doctor may be interested in determining if there is a relationship between air pollution and respiratory disease; a psychologist may want to know if there is a relationship between adulthood neurosis and mistreatment as an infant; a biologist may wish to discover if there is a relationship between DDT and the thickness of eggshells of certain birds. (Of course, we may all be interested in such questions. It is the experimenter, however, who conducts the scientific analyses.)

In similar fashion, we may be interested in using information from a set of data to predict other events. An economist may wish to predict general trends in the economy based on results of tests in specific areas; a high school counselor may want to predict a student's success or failure in college based on his high school grades or aptitude test scores; an engineer may wish to predict the strength of a new type of metal based on the proportion of one of its components. The techniques we shall discuss in this chapter, the *least-squares line* and the *correlation coefficient,* are the two most commonly used tools for the prediction and comparison of paired data. Generalizations of these techniques are used for related sets of data which occur in larger blocks than pairs. However, we shall not discuss these more complicated techniques.

The type of relationship which is most often studied, partially because of its relative simplicity, is the *linear relationship.* Two variables X and Y are linearly related if we can accurately predict one from the other by the use of a nonvertical straight line. As you may remember from high school, a straight line can be expressed by an equation of the form $Y = aX + b$, where a is the slope of the line and b is the Y intercept. In Fig. 13-1 we have drawn the graphs of some straight lines along with their equations.

To say that the variable Y is a linear function of X means that there are numbers a, b such that if we know $X = x$ we can determine Y by the formula $Y = ax + b$. This is called a *deterministic relationship:* If we know X, we can then determine Y. Of course, in real life there are few deterministic relationships. It is more often the case that a random factor enters in. We may still believe that a linear or near linear relationship exists between two variables, but now we wish to account for the random factor. We do this by modifying our formula a bit and saying that X and Y are related by the formula $Y = aX + b + \epsilon$, where ϵ is the random error term. This formula gives rise to what is called the *elementary linear model.*

> **Definition.** Two random variables X, Y are said to fit an elementary linear model, or have a linear relationship, if there exist numbers a, b such that $Y = aX + b + \epsilon$, where ϵ is a *random error term.*

Often assumptions are made about the distribution of the error term ϵ, usually that ϵ is normally distributed with a mean of zero. We shall not study the error term but shall discuss one main aspect of this model, namely, how to estimate the numbers a, b based on given data. We shall also discuss a measure of the linear relationship between two variables, called the *correlation coefficient.* We begin with the following example.

Example 13-1. Some data were gathered from college graduates in an attempt to determine the relationship (if any) between high school grades and college grades. In Fig. 13-2, the X value is the student's grade-point average in his final 2 years of high school and the corre-

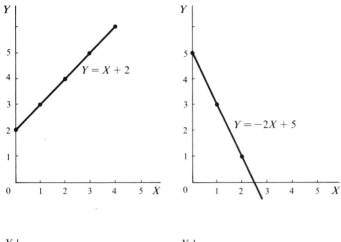

Figure 13-1   Graphs and equations of straight lines.

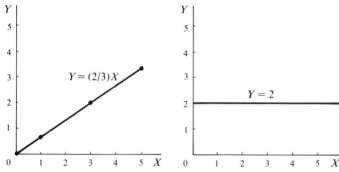

X	Y
2	1.5
1	2
3	2
3	3
4	3.5
4	4
4	4

Figure 13-2

sponding Y value is his grade-point average in the first 2 years of college. In Fig. 13-2, we give a pictorial display of this information in a *scatter diagram,* a graph of the points (x,y), the X value being the X coordinate, the Y value being the Y coordinate. Points (x, y) which appear more than once in our data are appropriately labeled on the scatter diagram. Based on this information, we wish to find a reasonable method of predicting a student's grades for the first 2 years of college, based on his high school grades.

We emphasize that a scatter diagram itself is an extremely important device for detecting whether or not there is a relationship between two variables. One may stare unsuccessfully at columns of numbers for long periods of time, futilely trying to see if a relationship exists, yet instantly see a relationship when the data are graphed in a scatter diagram. The usefulness of the scatter diagram becomes even more apparent when large amounts of data are collected (a situation which we have avoided for clarity of presentation).

The least-squares line

Suppose now, that we could draw a straight line on the scatter diagram in such a way that every pair of data points would be on the line. (We obviously can't do this in Fig. 13-2.) If we could, we would be justified in saying that there was a *linear relationship* between X and Y. If the line sloped upward, we could say X and Y were *positively correlated* (as X grows large, so does Y); if the line sloped downward, we would say X and Y are *negatively correlated* (as X becomes large, Y becomes small). We could then use this line as a *predictor;* if we obtained a new value x, we would predict Y to be the value y which would make the point (x, y) lie on our line. Of course, requiring that there be a straight line which passes through each point on the scatter diagram puts us out of business; such situations rarely occur. We could introduce an error term, mentioned previously, but this will not tell us how to determine a good straight line to fit our data. Instead, we use the following approach.

Let's use a notion we have used previously—the idea of squared distance as a measure of closeness. We shall say a straight line $L(x)$ is the best linear predictor of Y for the given data if the average squared error between the predicted value of Y and the real value of Y is minimized (with respect to all other straight lines). In terms of the scatter diagram, the best line is the line which best fits the points on the scatter diagram in the sense that the average squared vertical distance from the points to the line is minimized. We shall call this line the *least-squares line,* or *regression line.* In theory, for each scatter diagram, or set of paired data, we can obtain a best line. Figure 13-3 shows, for an arbitrary line, the errors whose squares we are trying to minimize to obtain the best line for the scatter diagram of the example.

Recall that a straight line can be written in the form $L(x) = ax + b$, where a is the slope of the line and b is the Y coordinate of the point where the line hits the Y axis (sometimes called the Y *intercept*). In order to define a particular line, then, we need know only the appropriate numbers a, b; this is especially true for the least-squares line. It turns out that the equation for the least-squares line $L(x)$ is not hard to compute.

> **Definition.** The least-squares line based on X to predict Y for a particular set of paired data (X, Y) is obtained as follows:
>
> $L(x) = ax + b$
>
> $$a = \frac{n\Sigma XY - (\Sigma X)(\Sigma Y)}{n\Sigma X^2 - (\Sigma X)^2} \qquad (13\text{-}1)$$
>
> $b = \bar{Y} - a\bar{X}$

Now, we shall compute a, b for Example 13-1; our computations are displayed in Fig. 13-4. Using Eqs. (13-1) we obtain

$$a = \frac{n\Sigma XY - (\Sigma X)(\Sigma Y)}{n\Sigma X^2 - (\Sigma X)^2}$$

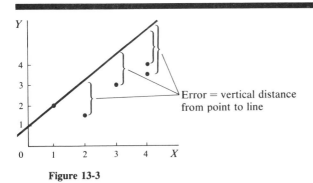

Figure 13-3

X	Y	XY	X^2	Y^2	
2	1.5	3	4	2.25	$\bar{X} = 3$
1	2	2	1	4	$\bar{Y} = 2.86$
3	2	6	9	4	$n = 7$
3	3	9	9	9	
4	3.5	14	16	12.25	
4	4	16	16	16	
4	4	16	16	16	
$\Sigma X = 21$	$\Sigma Y = 20$	$\Sigma XY = 66$	$\Sigma X^2 = 71$	$\Sigma Y^2 = 63.50$	

Figure 13-4

$$= \frac{7(66) - (21)(20)}{7(71) - (21)^2}$$

$$= .75$$

$$b = \bar{Y} - a\bar{X}$$

$$= 2.86 - .75(3)$$

$$= .61$$

and the least-squares line is $L(x) = .75x + .61$. In Fig. 13-5 we graph $L(x)$ on the scatter diagram. An easy way to graph a straight line is to plot two points on it and then draw the line which passes through these points. By computing $L(0) = .75(0) + .61 = .61$ and $L(1) = .75(1) + .61 = 1.36$, we see that $(0,.61)$ and $(1,1.36)$ are on the line $L(x)$. We draw the straight line connecting these two points and obtain the graph of $L(x)$.

Suppose now that we want to predict how a student will do in college if his grade-point average for the final 2 years of high school is 3.3. Using the least-squares line we compute $L(3.3) = .75(3.3) + .61 = 3.09$. If his high school grades are 2.0, we predict that his college grades will be $L(2.0) = (.75)(2.0) + .61 = 2.11$.

Notice that the accuracy of $L(x)$ as a predictor for the given data depends on the scatter diagram. If there is a definite linear relationship between the data, we would be more likely to regard the least-squares line as a good predictor. Also, the use of the least-squares line should be restricted to observations lying within the range of the sample data. For example, if the least-squares line is based on data which have X values ranging from, say, 1 to 4, we would not use it to predict Y for an

X value of 20. A linear relationship within a given range of data does not imply anything about a wider range.

The correlation coefficient

We shall now define a measure of the amount of linear relationship between X and Y, but first, we shall establish some terminology. We say that X and Y are *positively correlated* if when the values of X grow large, so, on the average, do the values of Y in a linear manner. We say that (X,Y) are *negatively correlated* if when the values of X grow large, the values of Y become small in a linear manner. In either case, X and Y are said to be *correlated,* and in such situations the least-squares line would be a good predictor. If no such pattern exists, we say that X and Y are *uncorrelated.* Observe that in Fig. 13-6a, X and Y seem to be positively correlated, in Fig. 13-6b, X and Y seem to be negatively correlated, and in Fig. 13-6c, X and Y seem to be uncorrelated.

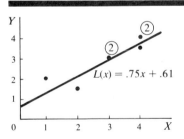

$L(x) = .75x + .61$

Figure 13-5

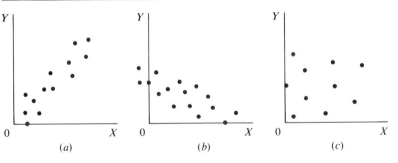

(a)   (b)   (c)

Figure 13-6

We make these ideas precise with the *correlation coefficient r,* a number which measures the degree of correlation of paired data. We shall discuss only the correlation coefficient based on sample data. The population correlation coefficient is defined in an analogous way. The formula for computing the sample correlation coefficient *r* is as follows:

$$r = \frac{n\Sigma XY - (\Sigma X)(\Sigma Y)}{\sqrt{n\Sigma X^2 - (\Sigma X)^2} \ \sqrt{n\Sigma Y^2 - (\Sigma Y)^2}}$$

Here are some facts about *r:*

1   *r* is the same no matter which set of data we label *X* and which set we label *Y*.

2   *r* is always between -1 and 1. If *r* is close to -1, *X* and *Y* are negatively correlated. If *r* is close to 1, *X* and *Y* are positively correlated. If $r = 0$, *X* and *Y* are uncorrelated.

3   If *X* and *Y* are highly correlated, that is, if *r* is close to $+1$ or -1, then the least-squares line is a good predictor for data in the range of *X* on the scatter diagram. If *r* is close to 0, the least-squares line is not a good predictor.

Notice that if $r = 0$, then $a = 0$ also. Substituting into the formula for $L(x)$, we see that if $r = 0$, then $L(x) = 0 + \bar{Y} = \bar{Y}$ (the horizontal line with height $\bar{Y}$). In other words, if *X* and *Y* are uncorrelated, information about *X* gives no information about *Y* which can be used in a linear way and we always predict *Y* to be the sample mean $\bar{Y}$, no matter that value *X* we observe.

One interpretation of *r* is the following: The square of the correlation coefficient tells us how much better (on the average) the least-squares line is in predicting *Y* than if we had no information from *X* at all (and just guessed *Y* to be $\bar{Y}$ in all cases). In fact, r^2 is the proportion of squared error we are able to eliminate by using $L(x)$ as a predictor of *Y*. For example, if we compute $r = .9$, then $r^2 = .81$ and $L(x)$ is 81% better than $\bar{Y}$ as a predictor for the given data.

Let us now compute *r* for Example 13-1. Using Fig. 13-4 and Eq. (13-2), we compute

$$r = \frac{7(66) - 21(20)}{\sqrt{7(71) - (21)^2} \ \sqrt{7(63.5) - (20)^2}}$$

$$= .84$$

We conclude that *X* and *Y* are positively correlated.

Example 13-2. Compute $L(x)$ and r for the data in Fig. 13-7.

SOLUTION: Using Eqs. (13-1) and (13-2) and Fig. 13-7, verify that $a = 0$, $b = 1$, and so $L(x) = 0(x) + 1 = 1$ and $r = 0$. These sample data indicate that X and Y are uncorrelated; the least-squares line is the constant $\bar{Y}$ and does not use X at all.

Example 13-3. Compute the least-squares line and correlation coefficient for the data in Fig. 13-8.

SOLUTION: Using Eqs. (13-1) and (13-2) and Fig. 13-8, verify that $a = -1$, $b = 3$, and so $L(x) = -x + 3$. Also $r = -1$, indicating that X and Y are negatively correlated (in fact, all the data lie on the least-squares line).

Example 13-4. A study was done to determine the relationship be-

X	Y
0	0
0	2
1	1
2	3
2	-1

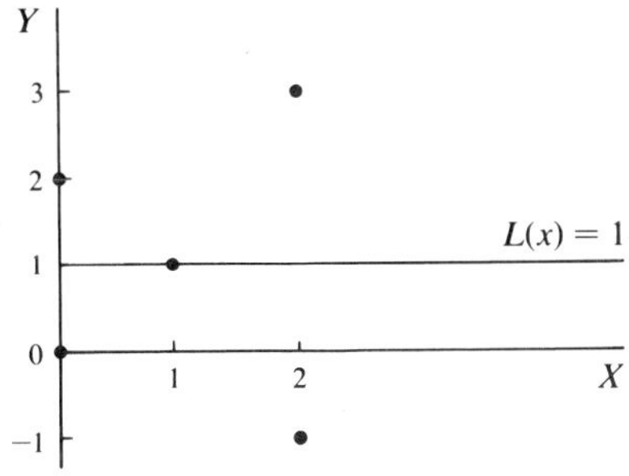

Figure 13-7

X	Y
0	3
1	2
2	1
3	0

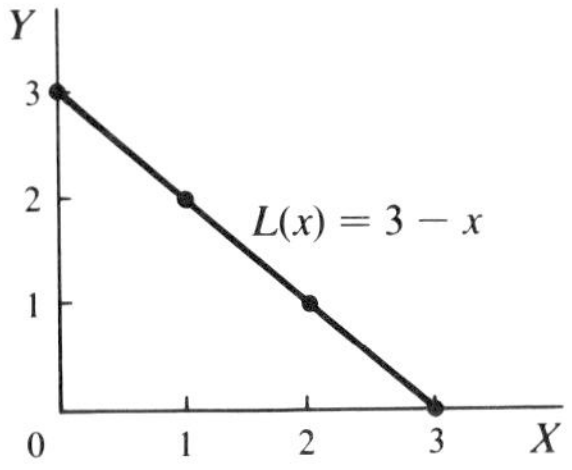

Figure 13-8

tween the presence of a certain food-coloring agent in processed food and the rate of hyperactivity in a group of hyperactive children. Figure 13-9 presents the results of this study, X denoting the amount of food coloring and Y the (average) amount of hyperactive behavior in the test group measured on a suitable scale. Compute $L(x)$ and r and discuss the relationship between X and Y.

SOLUTION: Verify, using Eqs. (13-1) and (13-2) and Fig. 13-9, that $a = 1$, $b = 0$, $L(x) = x$, and $r = .69$. Thus, X and Y are positively correlated. Based on this sample, there seems to be a strong correlation between X and Y.

In this situation we cannot help but suspect a cause-and-effect relationship, namely, that the food additive in question causes or increases hyperactive behavior. Of course our analysis should not stop here. We might wish to conduct a similar study with nonhyperactive children or to see in much greater detail what happens when a diet that is free of this substance is administered to hyperactive children.

MEASURING THE RELIABILITY OF r

The use of the sample correlation coefficient to determine whether or not two variables are correlated is a statistical procedure and is thus subject to random error. In order to obtain the population correlation between X and Y, of course we would have to compute the correlation coefficient for the entire set of population data instead of for just the sample data. Since we usually can't do this, it would thus be of interest to be able to measure, in some way, the reliability of r.

In many situations, we can infer whether or not X and Y are correlated with a test of hypotheses as follows:

H_0:   Population correlation coefficient $= 0$

H_a: Population correlation coefficient $\neq 0$

The test would be performed using r as the test statistic. If r is close to zero, we would accept H_0: true correlation $= 0$. Otherwise, we would reject H_0, saying that there is a correlation between X and Y. If the variables under study are normally distributed, tables are available which give rejection regions for tests of this type. Such a table is given in the Appendix. In looking at this table, notice that as the sample size grows larger, r becomes more reliable in the sense that not as large a deviation from zero is needed to conclude that X and Y are correlated. For small sample sizes, such as in the examples given previously, we would be hesitant to make any inferences about population correlation.

For example, for a sample of size 5, testing at a significance level of $\alpha = .05$, you would have to observe an r value of at least .88 (or $\leq -.88$) to infer any population correlation.

Example 13-5. A researcher wishes to determine if there is a correlation between weight at birth and weight as an adult. He compares the weights at birth and as an adult for 25 people and computes $r = .27$. Assuming these weights to be normally distributed, at a significance level of $\alpha = .05$, what should the researcher conclude about the population correlation?

SOLUTION: The researcher wishes to test the hypotheses

H_0:   Population correlation $= 0$

H_a: Population correlation $\neq 0$

Turning to the Appendix, we see that when $n = 25$ for a two-tailed test at a significance level of $\alpha = .05$, the rejection region starts at $r = .40$. This means that we reject H_0 whenever $r \geq .40$ or $r \leq -.40$. Since we observe $r = .27$, we conclude that there is not significant evidence to indicate that X and Y are correlated over the entire population. Thus, the researcher should conclude that weight at birth and weight as an adult are not necessarily correlated.

CONFIDENCE INTERVALS BASED ON r

Instead of testing hypotheses about the true correlation based on the sample correlation coefficient, we may wish to obtain a confidence interval for Y based on the observed value of X when $L(x)$ is used as a predictor. In cases where X and Y are normally distributed, the

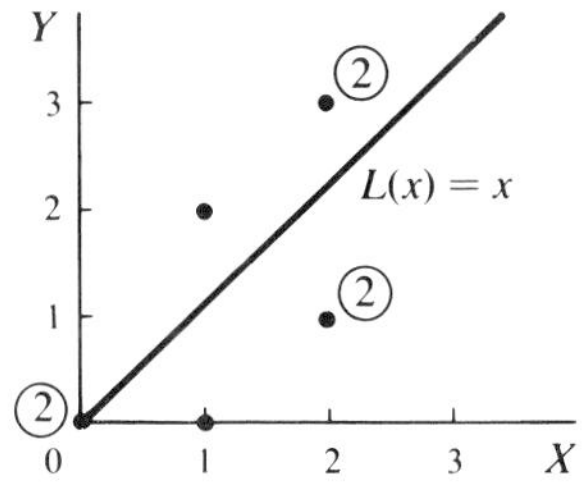

Figure 13-9   Hyperactivity and food additives.

calculations are not difficult and are similar to those for the confidence intervals studied in Chap. 5. In such cases, if we observe X, then the 95% confidence interval for Y, using $L(x)$ as a prediction, is given by

> *95% confidence interval for Y given X is L(x) ±*
> $1.96\sqrt{1 - r^2}\, s_Y.$ (13-3)

Example 13-6. A high school counselor wishes to determine if there is a relationship between students' mathematical aptitudes and their overall college grade-point averages. She compares scores on a mathematics aptitude test given to a group of high school seniors with their college grades in later years and computes $r = .8$, $L(x) = .6x + .8$, $s_Y = 1.0$. Suppose the counselor then gives the aptitude test to another high school senior, who obtains a score of $X = 4.0$. Based on this score, and assuming these scores and college grades are normally distributed, what is a 95% confidence interval for Y based on $L(4.0)$?

SOLUTION: We use formula (13-3) and obtain the 95% confidence interval

$L(4.0) \pm 1.96(1 - .64)(1.0) = 3.2 \pm .71 = (2.49, 3.91)$

Based on these data, we are 95% certain that a student receiving an aptitude test score of a 4.0 will have college grades between 2.49 and 3.91.

Final note

Often, the least-squares line and correlation coefficient are misused or misconstrued. If X and Y are positively correlated, we should *not* conclude that X is in some way causing Y to behave the way it does or vice versa. Beware of arguments like the following: "There is a positive correlation between the number of flowers in bloom in a particular garden and the number of hours spent outside by members of the family. Therefore, staying outside makes flowers bloom." (Explain this.) We do not use r by itself to make any statements concerning cause and effect.

Summary

We have learned a method of comparing paired data and the best linear method of prediction in certain situations. We have learned to compute the least-squares line, the line which minimizes the average squared error in predicting Y based on X. If X and Y are correlated, we can effectively use the least-squares line as a method for predicting Y when we are given new X values. The sample correlation coefficient r is a measure of the correlation, or degree of linear relationship, between X and Y. When r is close to 1, X and Y are positively correlated; when r is close to -1, X and Y are negatively correlated. When $r = 0$, X and Y are uncorrelated and X furnishes no information that can be useful in predicting Y in a linear way.

We have also seen how to decide if sample data indicate a correlation in the population. As the sample size grows large, the sample correlation coefficient becomes more reliable as a measure of the population correlation.

Important terms

Best linear predictor

Least-squares line

Linear function

Negatively correlated

Positively correlated

Sample correlation coefficient r

Scatter diagram

Uncorrelated

Problems

13-1 Use the following data to make a scatter diagram. Then compute the least-squares line of X to predict Y and graph it on the scatter diagram. Also, compute the sample correlation coefficient r. Are X and Y positively correlated? Negatively correlated? Uncorrelated? What can you conclude about this population correlation based on the sample testing at a significance level of $\alpha = .05$?

X	Y
0	1
1	1
2	3
2	2
3	1
4	4
4	3

13-2 A study is made to determine whether there is a correlation between lawyers' fees and lengths of jail sentences for certain
serious crimes. A random sample of 150 convicted criminals is
obtained from prison records, and interviews are made with the
prisoners to determine their lawyers' fees. Letting X represent
lawyers' fees and Y represent lengths of sentences, the following
data are obtained (X given in thousands of dollars, Y in years):

$$\Sigma XY = 600 \qquad \Sigma X = 200 \qquad \Sigma Y = 305 \qquad \Sigma X^2 = 430$$
$$\Sigma Y^2 = 1{,}040$$

Compute the correlation coefficient r and interpret your result.

13-3 A study is done with a large group of mothers to determine if
there is a linear relationship between number of packs of cigarettes smoked per day during pregnancy and weights of newborn
babies. (We shall present only a small number of observations in
order to make your task easier.) Letting $X =$ no. of packs of cigarettes smoked per day during pregnancy and $Y =$ weight of newborn baby, draw a scatter diagram, computing and graphing the
least-squares line of X to predict Y. Also, compute r. Interpret
your results. Testing at a significance level of $\alpha = .05$, what
would you conclude about the population correlation? Does this
suggest the need for more data?

X	Y
0	8.2
0	8.0
.5	7.9
1.1	7.6
1.5	7.4
2.3	7.1
2.5	7.3
3.0	6.9

13-4 An economist wishes to be able to predict the market value of a house 3 years from now based on its current market value. Believing that market values over the next 3 years will behave roughly the same as they have in the past 3 years, the economist takes a sample of houses and for each house records $X =$ its market value 3 years ago and $Y =$ its market value today. Based on the following data, draw the scatter diagram, computing and graphing the least-squares line.

Market value of houses 3 years ago and today (in units of $10,000)

X	Y
3.2	4.1
2.6	3.1
2.8	3.2
3.8	4.7
4.2	5.0
4.3	4.9

Based on the least-squares line you have computed, predict the market value of a house 3 years from now if it is valued at $31,500 today.

13-5 A bored airlines' ticket agent decides to see if he can predict the total weight of a passenger's luggage based on the one-way distance of his trip. In order to do this he randomly selects seven passengers from various flights and for each records $X =$ the one-way distance (in thousands of miles) and $Y =$ the total luggage weight (in 10-pound units). He then computes the least-squares line to use as a predictor. This is what you should now do. Also compute r. Is the least-squares line a good predictor based on your results?

X	Y
.6	3.1
.8	4.1
1.2	2.1
2.5	5.7
2.8	6.2
.3	4.0
.7	5.1

13-6 Make up some data for which (*a*) $r = 0$, (*b*) $r = -1$, and (*c*) $r = 2/3$.

True-false test

Circle the correct letter.

T F   1. If you are able to determine Y exactly, based on knowledge of X, then $r = 1$ or -1.

T F   2. If $r = 1$ or $r = -1$, then you are able to determine Y exactly based on knowledge of X.

T F   3. If $r > 0$, then as X increases, Y tends to increase.

T F   4. If r is close to 1 or -1, then there is a cause-and-effect relationship between X and Y.

T F   5. If the least-squares line is computed with data where X ranges from 0 to 27, you should not use it to predict Y for an X value of 112.

T F   6. The correlation coefficient will detect any relationship between X and Y.

T F   7. If all the X values are negative and all the Y value are positive, then r must be negative.

T F   8. Suppose the least-squares line is $L(x) = (2/3)x + 6$ and you have predicted $Y = 4$. This means that X must have been 8 2/3.

T F   9. Suppose you know that the points (3,5) and (1,4) are on the least-squares line. With this knowledge it is possible to determine the least-squares line exactly.

T F   10. If X and Y are negatively correlated and (0,0) is on the least-squares line, then if $X = 1$ is observed, the predicted value of Y must be negative.

Expectation 14

In this chapter and in Chaps. 15 to 18 we shall explore more deeply some of the mathematical aspects of probability, discussing also (in Chap. 17) a different approach to certain problems of statistical inference. In this chapter the average, or expectation, of a random variable is discussed.

Recall from Chap. 3 that a random variable is any assignment of numbers to the outcomes of a random experiment. We spoke also of the distribution of a random variable, namely, a list of the possible values the variable can assume along with the probabilities of these values. Structuring probability models in this way gave us an easy language with which to discuss many of the situations which arose in later chapters. Some of the random variables which we encountered in the first four chapters were the binomial variable S, the sample mean $\bar{X}$, and the sample standard deviation s. Now we shall discuss the mean of a random variable.

Just as the mean of a set of data is the average of the data values, the mean of a random variable is the average of the values that the variable assumes, weighted with respect to the probabilities of these values. Consider, now, the following example.

Example 14-1. You have $1 to invest and two different ways to invest it. If you invest in oil stock X, you will get back $25 if oil is found; otherwise you will lose your dollar. You know that there is a .1 chance of oil being found. On the other hand, if you invest in safe securities Y, with probability .9, you will get back $2 and with probability .1 you will get back $1.50. Which is a better investment?

> SOLUTION: One possible answer is "It depends." If oil is found, you would have done better with investment X; otherwise you would have done better with investment Y. This avoids the issue, however, as it doesn't help you in deciding which investment to make.
>
> One way to decide would be to make the investment which gives you the larger average gain. If you make investment X, you will gain $25 approximately .1 of the time and gain −$1 (lose your dollar) approximately .9 of the time. Your average gain will be ($25.00 × .1) − ($1.00 × .9) = $1.60. Using the same reasoning, investment Y will yield an average gain of ($2.00 × .9) + ($1.50 × .1) = $1.95. Thus, on the basis of average gain, investment Y is better.
>
> Here we have found the *mean of a random variable* by multiplying each value of the variable by its probability and then adding the results (this is the same as a weighted average). The distributions of X and Y are as follows:

X Value	Probability	Y Value	Probability
$25.00	.1	$2.00	.9
−1.00	.9	1.50	.1

In the previous chapters we denoted means by $\bar{X}$ (sample mean) and μ (population mean and mean of sampling variables). In probability it is customary to call the mean of a random variable its *expectation*. When discussing a particular variable X, we denote its expectation by $E(X)$. In the previous example, $E(X) = ($25.00 × .1) − ($1.00 × .9) = $1.60; E(Y) = ($2.00 × .9) + ($1.50 × .1) = $1.95.

Definition. If X is a random variable, then $E(X)$, the expectation, or mean, of X, is computed as follows: If X assumes values x_1, $x_2, \ldots, x_n$ with probabilities $p_1, p_2, \ldots, p_n$, respectively, then

$$E(X) = x_1 p_1 + \cdots + x_n p_n \qquad (14\text{-}1)$$

A famous theorem in probability is called the *law of large numbers:*

Law of large numbers. If repeated, independent observations are made of the same random variable X, then the sample average of the outcomes, $\bar{X}$, gets closer and closer to the true average, or expectation, $E(X)$.

This law gives a justification for defining $E(X)$ as we do since it shows that $E(X)$ is, in fact, the long-run average of observations of the variable X.

Example 14-2. A box contains one card marked 1, two cards marked 2, and three cards marked 3. One card is drawn at random from the box, with $Z =$ the number of the card selected. Using Eq. (14-1), compute the expectation of Z.

SOLUTION:

$$E(Z) = \left(1 \times \frac{1}{6}\right) + \left(2 \times \frac{2}{6}\right) + \left(3 \times \frac{3}{6}\right)$$

$$= \frac{1 + 2 + 2 + 3 + 3 + 3}{6} = 2\frac{1}{3}$$

Notice that $E(Z)$ is the numerical average of the numbers on the cards.

Recall, again, that the distribution of a random variable is a list of the values that the variable assumes, along with the probabilities of these values. Since the expectation is the weighted average of the values with respect to the probabilities, if we have the distribution, then the expectation is easy to compute.

Example 14-3. In a large group of construction workers, 50% earn a wage of \$4.00 per hour, 30% earn \$4.50 per hour, and 20% earn \$5.00 per hour. A person is selected at random from this group, with X denoting the selected person's hourly wage. Find $E(X)$.

SOLUTION: We list the distribution of X:

X Value	Probability
\$4.00	.5
4.50	.3
5.00	.2

Using Eq. (14-1), we obtain

$$E(X) = (\$4.00 \times .5) + (\$4.50 \times .3) + (\$5.00 \times .2) = \$4.35$$

The expected wage of the selected person is \$4.35.

Example 14-3 illustrates the similarity between the population mean and the expectation. If we let $\mu =$ the population mean of the group of construction workers in the example, then $\mu = \$4.35$. In fact, if we let $X =$ the result of one random selection from a population under study, then $E(X) = \mu$, the population mean.

Example 14-4. At a large rock concert, the police are worried about the heavy use of drugs, one estimate being that 60% of those attending are "stoned." A random sample of size 3 is taken, with S representing the number of stoned people in the sample. What is the distribution of S? What is $E(S)$?

SOLUTION: You should recognize S to be a binomial variable. Letting $p = P$(a stoned person is selected) and using the binomial formulas (from Chap. 4), we compute $\mu = np = (3)(.6) = 1.8$. If 60% of the population is stoned, we would expect about 60% of the sample to be stoned; in this case, 60% of $3 = (.6)(3) = 1.8$. We shall see if this agrees with our definition of expectation. Remember that μ should be the same as $E(S)$.

We first determine the distribution of S. Since three people are selected, S can assume values either 0, 1, 2, or 3. There is only one way S can be 0, namely, the outcome NNN in which no one in the sample is stoned. Since the crowd is large, we can assume the selections are independent; and using the multiplication rule (from Chap. 3) and the fact that P(selecting someone who is not stoned) $= 1 - .6 = .4$, we see $P(S = 0) = P(NNN) = .4 \times .4 \times .4 = .064$. If we let Y represent a stoned person being selected, we see that there are three ways that one out of the three people selected is stoned: $(S = 1) = (YNN, NYN, NNY)$. Using the multiplication rule again, we see that each of these outcomes has probability $.6 \times .4 \times .4 = .096$. Since there are three such outcomes, we use the addition rule of probability to compute $P(S = 1) = 3 \times .096 = .288$. Likewise, we see $(S = 2) = (YYN, YNY, NYY)$, each of these outcomes having probability $.6 \times .6 \times .4 = .144$. Continuing as above, we find $P(S = 2) = 3 \times .144 = .432$. Finally, we obtain $(S = 3) = (YYY)$, and so $P(S = 3) = .6^3 = .216$. We thus determine the distribution of S,

Distribution of S

S Value	Probability
0	.064
1	.288
2	.432
3	.216

Figure 14-1

which we list in Fig. 14-1. (If you have had trouble following this discussion, review the section of Chap. 3 dealing with independent sampling.)

To compute $E(S)$, we follow Eq. (14-1):

$$E(S) = (0 \times .064) + (1 \times .288) + (2 \times .432) + (3 \times .216) = 1.8$$

which agrees with the formula $\mu = np = (3)(.6) = 1.8$.

Problems

14-1 A variable X has the following distribution:

X Value	Probability
2	.5
−1	.25
4	.25

Compute $E(X)$.

14-2 A machine which gives change for a one-dollar bill will give $1.00 in change with probability .9, $1.50 in change with probability .05, and $.75 in change with probability .05. If you put $1.00 into this machine, how much should you expect to get back?

14-3 A population contains .6 who are "moderate" drinkers. If two people are selected at random from this population, how many moderate drinkers would you expect to be selected?

14-4 The price of eggs over a 1-year period was $.90 per dozen .3 of the time, $.80 per dozen .5 of the time, and $1.00 per dozen .2 of the time. During this period, what was the average price of eggs?

The addition rule for expectation

We now discuss two important properties of expectation. For any two variables X, Y, we have

$$E(X + Y) = E(X) + E(Y)$$

In fact, for any variables X_1, X_2, ..., X_n, we have

$$E(X_1 + X_2 + \cdots + X_n) = E(X_1) + E(X_2) + \cdots + E(X_n) \qquad (14\text{-}2)$$

This is called the *addition rule for expectation*. Also, if c is any number, then

$$E(cX) = cE(X) \qquad\qquad\qquad\qquad (14\text{-}3)$$

Equation (14-3) can be easily verified by studying the definition of expectation. If X assumes values x_1, x_2, ..., x_n with probabilities p_1, p_2, ..., p_n, then cX assumes values cx_1, cx_2, ..., cx_n with probabilities $p_1, p_2, \ldots, p_n$. Thus

$$E(cX) = cx_1 p_1 + cx_2 p_2 + \cdots + cx_n p_n$$
$$= c(x_1 p_1 + x_2 p_2 + \cdots + x_n p_n)$$
$$= cE(X)$$

to yield Eq. (14-3). Equation (14-2), the addition rule for expectation, can also be verified without difficulty. We leave it to you as an exercise.

Clever use of Eqs. (14-2) and (14-3) enables us to easily compute the expectation of variables with distributions that are very difficult or which may otherwise involve difficult computations. The next three examples illustrate this.

Example 14-4 revisited. We return to the rock concert. We let X_1 = the number of "stoned" people on the first selection ($X_1 = 1$ if a stoned person is selected; $X_1 = 0$ if not), X_2 = the number of stoned people on the second selection, and X_3 = the number of stoned people on the third selection. See that $S = X_1 + X_2 + X_3$.

SOLUTION: We see that $S = X_1 + X_2 + X_3$. Each of the variables X_1, X_2, X_3 has the same distribution:

X Value	Probability
0	.4
1	.6

Using Eq. (14-1), we see $E(X_1) = E(X_2) = E(X_3) = (0 \times .4) + (1 \times .6) = .6$. Now using Eq. (14-2), the addition rule, we obtain $E(S) = E(X_1 + X_2 + X_3) = E(X_1) + E(X_2) + E(X_3) = .6 + .6 + .6 = 1.8$, as desired.

A situation where we represent a variable by a sum of variables, each variable in the sum indicating whether or not a certain event has taken place, is called the *method of indicators*. In our example, X_1 *indicated* whether or not a stoned person was the first person selected and X_2 and X_3 indicated the analogous facts about the second and third selections.

Example 14-3 revisited. We return to the group of construction workers. Now, instead of selecting 1 worker at random, we select 50 at random. Letting $\bar{X}$ = the sample mean of the 50 selected workers, find $E(\bar{X})$.

SOLUTION: First, be certain you agree that $\bar{X}$ is a random variable; it varies according to the results of our sample. If we let X_1 = the hourly wage of the first person selected, X_2 = the hourly wage of the second person selected, $\ldots$, X_{50} = the hourly wage of the fiftieth person selected, then

$$\bar{X} = \frac{X_1 + X_2 + X_3 + X_4 + \cdots X_{50}}{50} = \frac{1}{50}(X_1 + \cdots + X_{50})$$

We have already shown that the mean for a particular selection is $4.35 (as is the population mean μ). We now use Eqs. (14-2) and (14-3) to obtain

$$E(\bar{X}) = E\left[\frac{1}{50}(X_1 + \cdots + X_{50})\right]$$

$$= \frac{1}{50}E(X_1 + \cdots + X_{50})$$

$$= \frac{1}{50}[E(X_1) + E(X_2) + \cdots + E(X_{50})]$$

$$= \frac{1}{50}(\$4.35 + \cdots + \$4.35)$$

$$= \frac{1}{50}(50)(\$4.35)$$

$$= \$4.35$$

In other words, the average, or expectation, of the sample mean is, in fact, the population mean. This fact provides a justification for using $\bar{X}$ as an estimate of μ in situations where we don't know μ but have only sample data. We call such an estimate an *unbiased estimate*.

> **Definition.** If we are using a random variable X to estimate a population parameter θ, and if $E(X) = \theta$, we say that X is an unbiased estimate of θ. For example, the sample mean $\bar{X}$ is always an unbiased estimate of the population mean μ: $E(\bar{X}) = \mu$.

Example 14-5. A fair coin is tossed 10 times. What is the expected number of heads?

> SOLUTION: If we let S = the number of heads in the 10 tosses, we wish to obtain $E(S)$. Intuitively, the anwer is 5; however, to actually compute $E(S)$ by Eq. (14-1), we need to know the distribution of S. Even if we could compute the distribution of S, however, we would have to perform some formidable calculations to obtain $E(S)$. Fortunately, we can apply Eq. (14-2) using the method of indicators.
>
> Let X_1 = the number of heads on the first toss, X_2 = the number of heads on the second toss, $\ldots$, X_{10} = the number of heads on the tenth toss. Another way of saying this is let X_i = the number of heads on the ith toss, $i = 1, 2, \ldots, 10$. We observe that each X_i has the same distribution and expectation:

X_i Value	Probability
0	1/2
1	1/2

$$E(X_i) = \left(0 \times \frac{1}{2}\right) + \left(1 \times \frac{1}{2}\right)$$
$$= \frac{1}{2}$$

Notice that $S = X_1 + X_2 + \cdots + X_{10}$. Therefore,

$$E(S) = E(X_1 + X_2 + \cdots + X_{10})$$
$$= E(X_1) + E(X_2) + \cdots + E(X_{10})$$
$$= 10 \times 1/2$$
$$= 5$$

Example 14-6. (This example is tricky.) You have 7 letters to mail and 7 mailboxes in which you can mail them. Suppose you mail the letters at random; i.e., for each letter, you select a mailbox at random, independently of the other selections, and mail the letter in the selected mailbox. If you mail all the letters in this manner, what is the expected number of unused mailboxes?

SOLUTION: If we let $Y =$ the number of unused mailboxes (after all the letters are mailed), the problem is to find $E(Y)$. Before proceeding, try to obtain the distribution of Y; for example, try to obtain $P(Y = 3)$, the probability that exactly 3 mailboxes are unused. You will quickly discover that this is a most difficult task. Fortunately, Eq. (14-2) will enable us to compute $E(Y)$ without knowing these probabilities.

To begin, we make the notation easier by numbering the mailboxes from 1 to 7. Now we let $Y_1 = 1$ if no letters are ever mailed in mailbox 1; $Y_1 = 0$ otherwise (for example, $Y_1 = 0$ if any letters are mailed in mailbox number 1). We let $Y_2 = 1$ if no letters are mailed in mailbox number 2; $Y_2 = 0$ otherwise; etcetera. ($Y_i = 1$ if no letters are mailed in mailbox i; $Y_i = 0$ otherwise.) After careful thought, you should see that $Y = Y_1 + Y_2 + \cdots + Y_7$. Also, since the mailboxes are selected independently, each Y_i has the same distribution and hence the same expectation.

Now we shall compute the distribution, say, of Y_1. Notice that $Y_1 = 1$ when no letters are mailed in mailbox 1, that is, when all the letters are mailed in mailboxes other than mailbox 1. The probability of a specific letter being mailed in a mailbox other than number 1 is 6/7. Therefore, the probability that none of the letters are mailed in mailbox 1 is $(6/7)^7$. Thus, $P(Y_1 = 1) = (6/7)^7$, and $P(Y_1 = 0) = 1 - (6/7)^7$. The same is true for the other mailboxes, and the distribution and expectation of each Y_i is as follows:

Y_i Value	Probability
0	$1 - (6/7)^7$
1	$(6/7)^7$

$$E(Y_1) = \left\{ 0 \times \left[1 - \left(\frac{6}{7}\right)^7 \right] \right\} + \left[1 \times \left(\frac{6}{7}\right)^7 \right]$$

$$= \left(\frac{6}{7}\right)^7 = .3399$$

We now use Eq. (14-2) to obtain

$$E(Y) = E(Y_1) + E(Y_2) + \cdots + E(Y_7)$$
$$= 7 \times (6/7)^7$$
$$= 2.379$$

In other words, 2.379 mailboxes are unused, on the average.

Example 14-7. A deck of five cards consists of two cards marked 3, two cards marked 2, and one card marked 1. One card is to be selected at random, and you have to predict the number on the card selected, losing the square of your error; that is, if you guess t and the outcome is X, you lose $\$(X - t)^2$. What is your best guess?

SOLUTION: First we have to decide what is meant by "best." If you predict 3, and a card marked 3 is selected you don't lose anything; however, if a card marked 1 is drawn, you lose $(3 - 1)^2 = \$4$. In some cases, then, a 3 guess is better than a 1 guess and vice versa. Clearly you can't have a prediction which is the best in all possible cases. However, you may have a prediction which is best "on the average." We shall call a prediction "best" if using it minimizes the expected loss. If we let $X = $ the number on the card selected, and if t is your guess, the loss is then the variable $(X - t)^2$ and the expected loss is $E(X - t)^2$. The problem, then, is to find the number t which minimizes $E(X - t)^2$.

We can solve this problem by using some elementary algebra and then Eqs. (14-2) and (14-3):

$$E(X - t)^2 = E(X^2 - 2tX + t^2)$$
$$= E(X^2) - 2tE(X) + t^2$$
$$= [E(X) - t]^2 + E(X^2) - [E(X)]^2$$

The only place t appears in the final expression is in $[E(X) - t]^2$. This is minimized when $t = E(X)$. Notice that nowhere have we used specific information about X; we have used only algebraic properties of expectation. We have thus proved the following interesting result: If X is a random variable and you have to predict X (with a single guess), losing the square of your error, the guess which minimizes your expected loss is $E(X)$.

In this example, X has the following distribution and expectation:

X Value	Probability
1	1/5
2	2/5
3	2/5

$$E(X) = 1 \times \frac{1}{5} + 2 \times \frac{2}{5} + 3 \times \frac{2}{5}$$

$$= \frac{11}{5}$$

Thus, your best guess is 11/5. Using this guess your loss is the variable $L = (X - 11/5)^2$ with the distribution

X Value	(X − 11/5)	L = (X − 11/5)²	Probability
1	−1.2	1.44	1/5
2	−.2	.04	2/5
3	.8	.64	2/5

Check that $E(L) = .56$; if you use 11/5 to predict X, you should expect to lose $.56 on the average. As an exercise, suppose you guess 2 instead of 11/5. How much is your expected loss?

Example 14-8. Which of the following games would you rather play?

Game X. A fair coin is tossed. If it comes up heads, you win $500. If it comes up tails, you lose $500.

Game Y. A coin with $P(\text{heads}) = .1$ is tossed. If it comes up heads, you win $530. If it comes up tails, you lose $60.

Game Z. A coin with $P(\text{heads}) = .01$ is tossed. If it comes up heads, you lose $9,800. If it comes up tails, you win $100.

SOLUTION: Before deciding which game is best, we recall the famous theorem known as the *law of large numbers*.

In this case the law of large numbers implies that if we play a gambling game over and over again a large number of times, the average winnings per game will approach the true expected winnings of the game. We can use this fact to help decide whether or not to play a game a large number of times or which of several games to play.

Using Eq. (14-1), we see that

$$E(X) = (.5)(500) + (.5)(-500) = 0$$

$$E(Y) = (.1)(530) + (.9)(-60) = -\$1$$

$$E(Z) = (.01)(-9,800) + (.99)(100) = \$1$$

The law of large numbers asserts that if we play game X a large number of times, we shall break even on the average; if we play game Y a large number of times, we shall lose an average of $1 per game; if we play game Z, we shall win an average of $1 per game. Thus, if we are to play the games repeatedly, game Z would definitely be preferable since $E(Z) > E(X) > E(Y)$. In the long run, we would do better by playing game Z.

On the other hand, if we play the selected game only once, the law of large numbers would not apply and we might choose another game or decide not to play at all. This decision would have to be based on our personal feelings about gambling, our personal value of money, or the like. (We shall discuss how our opinions may affect our decision-making procedures in Chap. 17.)

Summary

We have defined and discussed the expectation of a random variable. We have seen how to compute expectations of variables using the definition in Eq. (14-1) and the two useful equations (14-2) and (14-3). We have shown how the notion of expectation can be used in making decisions about which of certain courses of action to take or in formulating strategies for games. We have stated the law of large numbers and have shown how it provides a justification for our definition of expectation.

Important terms

Distribution

Expectation

Law of large numbers

Method of indicators

Random variable

Additional Problems

(Starred problems are more difficult.)

14-5 A certain commuter train takes 1 hour about .3 of the time, 1.5 hours about .2 of the time, and 2 hours about .5 of the time. What is the average length of time a commuter spends on this train?

14-6 In a certain gambling game, you will win $1 with probability 1/8, $2 with probability 1/4, and $3 with probability 1/2. However, you will lose $15 with probability 1/8. Would you like to play this game? What would your average winnings be?

14-7 On an assembly line there is a 1/10 chance of producing a defective item. There is a safety check made whenever a defective item is produced or after every fifth item, whichever happens first. How many items are produced, on the average, before a safety check is made?

14-8 A committee of size 3 is selected at random from a group of 5 men and 3 women. What is the expected number of women on the committee? The expected number of men?

14-9 From a box containing 5 tickets marked $1.00, 3 tickets marked $2.00, 1 ticket marked $5.00, and 1 ticket marked $.05, 1 ticket is drawn at random and you win the amount on the ticket. How much should you expect to win? Suppose, instead of drawing 1 ticket, 10 tickets are drawn (with replacement) and you win the sum of the amount on the tickets. How much should you expect to win?

14-10 In a certain population of voters, 60% will vote for candidate A. If 200 voters are selected at random, how many would you expect to vote for candidate A?

14-11 You are moving to another city, and the moving company is going to charge you $400. However, with probability 1/3, the movers will run into inclement weather and charge you a $50 "inclement weather fee." Also, with probability 3/4, your antique urn will be damaged, costing $75 to be repaired. Finally, with probability 1/2, the movers are extortionists and will charge you a $100 "protection" fee. How much should you expect to spend on the move?

14-12 The average number of cars owned per family in a large city is an unknown number μ. If you select 40 families at random, letting $\bar{X}$ = the sample mean, what is $E(\bar{X})$?

14-13 A die is rolled repeatedly until the sum of the faces exceeds 5. How many rolls would you expect to be made?

14-14 If you submit your data and computer program at the computer center, your data will be processed in 8 hours if all goes well. However, with probability 1/3 things will be busy and there will be a 2-hour delay. Also, with probability 1/2 the computer operator will be in a bad mood and there will be a 1-hour delay because of that. How long should you expect to wait before your data are processed?

14-15 A certain type of house (type A) takes 6 months to build; type B takes 1 year to build; and type C takes 1 1/2 years to build. A builder builds type-A houses about 1/4 of the time, type-B houses about 1/2 of the time, and type-C houses about 1/4 of the time. On the average, how long will it take him (in total time spent) to build three houses?

14-16 A man has 12 keys in his pocket, and only 1 key fits a certain lock. If he selects the keys at random, discarding the selected key if it doesn't fit the lock, how many selections should he expect to make before he selects the correct key?

14-17 a. Make up a variable X for which $E(X) = 7$ and $P(X = 9) = 1/3$.

b. Make up a variable Y for which $E(Y) = 0$ and $P(Y \leq -2) = 1/2$.

c. Make up a variable Z for which $E(Z) = 8$ and $P(Z = 9) = 1/3$ and $P(Z = -1) = 1/3$.

*14-18 You are to sample with replacement from a large collection of machines until you select the first defective one. If 1/4 of the machines are defective, how large should you expect your sample to be? (*Hint*: If you obtain a nondefective machine on the first selection, you're in the same situation as when you started.)

*14-19 A miner is lost in a mine after a power failure. There are three tunnels in front of him, one leading to safety in 1 hour, one in which he will wander around for 3 hours before coming back to the starting point, and one in which he will wander around for 2 hours before coming back to the starting point. If each time he gets back to the starting point he selects a tunnel at random and

enters that tunnel, how long should he expect it will take to get to safety? (*Hint*: Once the miner returns to the starting place, he is really starting all over again.)

*14-20 A magician who claims to have ESP powers has you place 10 cards numbered from 1 to 10 face down on the table. He then will try to guess the number of each card. If he is guessing purely at random, how many numbers should you expect him to guess correctly? What if he does the same thing with 50 cards? [*Hint*: Use Eq. (14-2).]

*14-21 *The Petersburg paradox.* You play the following game. A fair coin is tossed until the first head occurs. If this happens on the *n*th toss, you receive 2^n. How much money should you expect to win if you play this game? How much would you be willing to pay as an entrance fee to play this game if you could play as many times as you like? What if you could play only once?

True-false test

Circle the correct letter.

T   F   1. To find $E(X)$, you must know the distribution of X.

T   F   2. In a certain gambling game, your expected "winnings" are $E(W) = -$1 million. It is therefore impossible for you to win at this game.

T   F   3. According to the law of large numbers, if you play roulette at Las Vegas long enough, you are bound to go broke.

T   F   4. A random variable X has $E(X) = 4$. It would thus be very unlikely to observe $X = 1,000$.

T   F   5. If the population mean equals μ, then $E(\bar{X}) = n\mu$, where $n =$ the sample size.

T   F   6. If $E(X) = 5$, then there is some chance that $X = 5$.

T   F   7. $E(X^2) = [E(X)]^2$

T   F   8. Your expected winnings for game A are $2, and your expected winnings for game B are $3. It would be a good strategy to play each game some of the time in case you are lucky in one game but not in the other.

T F   9. A certain variable X has $E(X) = 20$. There must be some chance that $X \geq 20$.

T F   10. You know that $E(X + Y) = 12$. Suppose you observe $X = 7$. Then Y will have to be 5.

Variance

As we saw in the last chapter, the expectation of a random variable measures its average value. Another important summarizing characteristic is the *variance*, which measures the dispersion, or spread, of the values. The variance is itself an average, measuring the average squared deviation of a variable from its mean. Just by knowing the expectation and variance of a random variable we can usually obtain a pretty good idea about the shape of its distribution.

In Fig. 15-1, the histograms for the random variables X_1 and X_2 each have the same mean, but the variance of X_1 is smaller than the variance of X_2. The variance of X_2 equals the variance of X_3, but the two means are different. The formula for the variance of a random variable X, written Var X, is given by

$$\text{Var } X = E[X - E(X)]^2 \tag{15-1a}$$

An equivalent formula, which is easier for computations, is

$$\text{Var } X = E(X^2) - [E(X)]^2 \tag{15-1b}$$

This equation can be derived using Eqs. (14-2) and (14-3) as follows:

$$E[X - E(X)]^2 = E(X^2) - 2E(X)E(X) + [E(X)]^2$$
$$= E(X^2) - [E(X)]^2$$

Var X is given in squared X units. For example, if X is measured in feet, Var X is expressed in units of feet squared; if X is measured in seconds, Var X is expressed in seconds squared. In order to convert Var X back to the units of X, the square root is taken. This measure is called the *standard deviation of X* and is abbreviated by SD X; that is,

$$\text{SD } X = \sqrt{\text{Var } X} \qquad\qquad (15\text{-}2)$$

Sometimes, as when denoting population standard deviations, we use σ instead of SD. If we let X denote the result of a sample of size one from a given population, then SD $X = \sigma$ and Var $X = \sigma^2$.

The following examples will illustrate how to use Var X.

Example 15-1. Let X be a random variable with the following distribution:

X Value	Probability
0	1/4
2	1/2
3	1/4

Find Var X.

SOLUTION: To be systematic, it helps to use the following program:

(a) Compute $E(X)$. Here, $E(X) = 0(1/4) + 2(2/4) + 3(1/4) = 7/4$.

(b) Add the column headed by X^2 to the distribution table:

Probability	X	X²
1/4	0	0
2/4	2	4
1/4	3	9

(c) Compute $E(X^2)$. $E(X^2) = 0(1/4) + 4(2/4) + 9(1/4) = 17/4$.

(d) Finally,

$$\text{Var } X = E(X^2) - [E(X)]^2 = 17/4 - (7/4)^2 = 19/16 \approx 1.19$$

Also,

$$\text{SD } X = \sqrt{\frac{19}{16}} \approx \sqrt{1.19} \approx 1.09$$

Example 15-2. Peter has the opportunity to play the following game. He will toss a fair coin until he gets a head or three tosses have been made, whichever comes first. What is the variance of the number of tosses he will make?

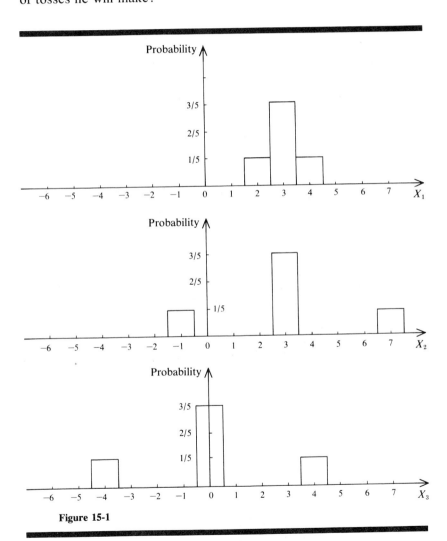

Figure 15-1

SOLUTION: Let X = number of tosses made. The distribution of X is

Probability	X	X^2
1/2	1	1
1/4	2	4
1/4	3	9

Thus,

$$E(X) = 1\left(\frac{1}{2}\right) + 2\left(\frac{1}{4}\right) + 3\left(\frac{1}{4}\right) = \frac{7}{4}$$

$$E(X^2) = 1\left(\frac{1}{2}\right) + 4\left(\frac{1}{4}\right) + 9\left(\frac{1}{4}\right) = \frac{15}{4}$$

$$\text{Var } X = E(X^2) - [E(X)]^2 = \frac{15}{4} - \left(\frac{7}{4}\right)^2 = \frac{60 - 49}{16} = \frac{11}{16} = .69$$

Also,

$$SD\ X = \sqrt{\text{Var } X} = \sqrt{\frac{11}{16}} = \sqrt{.69} = .83$$

Example 15-3. In Example 14-7, a deck of five cards contained two cards marked 3, two cards marked 2, and one card marked 1. One card was to be selected at random, and you were asked to predict the number on the selected card, losing the square of your error. We let X denote the number on the selected card. It was shown that the best predictor of X is $E(X)$; that is, the number t which minimizes the average loss $E(X - t)^2$ is $t = E(X) = 11/5$. It was also shown that if you used $E(X) = 11/5$ as your predictor, then the expected loss L is $E[X - E(X)]^2 = \$.56$. Notice that this is just the formula for Var X. In other words, Var X is the expected squared error loss if $E(X)$ is used to predict X.

Let's modify the problem a bit. Suppose you could find out if $X = 1$ or $X > 1$ before making your prediction. Using this information, what would be the best predictor? Also, how much would you be willing to pay to obtain this additional information?

SOLUTION: Clearly, if you are told $X = 1$, then since you have to guess X, you should guess 1. If you are told $X > 1$, then you know the chosen card has an equal chance of being 2 or 3 and so it would be reasonable to guess the average of these numbers:

2.5. This minimizes your average loss, assuming you were told $X > 1$. The best predictor is then given by the following rule: If you are told $X = 1$, then predict $X = 1$; if you are told $X > 1$, then predict $X = 2.5$. Let $L = $ the loss $= (X - $ your prediction$)^2$. Then the distribution of X and L is given by

Probability	X	L
1/5	1	$(1 - 1)^2 = 0$
2/5	2	$(2 - 2.5)^2 = \$.25$
2/5	3	$(3 - 2.5)^2 = \$.25$

Thus $E(L) = 0(1/5) + \$.25(2/5) + \$.25(2/5) = 1/5 = \$.20$.

If you are told whether $X = 1$ or $X > 1$, you can reduce your expected loss from \$.56 to \$.20, a saving of \$.36. Hence, you should be willing to pay up to \$.36 for this information about X and still reduce your loss on the average.

A few simple formulas will often aid in computing variances. If X is a random variable and c is a constant, then

$$\text{Var } (X + c) = \text{Var } (X) \qquad (15\text{-}3a)$$

$$\text{Var } cX = c^2 \text{ Var } X \qquad (15\text{-}3b)$$

If $X_1, X_2, \ldots, X_n$ are independent, then

$$\text{Var } (X_1 + \cdots + X_n) = \text{Var } X_1 + \cdots + \text{Var } X_n \qquad (15\text{-}4)$$

Example 15-4. If a student randomly guesses the answer on each of 50 questions of a multiple-choice test with 4 answers per question, what is the variance of X, the number of correct answers?

SOLUTION: We use the method of indicators, as discussed in Chap. 14. We let $X_1 = 1$ if the first answer is correct, and $X_1 = 0$ if it is incorrect; we let $X_2 = 1$ if the second answer is correct, and $X_2 = 0$ if not; etcetera. Then $X = X_1 + X_2 + \cdots + X_{50}$. Since the questions are answered independently, Var $X = $ Var $X_1 + $ Var $X_2 + \cdots + $ Var X_{50}. Since $X_1, X_2, \ldots, X_{50}$ have the same distribution, they have the same variance and hence Var $X = 50$ Var X_1. The distribution of X_1, X_1^2 is given by

Probability	X_1	X_1^2
3/4	0	0
1/4	1	1

Thus,

$$E(X_1) = 0\left(\frac{3}{4}\right) + 1\left(\frac{1}{4}\right) = \frac{1}{4}$$

$$E(X_1^2) = 0\left(\frac{3}{4}\right) + 1\left(\frac{1}{4}\right) = \frac{1}{4}$$

$$\text{Var } X_1 = \frac{1}{4} - \left(\frac{1}{4}\right)^2 = \frac{3}{16}$$

Finally,

$$\text{Var } X = 50\left(\frac{3}{16}\right) = 9.375$$

Example 15-5. In a large group of transistors, 20% will last 6 hours, 30% will last 7 hours, and 50% will last 4 hours. If 10 transistors are selected at random, what is the variance of their average lifetime.

SOLUTION: We use a method similar to the method of indicators. We let X_1 = lifetime of first transistor chosen, X_2 = lifetime of second transistor, ..., X_{10} = lifetime of tenth transistor. Let

$$\bar{X} = \frac{X_1 + X_2 + \cdots + X_{10}}{10}$$

denote the sample average. According to Eq. (15-4), Var $\bar{X}$ = Var $(X_1/10)$ + Var $(X_2/10)$ + $\cdots$ + Var $(X_{10}/10)$. Since X_1, $X_2, \ldots, X_{10}$ have the same distribution, Var $(X_1/10) = \cdots =$ Var $(X_{10}/10)$, and hence, using Eq. (15-3),

$$\text{Var } \bar{X} = 10 \text{ Var } \frac{X_1}{10} = \frac{10}{100} \text{ Var } X_1 = \frac{1}{10} \text{ Var } X_1$$

Now, Var X_1 is computed from the table:

Probability	X_1	X_1^2
.2	6	36
.3	7	49
.5	4	16

Thus,

$$E(X_1) = 6(.2) + 7(.3) + 4(.5) = 5.3$$

$$E(X_1^2) = 36(.2) + 49(.3) + 16(.5) = 29.9$$

$$\text{Var } X_1 = 29.9 - (5.3)^2 = 1.81$$

Finally,

$$\text{Var } \bar{X} = \frac{\text{Var } X_1}{10} = \frac{1.81}{10} = .181$$

Problems

15-1 Compute the variances of the following variables:

a.

X Value	Probability
0	1/4
1	1/4
3	1/2

b.

Y Value	Probability
−1	1/8
2	2/8
3	1/8
7	4/8

c.

Z Value	Probability
1,008	.3
1,010	.3
1,011	.2
1,012	.2

15-2 In a certain city 40% of the voters support proposition A. If 10 people are selected at random, with S representing the number of selected voters who support proposition A, find Var S.

15-3 In a certain gambling game you win $2 with probability 1/2 and lose $1 with probability 1/2. If you play this game six times, find the variance of your winnings.

15-4 A sample of size n is drawn from a population whose standard deviation is σ. What is the variance of the sample mean $\bar{X}$?

Chebyshev's inequality

As stated previously, Var X gives a measure of the dispersion of possible X values. If Var X is small, the values of X are likely to be clustered about its mean. This can be made more precise by an inequality called *Chebyshev's inequality.*

> **Chebyshev's inequality.** If X is a random variable and t is a constant, then
>
> $$P[|X - E(X)| \geq t] \leq \frac{\text{Var } X}{t^2}$$

The proof of Chebyshev's inequality is not hard, and we shall present it here. First, consider a random variable Y, which assumes nonnegative values, and a positive number t. Let $\tilde{Y} = t$ if $Y \geq t$, and $\tilde{Y} = 0$ if $Y < t$. Notice that no matter what the value of Y is, $Y \geq \tilde{Y}$. This implies that, on the average, Y will be larger than $\tilde{Y}$ and hence $E(Y) \geq E(\tilde{Y})$. Since

$$E(\tilde{Y}) = t[P(\tilde{Y} = t)] + 0[P(\tilde{Y} = 0)] = t[P(Y \geq t)]$$

we may conclude $P(Y \geq t) \leq E(Y)/t$. Applying this inequality to $[X - E(X)]^2$, a nonnegative variable, we obtain

$$P[|X - E(X)| \geq t] = P\{[X - E(X)]^2 \geq t^2\} \leq \frac{E[X - E(X)]^2}{t^2}$$

$$= \frac{\text{Var } X}{t^2}$$

Chebyshev's inequality may be used as a quick rule of thumb to obtain bounds on the probability of large deviations, expressed in units of standard deviations. Let X be a random variable having mean μ and standard deviation σ. Then, using Chebyshev's inequality,

$$P(X \text{ differs from } \mu \text{ by at least } 2\sigma) = P(|X - \mu| \geq 2\sigma) \leq \frac{\text{Var } X}{(2\sigma)^2}$$

$$= \frac{\sigma^2}{4\sigma^2} = \frac{1}{4}$$

In general,

$$P(X \text{ differs from } \mu \text{ by at least } k\sigma) = P(|X - \mu| \geq k\sigma)$$

$$\leq \frac{\text{Var } X}{(k\sigma)^2}$$

$$= \frac{1}{k^2}$$

Example 15-6. A population is found to have a mean $\mu = 10$ and a standard deviation $\sigma = 2$. If an element is selected at random from this population, how certain are you that the selected element is between 4 and 16?

SOLUTION: Since $\mu = 10$, $\sigma = 2$, the range of values from 4 to 16 are exactly the values which are within 3 standard deviations of the mean, that is, the values X for which $|X - \mu| \le 3\sigma$. Using Chebyshev's inequality, we obtain

$$P(|X - \mu| \ge 3\sigma) \le \frac{\sigma^2}{3^2\sigma^2} = \frac{1}{9}$$

implying

$$P(4 \le X \le 16) = P(|X - \mu| \le 3\sigma) \ge 1 - \frac{1}{9} = \frac{8}{9}$$

We conclude that with probability of at least 8/9, the selected element is between 4 and 16.

Example 15-7. A set of data has a mean $\mu = 23$ and a standard deviation $\sigma = 2$. What proportion of the data are within 4 standard deviations of the mean?

SOLUTION: Chebyshev's inequality states that with probability of at least $1 - 1/k^2$, an element selected at random is within k standard deviations of the mean. In terms of proportions of data, this means that a proportion of at least $1 - 1/k^2$ of a set of data is within k standard deviations of the mean. Thus, at least $1 - 1/4^2 = 15/16$ of any set of data are within 4 standard deviations of the mean. In this example, since the standard deviation $\sigma = 2$, at least 15/16 of the data are within 8 units of the mean. Since the mean $\mu = 23$, we conclude that at least 15/16 of the data are between 15 and 31.

Example 15-8. The average weight of a person in a certain large university is 150 pounds, with a standard deviation of 25 pounds. The campus elevators will hold about 2,000 pounds before the supporting cables will break. If 10 students are randomly selected and placed in an elevator, what can be said about the probability that the cables will break?

SOLUTION: Let $X = $ total weight of the 10 students. In order to apply Chebyshev's inequality to obtain an upper limit on the probability of the cables breaking, we need to know $E(X)$ and Var X. To find these, we let X_1 = weight of first person chosen, X_2 = weight of second person, $\ldots$, X_{10} = weight of tenth person. Then $X = X_1 + \cdots + X_{10}$. Also, $E(X) = E(X_1) + \cdots + E(X_{10}) = 10E(X_1) = 1,500$, and Var $X = $ Var $X_1 + \cdots + $ Var $X_{10} = 10$ Var $X_1 = 10(SD X_1)^2 = 10(25)^2 = 6,250$.

The cable will break if $X \geq 2,000$, which is the same as $X - 1,500 \geq 500$. If $X - 1,500 \geq 500$, then certainly $|X - 1,500| \geq 500$, and hence the event $(|X - 1,500| \geq 500)$ has higher probability than the event $(X - 1,500 \geq 500)$. Thus, using Chebyshev's inequality gives

$$P(\text{cables break}) = P(X - 1,500 \geq 500)$$

$$\leq P(|X - 1,500| \geq 500) = P(|X - E(X)| \geq 500)$$

$$\leq \frac{\text{Var } X}{(500)^2} = \frac{6,250}{(500)^2} = .025$$

That is, there are less than 25 chances in 1,000 that the cables will break.

Example 15-9. When there was conscription, the Army would not draft a man if his height either exceeded 78 inches or was less than 62 inches. Assuming that the average height of draft-age males was 70 inches with a variance of 9 inches, what can be said about the proportion of males who were excluded from the draft on the basis of height?

SOLUTION: Let $X = $ height of a draft-age male chosen at random. The man would be excluded if $X \geq 78$ or $X \leq 62$, that is, if $|X - 70| \geq 8$. From the given information $E(X) = 70$ and Var $X = 9$. Therefore, from Chebychev's inequality,

$$P(|X - 70| \geq 8) = P(|X - E(X)| \geq 8) \leq \frac{9}{(8)^2} = \frac{9}{64} = .14$$

Less than .14 of draft-age males would be excluded on the basis of height.

The probability bounds obtained by Chebyshev's inequality are very crude. It is an important tool only in applications where no assumptions can be made about the underlying probability distribution of X. For example, in Example 15-9 if we assume the heights are normally distributed (not an unreasonable assumption), then, using the techniques of Chap. 4, we may compute $P(|X - 70| \geq 8) = .0076$. In the next section we shall show how Chebyshev's inequality may be used to prove the law of large numbers.

Problems

15-5 A collection of data has a mean $\bar{X} = 10$ and a standard deviation $s = 2$. Using Chebyshev's inequality, find a lower limit for the proportion of data values between 1 and 19.

15-6 A random variable X has $E(X) = 10$ and Var $X = 3$. What can be said about $P(X \geq 15)$?

15-7 A collection of watches has an average price of \$73 and a standard deviation of \$6. What proportion of the watches cost between \$61 and \$85?

15-8 In a final examination in sociology, the average score was 81 and the standard deviation was 5. At least what proportion of the scores were between 71 and 91?

Law of large numbers

What would you expect to happen if a fair coin is tossed a large number of times? You should answer that the proportion of times the coin will land heads will probably be near 1/2. This fact is a special case of a very general law of probability called the *law of large numbers*, which we mentioned in Chap. 14. We shall state it again but with a more precise mathematical notation.

> **Law of large numbers.** Suppose a random experiment is repeated over and over again. If $X_1, X_2, \ldots, X_n$ represent the outcomes observed on the first n trials, then the sample mean $\bar{X} = (X_1 + \cdots + X_n)/n$ will get closer and closer to $E(X_1)$, the larger n becomes, with high probability.

This law applies to coin tossing in the following way. Let $X_1 = 1$ if the first toss is heads and $X_1 = 0$ if the first toss is tails; let $X_2 = 1$ if the second toss is heads and $X_2 = 0$ if the second toss is tails; etc. If the coin is fair, $E(X_1) = 0(1/2) + 1(1/2) = 1/2$. The quantity $X_1 + \cdots + X_n$ simply gives the number of heads in n tosses, and $(X_1 + \cdots + X_n)/n$ gives the proportion of times the coin lands heads.

According to the law of large numbers, this proportion will get closer and closer to $E(X_1) = 1/2$ as n becomes larger.

Using Chebyshev's inequality, we may give a simple proof of the law of large numbers. Let $X_1, X_2, \ldots, X_n$ be independent random variables having the same distribution, with expectation μ and variance σ^2. Let $\bar{X} = (X_1 + \cdots + X_n)/n$, the sample mean.
Then

$$E(\bar{X}) = \frac{E(X_1)}{n} + \cdots + \frac{E(X_n)}{n} = n\frac{E(X_1)}{n} = E(X_1) = \mu$$

The variance is

$$\text{Var } \bar{X} = \text{Var } \frac{X_1}{n} + \cdots + \text{Var } \frac{X_n}{n} = n \text{ Var } \frac{X_1}{n} = \frac{\text{Var } X_1}{n} = \frac{\sigma^2}{n}$$

using Eqs. (15-2) and (15-3). From Chebyshev's inequality, for any positive constant t

$$P(|\bar{X} - \mu| \geq t) \leq \frac{\text{Var } \bar{X}}{t^2} = \frac{\sigma^2}{nt^2}$$

As n becomes larger, the right-hand side σ^2/nt^2 becomes smaller and smaller. Therefore, $P(|\bar{X} - \mu| \geq t)$ is close to 0, or, equivalently, $P(|\bar{X} - \mu| \leq t)$ is close to 1. This last assertion means it is almost certain that the sample mean $\bar{X}$ is within t of μ when n is large. This holds for any t, no matter how small.

Example 15-10. A coin has unknown probability p of landing heads. To obtain some idea about p you could toss the coin and use the proportion of heads as an estimate of the true probability of heads p. How many tosses should be made in order that the estimate of p be within .05 of the true value 95% of the time.

SOLUTION: Let $X_1 = 1$ if heads occur on the first toss and $X_1 = 0$ if tails occur on the first toss; let $X_2 = 1$ if heads occurs on the second toss and $X_2 = 0$ if tails occurs on the second toss; etc.

Let $\bar{X} = (X_1 + \cdots + X_n)/n$. Then $\bar{X} =$ proportion of heads in the first n tosses, this being our estimate of p. The requirement is $P(|\bar{X} - p| \leq .05) \geq .95$, or, equivalently, $P(|\bar{X} - p| \geq .05) \leq .05$. Since

$$E(\bar{X}) = E\left(\frac{X_1}{n}\right) + \cdots + E\left(\frac{X_n}{n}\right) = nE\left(\frac{X_1}{n}\right) = E(X_1) = p$$

and

$$\text{Var } \bar{X} = \text{Var } \frac{X_1}{n} + \cdots + \text{Var } \frac{X_n}{n} = n \text{ Var } \frac{X_1}{n} = \frac{\text{Var } X_1}{n}$$

$$= \frac{p(1-p)}{n}$$

Chebyshev's inequality asserts

$$P(|\bar{X} - p| > .05) \leq \frac{p(1-p)}{n(.05)^2}$$

Now

$$\frac{p(1-p)}{n(.05)^2} \leq .05$$

if and only if

$$n \geq \frac{p(1-p)}{(.05)^3} = \frac{p(1-p)}{.000125}$$

Although p is unknown, for any number p between 0 and 1, $p(1 - p) \leq 1/4$. Hence, we are safe if $n = 1/4(.000125) = 2,000$.

Summary

The variance of a random variable X, written Var X, is a number which indicates the possible spread, or dispersion, of the values of X and is deformed by Var $X = E[X - E(X)]^2$. The following formulas are useful in computing the variance in complicated situations:

If c is a constant, then

Var $cX = c^2$ Var X and Var $(X + c) = $ Var X

If $X_1, X_2, \ldots, X_n$ are independent random variables, then

Var $(X_1 + \cdots + X_n) = $ Var $X_1 + \cdots + $ Var X_n

Chebyshev's inequality gives a method for obtaining bounds on the probability of wild deviations of X from its mean. It says, for any positive number t,

$$P(|X - E(X)| \geq t) \leq \frac{\text{Var } X}{t^2}$$

Using this inequality, the law of large numbers may be proved. The law of large numbers states:

> *Suppose a random experiment is repeated over and over again. If $X_1, X_2, \ldots, X_n$ represent the outcomes observed on n trials, then the sample mean $(X_1 + \cdots + X_n)/n$ will get closer and closer to $E(X_1)$, the larger n becomes, with high probability.*

Important terms

Chebyshev's inequality

Law of large numbers

Method of indicators

Variance

Additional problems

15-9 A collection of data has a mean $\bar{X} = 0$ and a standard deviation $s = 2$.

   a. Using Chebyshev's inequality, find an upper limit on the proportion of measurements lying outside the interval $(-4,4)$.

   b. Assuming the data are normally distributed with a mean of 0 and a standard deviation of 2 compute the probability that an observation selected at random will be outside the interval $(-4,4)$. Compare this probability with the proportion in (*a*).

15-10 A pair of dice are rolled and you must guess the sum of the dots which show up, losing the square of your error.

   a. What is your best guess and how much should you expect to lose?

   b. How much would you be willing to pay to see the face on one of the dice before making your guess?

15-11 Find a random variable X with a distribution such that $E(X) = 2$ and Var $X = 10$.

15-12 You have a balance and a collection of calibrated weights. Unfortunately, the balance makes random errors with a mean $\mu = 0$. Suppose you are given two objects and are asked to estimate the weight of each of the objects. You are allowed to use the balance twice. Two procedures are possible: (1) Weigh each object separately, taking the observed weights as the estimates. (2) Place both objects on the same side of the balance together, weighing their sums. Then place one object on each of the balances and observe the difference in weights.

    a. How can the second procedure be used to obtain estimates of the weights of the two individual objects?

    b. Which procedure is better? Why?

True-false test

Circle the correct letter.

T   F   1. If $E(X) = 0$ and Var $X = 100$, then $P(X \leq -10$ or $X \geq 10) > 0$.

T   F   2. Var $(X + 3) = $ Var $(X + 4)$.

T   F   3. If Var $X = 0$, then $X = 0$ with probability 1.

T   F   4. Suppose X and Y are independent, then SD$(X + Y)$ = SD X + SD Y.

T   F   5. If all the X values are negative, then Var X will be negative.

T   F   6. Let X be the outcome of a single measurement, and let $\bar{X}$ be the sample mean based on 12 measurements. Then Var $\bar{X} < $ Var X.

T   F   7. Suppose you must guess the value of a variable X, losing the square of your error. If you guess $E(X)$, then, on the average, you will lose Var X.

T   F   8. The law of large numbers says that as sample size becomes large, $\bar{X}$ becomes normally distributed.

T   F   9. If every value of a variable is multiplied by 2, then the variance will be multiplied by 4.

T   F   10. If X takes on only two values, then Var X takes on only two values.

More probability 16

In this chapter we shall discuss four commonly used probability distributions: the binomial, hypergeometric, geometric, and Poisson distributions. The presentation of these distributions is more from the perspective of probability theory than from a statistical point of view, and we assume that you have already read Chap. 3. A knowledge of the binomial distribution as presented in Chap. 4 will be helpful but not essential. In the first section we shall give a few powerful methods of counting combinations and permutations of events. These methods are necessary for a derivation of the formulas for the binomial and hypergeometric distributions, as well as being interesting in their own right.

Methods of counting

Anyone reading this knows how to count the number of letters in the alphabet. However, unless you are particularly clever or have already studied this material, you would have great trouble determining the number of different subcollections of 10 letters which can be formed from these 26 letters. If you try to list all such subcollections, after a while you probably would become confused about which subcollection to write down next, thus reaching an impasse. In this section we shall explain how to count such things in a simple manner.

To begin, consider the following problem:

How many ways can you rearrange (or permute) the letters A, B, C? In other words, how many different lists can you make with these letters, using each letter only once?

We can easily determine the number of different rearrangements from Fig. 16-1. Each possible list is obtained by following one of the paths from top to bottom. Since there are $3 \times 2 \times 1 = 6$ different paths, there are 6 different lists.

We can use the same reasoning to determine the number of different lists which can be formed from the 10 letters *A, B, C, D, E, F, G, H, I, J*. There are 10 letters to begin with. After the first letter has been chosen, there are 9 possible letters left for the second letter in the list; after the second letter has been chosen, there are 8 possible letters left for the third letter in the list; then 7, 6, 5, 4, 3, 2, and finally 1. Thus, there are $10 \times 9 \times 8 \times 7 \times 6 \times 5 \times 4 \times 3 \times 2 \times 1 = 3,628,800$ different lists. We shall abbreviate $10 \times 9 \times 8 \times 7 \times 6 \times 5 \times 4 \times 3 \times 2 \times 1$ by 10! (pronounced "ten factorial"). In general:

Suppose a list contains n different symbols. Then there are $n! = (n)(n - 1) \cdots (2)(1)$ ways to permute (or rearrange) the list. By convention, $0! = 1$ since there is essentially 1 way to arrange 0 symbols.

Example 16-1. Suppose a person has 6 keys and wishes to try them on his locked door. If none of the keys fits, how many different orders of the 6 keys are possible?

SOLUTION: There are $6! = 720$ different orders with which to try the keys.

Example 16-2. A line consists of 10 people who then form a new line, choosing their position randomly by drawing numbers out of a hat. What is the probability that they reform into the original order?

SOLUTION: There are $10! = 3,628,800$ different lines which may be formed by 10 people. Since they form the new line at random, each of the 10! lines is equally likely. Since there is only one way to reform the original line, the probability of reforming the original line is $1/10! = 1/3,628,800 = .00000027$.

Next, suppose we wish to know the number of different lists of 3 letters which can be formed from the 10 letters *A, B, C, D, E, F, G, H, I, J*. There are 10 possible letters to begin with. After the first letter has been chosen, there are 9 letters left to use for the second position; and finally there are 8 possible letters for the third position. Thus, there are $10 \times 9 \times 8 = 720$ different lists of 3 letters which may be formed. In general, the number of lists of size k which may be formed from n distinct symbols is $n \times (n-1) \times (n-2) \times \cdots \times (n-k+1)$.

Example 16-3. A lottery which sells 20 tickets is held 2 weeks in succession. If a first-, second-, third-, and fourth-place prize is awarded each time and the same 20 people participate in both lotteries, what is the probability that each of the 4 people who receive prizes the first week win the same prize the second week?

SOLUTION: If we label the 20 people using the first 20 letters of the alphabet, then we see that the number of ways the prizes may be awarded is the same as the number of different lists of 4 letters which may be formed from the first 20 letters of the alphabet. This number is $20 \times 19 \times 18 \times 17 = 116,280$. Since each of these 116,280 ways to award the four prizes is equally likely, and there is only one way to award each of the prizes from the second lottery to the same winners of the first lottery, the desired probability is $1/116,280 \approx .0000086$.

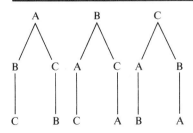

Figure 16-1

The previous examples involved *permutations* of objects or symbols. We were concerned with counting the number of different orderings of the symbols. A related problem is to find the number of different subcollections, or combinations, which can be formed from a group of different symbols. In this case we are concerned only with the particular symbols included in the subcollection, not the order in which they occur.

Example 16-4. How many subcollections containing 2 letters can be formed from the 5 letters *A, B, C, D, E*?

SOLUTION: We can simply list the subcollections, obtaining

AB	*AC*	*AD*	*AE*	*BC*
BD	*BE*	*CD*	*CE*	*DE*

There are thus 10 different subcollections of size 2 which can be formed from 5 different symbols.

Unfortunately, the approach used in Example 16-4 is unsatisfactory for larger collections since it would be too confusing and tedious to list all subcollections of a certain size. For example, how many subcollections of size 4 can be formed from the 10 letters *A, B, C, D, E, F, G, H, I, J*? Such a problem can be solved as follows: First, we determine the number of different lists of 4 letters, counting different permutations of the same 4 letters as different lists. Applying the reasoning we used in problems of permutations, we see that there are $10 \times 9 \times 8 \times 7 = 5,040$ different *ordered* lists of 4 letters which can be formed from the 10 letters we are considering.

Now, observe that in this collection of 5,040 lists we have, for example, counted *ABCD, BADC,* and *CDBA* as different lists. In fact, every unordered subcollection of 4 letters has been included 4! times since any group of 4 different letters can be permuted to make 4! different lists. This means that the 5,040 *ordered* lists of 4 letters which we have counted is 4! times the number of different *unordered* subcollections of 4 letters which can be formed from the 10 letters. Hence, there are

$$\frac{10 \times 9 \times 8 \times 7}{4 \times 3 \times 2 \times 1} = 210$$

different unordered subcollections, or combinations, of size 4 which may be formed from 10 different symbols. An equivalent way to write this number is 10!/4!6! since

$$\frac{10!}{4!6!} = \frac{10 \times 9 \times 8 \times 7 \times (6 \times 5 \times 4 \times 3 \times 2 \times 1)}{(4 \times 3 \times 2 \times 1)(6 \times 5 \times 4 \times 3 \times 2 \times 1)}$$

$$= \frac{10 \times 9 \times 8 \times 7}{4 \times 3 \times 2 \times 1}$$

by cancellation. This leads to the following general formula:

The number of different subcollections of size k which can be formed from a collection of n different objects or symbols is

$$\frac{n!}{k!\,(n-k)!} \tag{16-2}$$

The formula $n!/k!\,(n-k)!$ is commonly abbreviated by $\binom{n}{k}$, pronounced "*n* choose *k*." This formula may be used to solve the problem we raised in the beginning of this section. Formula (16-2) shows that there are $\binom{26}{10} = 5{,}311{,}735$ subcollections of size 10 which can be formed from the 26 letters of the alphabet.

Example 16-5. In a certain contest 7 people are tied for first place. Since there are only 3 prizes, the judge decides to choose 3 people at random to determine the winners. How many different groups of 3 winners are possible?

SOLUTION: Applying the formula in (16-2) with $n = 7$ and $k = 3$, we see there are $\binom{7}{3} = 7!/3!4! = 35$ possible groups of 3 winners.

Example 16-6. In an experiment to test a new pill as a cure for a disease, a doctor selects 8 patients who have the disease. He wants to administer the new pill to 4 patients, using the other 4 as controls (giving them a sugar pill). How many possible ways are there for him to divide the patients into 2 groups of 4?

SOLUTION: Dividing the patients into two groups of 4 is the same as choosing a group of 4 from the 8 and letting the remaining 4 be the other group of 4. Thus, using the formula in (16-2) with $n = 8$ and $k = 4$, we see that the doctor has $8!/4!4! = 70$ choices.

Problems

16-1 A woman owns 40 books, but only 30 will fit on her bookshelf. How many ways are there to arrange the books on the bookshelf?

16-2 A collection of 15 pieces of candy, including 6 chocolate bars, 4 peanut-butter cups, and 5 peppermint sticks, are placed randomly in the 15 slots of a vending machine. How many different arrangements are possible?

16-3 A computer-dating service has a pool of 20 women and 15 men who desire compatible mates. In order to cut costs and avoid the complicated method of matching people according to their psychological profiles, the dating service decides to randomly match the 15 men with 15 of the 20 women? How many different groups of 15 couples could be formed?

Binomial distribution

In Chap. 4 we showed how the binomial distribution could be used to compute probabilities associated with a binomial experiment. The importance of this distribution was evident throughout the first part of the book; we saw it arise over and over again in problems of hypothesis testing and estimation. In order to avoid any unnecessary mathematical discussion we did not give a formula for computing probabilities for the binomial distribution. Instead, we showed how to use the binomial tables to obtain exact probabilities when the sample sizes were small, and how to use the normal approximation in the case of large samples. Using the methods of counting discussed in the last section, we can now derive the formula for the binomial distribution. Although we still recommend obtaining binomial probabilities from tables and the normal approximation rather than directly from the formula, we hope this discussion will give you deeper insight into the random properties of binomial experiments.

Recall that the binomial distribution arises when a sample is drawn from a population divided into two groups, such as Democrat-Republican, male-female, long-short, and heads-tails. A typical problem is illustrated the following example.

Example 16-7. In a certain community 70% of the voters will vote for Mr. Smiley. A sample of 10 voters are selected at random (with replacement). What is the probability that 4 of the voters selected will vote for Mr. Smiley?

SOLUTION: Let us denote by S the number of selected voters who will vote for Mr. Smiley. Then S is a random variable; its value depends on which 10 voters are selected. Our problem can be stated in terms of S: Find $P(S = 4)$, where the event $(S = 4)$ means 4 out of the 10 voters selected will vote for Mr. Smiley.

The addition rule of probability tells us that if we can list the outcomes which make up the event $(S = 4)$, find the probability of each outcome, and then sum these probabilities, we shall have computed $P(S = 4)$. Let's look, then, at some of the outcomes which make up this event. One such outcome is *YYYYNNNNNN* (that is, the first 4 people we select will vote for Mr. Smiley, the next 6 will not.) Since the selections are made with replacement, they are independent and we can use the multiplication rule to obtain

$$P(YYYYNNNNNN) = (.7)(.7)(.7)(.7)(.3)(.3) \cdots (.3)$$

$$= (.7)^4(.3)^6$$

Another outcome in the event $(S = 4)$ is *YNYYYNNNNN*. Using the multiplication rule again, we see

$$P(YNYYYNNNNN) = (.7)(.3)(.7)(.7)(.7)(.3)(.3) \cdots (.3)$$

$$= (.7)^4(.3)^6$$

A moment of consideration should convince you that the probability of any outcome in the event $(S = 4)$ is $(.7)^4(.3)^6$. We need find out only how many outcomes there are in the event $(S = 4)$ and then add $(.7)^4(.3)^6$ that many times, according to the addition rule, and we shall have determined $P(S = 4)$.

The number of outcomes in $(S = 4)$ is the same as the number of different lists which can be formed from the group of 10 symbols including 4 Y's and 6 N's. Any such list can be specified by giving the positions of the 4 Y's, the 6 N's being assigned to the remaining 6 positions. Accordingly, the number of different lists of 10 symbols including 4 Y's and 6 N's is the same as the number of subcollections of 4 positions which may be formed from the 10 positions in the list. This number is seen to be $10!/4!6! = 210$ if we use formula (16-2) with $n = 10$ and $k = 4$ and consider the 10 different positions as 10 different objects. Therefore

$$P(S = 4) = \frac{10!}{4!6!} (.7)^4(.3)^6 = .037$$

The reasoning used in Example 16-7 gives a derivation of the formula for the binomial distribution. The general situation is summarized as follows:

> *Suppose you have a sequence of n independent trials of a random experiment and suppose there are only two possible results for each trial, say, success and failure. If we denote the total number of successes in the n trials by S and use p for the probability of success in a trial, then*

$$P(S = k) = \binom{n}{k} p^k (1 - p)^{n-k} \tag{16-3}$$

> *for any integer k between 0 and n. This equation is called the* binomial formula *with paramters n and p.*

The binomial formula (16-3) may be used to obtain $P(S = k)$ for any number of trials n, any success probability p, and any number of successes k. The binomial tables were computed from the binomial formula for selected values of n and p. In Chap. 4, several examples were given to illustrate the use of the binomial tables and the normal approximation to obtain various probabilities associated with binomial experiments. The following examples show how to use the binomial formula.

Example 16-8. Two ice-cream vendors, Joe and Joan, are competing for sales. If they are stationed near each other, and each of 10 customers chooses 1 vendor at random independently of the other customers, find $P(7$ of the 10 customers buy from Joe).

SOLUTION: Let $S =$ number of customers who buy from Joe. Then we are interested in $P(S = 7)$. The variable S has a binomial distribution with parameters $n = 10$ and $p = 1/2$. According to the binomial formula (16-3),

$$P(S = 7) = \binom{10}{7} \left(\frac{1}{2}\right)^7 \left(\frac{1}{2}\right)^3 = .117$$

Example 16-9. A newspaper delivery boy throws his papers toward his customer's doorstep while riding his bicycle down the street. Suppose that, on any throw, the paper will land on the porch with probability .6. A customer will give the boy a $1 tip if the paper lands on the porch on at least 5 out of the 7 days.

(*a*) Find P(the customer gives a tip).

(b) If the delivery boy has 5 customers on his route, find P(at least 4 customers give a tip).

SOLUTION:

(a) Consider a particular customer. Let S = number of days of the week when the customer's paper lands on his porch. Then S has a binomial distribution with success probability $p = .6$ and number of trials $n = 7$. Hence,

P(the customer gives a tip) $= P(S \geq 5) = P(S = 5) + P(S = 6)$
$$+ P(S = 7)$$
$$= \binom{7}{5} (.6)^5 (.4)^2 + \binom{7}{6} (.6)^6 (.4)^1$$
$$+ \binom{7}{7} (.6)^7 (.4)^0$$
$$= .261 + .131 + .028$$
$$= .420$$

(b) Let S' = number of customers who give a tip. If we consider the results from each customer as trials in a binomial experiment, then it becomes clear that the variable S' is a binomial distribution with parameters $n = 5$ (since there are 5 customers) and $p = P$(customer gives tip) $= .420$. Then

P(at least 4 customers give a tip) $= P(S' \geq 4)$
$$= \binom{5}{4} (.42)^4 (.58)^1$$
$$+ \binom{5}{5} (.42)^5 (.58)^0$$
$$= .1$$

EXPECTATION AND VARIANCE OF A BINOMIAL VARIABLE

The expectation and variance of a variable S which has a binomial distribution with parameters n and p is given by the equations

$$E(S) = np \quad \text{and} \quad \text{Var } S = np(1 - p) \tag{16-4}$$

To derive these equations directly from the definitions requires tricky algebraic manipulations. However, there is a technique called the *method of indicators,* as discussed in Chap. 14, which can be used to obtain a simple derivation. This method is based on the fact that S may be expressed as the sum of n independent random variables having the same distribution which assume only the values of 0 and 1.

To see this, let $X_1 = 1$ if a success occurs on the first trial and $X_1 = 0$ if a failure occurs; let $X_2 = 1$ if a success occurs on the second trial and $X_2 = 0$ if a failure occurs; etc. Thus, the variables $X_1, X_2, \ldots,$ X_n "indicate" the outcome for each of the n trials, having the value 1 for success and 0 for failure. Their sum $X_1 + \cdots + X_n$ gives the number of successes in the n trials since a 1 is added for each success. Thus, $S = X_1 + \cdots + X_n$.

Using Eq. (14-2) for computing the expectation of sums, we obtain $E(S) = E(X_1) + E(X_2) + \cdots + E(X_n)$. But $X_1, X_2, \ldots, X_n$ have the same distribution, and so $E(X_1) = E(X_2) = \cdots = E(X_n)$, implying $E(S) = nE(X_1)$. Now, $E(X_1) = (0)P(X_1 = 0) + (1)\ P(X_1 = 1) = 0 + p$ $= p$, and hence $E(S) = np$ as claimed. (Example 14-4 is another illustration of this method.)

The same kind of argument applies to the variance: $\text{Var } S = \text{Var } X_1 + \cdots + \text{Var } X_n$ according to Eqs. (15-3) for the variance of a sum of independent random variables; and $\text{Var } X_1 = \cdots = \text{Var } X_n$ since $X_1, \ldots, X_n$ have the same distribution. Thus, $\text{Var } S = n \text{ Var } X_1$. Now,

$$\text{Var } X_1 = E(X_1{}^2) - [E(X_1)]^2$$

$$= (1)^2\ P(X_1 = 1) + (0)^2\ P(X_1 = 0) - p^2 = p(1 - p)$$

leading to

$$\text{Var } S = np(1 - p)$$

Also, SD $(S) =$ standard deviation $S = \sqrt{np(1 - p)}$. (Notice that the formulas we have obtained for the mean and standard deviation of a binomial variable agree with the ones stated in Chap. 4.)

Example 16-9 revisited. In order to find the expectation of S', the number of tips the delivery boy will receive each week, we apply Eqs. (16-4) with $n = 5$ and $p = .420$:

$$E(S') = 5(.420) = 2.10$$

In other words, the delivery boy can expect to receive $2.10 each week, on the average. Also $\text{Var } S' = 5(.420)\ (.58) = 1.218$ and hence $\text{SD}(S') = \sqrt{1.218} = 1.1$.

Problems

16-4 What is the probability that a .300 hitter will get at least 6 hits in 10 turns at bat?

16-5 A sample of 15 balls are randomly placed in 3 boxes marked *A, B, C.*

    a. What is the probability that 8 balls are placed in box *A*?

    b. What is the expected number of balls in box *A?*

16-6 In a certain large elementary school, 40% of the children have colds. A group of 5 children is selected at random.

    a. What is the probability that at least 2 children in the sample have colds?

    b. What is the expected number of children selected who have colds?

Hypergeometric distribution

Perhaps the simplest type of sampling situation occurs when a random sample is drawn from a population divided into two categories. Let us call the categories 1 and 2. The outcome of such a sample may be reported by simply giving the number of sample members in category 1, the remaining sample members being from category 2. As we have seen in the last section, if the sampling is done with replacement, then the number of sample members from category 1 will have a binomial distribution (category 1 is "success," and category 2 is "failure"). However, if the sample is drawn without replacement, the draws will not be independent since the probability of choosing a member from category 1 on any draw after the first one will depend on the outcomes of previous draws. In this case, the number of sample members from category 1 has what is called a *hypergeometric distribution.*

    Example 16-10. A typical example of the hypergeometric distribution arises in the area of quality control. Suppose a lot of 100 items contains 25 defectives. If a random sample of 10 items is chosen without replacement, what is the probability that 4 are defective and 6 are nondefective?

    SOLUTION: If D denotes the number of defectives in our sample, we wish to find $P(D = 4)$. This probability is computed by using the counting methods discussed in the first section. There are $\binom{100}{10}$ possible samples of 10 items which could be selected from the 100 items in the lot. Since we are selecting at random, each of these possible samples is equally likely and therefore has probability $1/\binom{100}{10}$ of being drawn. From the addition rule, $P(D = 4)$ equals the sum of the probabilities of the outcomes having 4 defectives. This means $P(D = 4) =$ (number of possible samples with 4 defectives) $\times 1/\binom{100}{10}$.

To find the number of samples with 4 defectives we consider how such a sample must be chosen: 4 defectives must be chosen from the 25 defectives in the lot, and 6 nondefectives must be chosen from the 75 nondefectives. There are $\binom{25}{4}$ ways to choose 4 defectives from 25 defectives and $\binom{75}{6}$ ways to choose 6 nondefectives from 75 nondefectives. Accordingly, there are $\binom{25}{4}\binom{75}{6}$ ways to choose a sample of 10 items containing exactly 4 defectives; hence

$$P(D = 4) = \frac{\binom{25}{4}\binom{75}{6}}{\binom{100}{10}}$$

In more general terms, suppose a group of N items can be divided into two categories, which we call categories 1 and 2. Assume there are N_1 items from category 1 and N_2 items from category 2. Suppose a random sample of size n is drawn without replacement. Let $D =$ number of items in the sample from categorgy 1. Then

$$P(D = k) = \frac{\binom{N_1}{k}\binom{N_2}{n-k}}{\binom{N}{n}} \tag{16-5}$$

for any integer k between 0 and n. Of course, $P(D = k) = 0$ if $k \geq N_1$ or $n - k \geq N_2$. The probability distribution of D is called the hypergeometric distribution.

Example 16-11. A box contains 15 shoes of the same size and style, including 10 right shoes and 5 left shoes. If 8 shoes are drawn at random without replacement from this box, what is the probability that there are exactly 4 pairs of right and left shoes?

SOLUTION: Since the draws are made without replacement, D = number of right shoes selected has a hypergeometric distribution. Using Eq. (16-5) with N_1 = number of right shoes in the box = 10, N_2 = number of left shoes in the box = 5, n = number of draws = 8, we obtain

$P(4 \text{ pairs}) = P(4 \text{ right shoes and 4 left shoes are selected})$

$$= P[D = 4] = \frac{\binom{10}{4}\binom{5}{4}}{\binom{15}{8}} = .163$$

Example 16-12. Suppose that a jury of 12 people is selected at random from a pool of 50 people, including 30 women and 20 men. After the selection is made, the resulting jury includes 2 women and 10 men. The defense lawyer protests, claiming that the low proportion of women selected on the jury, $2/12 = .167$, compared to the proportion of women in the pool, $30/50 = .60$, indicates that the selection was not made at random. Is the defense attorney's protest justifiable?

SOLUTION: To evaluate the defense lawyer's protest, we consider how improbable it is to select a jury with so few women if the selections are actually made at random. Thus, letting D = number of women on a randomly selected jury, we want to find $P(D \le 2)$. Using Eq. (16-5) with N_1 = number of women in pool = 30, N_2 = number of men in pool = 20, and n = sample size = 12, we obtain,

$P(D \le 2) = P(D = 0) + P(D = 1) + P(D = 2)$

$$= \frac{\binom{30}{0}\binom{20}{12}}{\binom{50}{12}} + \frac{\binom{30}{1}\binom{20}{11}}{\binom{50}{12}} + \frac{\binom{30}{2}\binom{20}{10}}{\binom{50}{12}}$$

$$= .0007$$

Thus, if the jury was actually selected at random, there are only 7 chances in 10,000 of obtaining such a low proportion of women. Since this occurrence is so unlikely, it casts grave doubt upon whether the method of jury selection was random.

When the sample size n is small in relation to N_1 and N_2, it makes little difference in computing probabilities whether sampling is done with or without replacement. This is because it is unlikely that a sample drawn with replacement will include any duplication among the selections. It follows that in the setup of Eq. (16-5), the binomial distribution which D would have if sampling were done with replacement is nearly the same as the hypergeometric distribution which arises when sampling is done without replacement if n/N_1 and n/N_2 are small. Hence, just as for the binomial distribution, we may approximate the hypergeometric distribution by the normal distribution when n is large, but small in relation to N_1 and N_2. An application of this situation is given in the next example.

Example 16-13. A United States Congressional election was held between candidates A and B in which the votes cast were 15,000 and 14,975, respectively. Unfortunately for A, there were 60 "irregular" votes counted among the total of 29,975 votes, putting A's victory in doubt. If we make no presumption about the number of irregular votes each candidate received, is there much reason to believe that a reelection should be held?

SOLUTION: Making no assumption as to which votes went to whom, we need to determine the probability that B would end up with more votes than A if 60 votes were randomly selected from the total of 29,975 votes. That is, if we let D = number of irregular votes for candidate A, then $15,000 - D$ and $14,975 - (60 - D)$ are the respective totals after removal of irregular votes; hence we wish to find $P[(15,000 - D) < 14,975 - (60 - D)]$. By algebra, this probability equals $P(D > 42.5)$.

Now D has a hypergeometric distribution with $N_1 = 15,000$, $N_2 = 14,975$, and $n = 60$. Since $n/N_1 = 60/15,000 \approx .004$ and $n/N_2 = 60/14,975 \approx .004$ are both small, the distribution of D is approximated by the binomial distribution with parameters $n = 60$ and $p = N_1/(N_1+N_2) = 15,000/29,975 \approx .50$. Now, using the normal approximation to the binomial, we obtain

$$P(D > 42.5) \approx P\left[Z > \frac{42.5 - np}{\sqrt{np\,(1 - p)}}\right]$$

$$P(D > 42.5) = P\left[Z > \frac{42.5 - 60(.5)}{\sqrt{60(.5)(.5)}}\right]$$

$$= P(Z > 3.23)$$

$$= .001$$

Since this probability is only 1 chance in 1,000, there is little reason to believe a new election should be held unless we have more information about the person for whom the irregular votes were cast.

Geometric distribution

As we have seen, in situations where a fixed number of independent trials of a random experiment having only two outcomes are made, the probability of observing a certain number of outcomes of one type may be computed from the binomial distribution. Sometimes we are interested, not in the number of successes which occur, but rather in the number of trials necessary to obtain the first success. A typical example is the following.

Example 16-14. A personnel officer knows that about 20% of the applicants for a certain position are suitable for the job. What is the probability that the fourth person interviewed will be the first one who is suitable?

SOLUTION: When an applicant is selected at random and interviewed, there is a probability .2 that he is suitable and a probability $1 - .2 = .8$ that he is not suitable. Using the multiplication rule, we obtain P(fourth applicant interviewed is the first one suitable) $= (.8)(.8)(.8)(.2) = .1024$.
In more general terms:

Suppose a certain random experiment has only two possible outcomes, which we call success and failure, and suppose independent trials of this experiment are performed until the first success occurs. If we let $p = P$ (success), then the distribution of T = number of trials until first success is given by

$$P(T = k) = (1 - p)^{k-1}p \tag{16-6}$$

for any positive integer k.

We now derive $E(T)$ and show that $E(T) = 1/p$ by the following argument: Observe that the first trial will be success with probability p, in which case $T = 1$. On the other hand, the first trial will be a failure with probability $1 - p$, in which case the trials will continue just as if we were starting over. Hence, given that the first trial is a failure, an event having probability $1 - p$, the expected number of trials until the first success is $1 + E(T)$ (since beginning with the second trial we are "starting over"). Thus

$$E(T) = (1)\, P(T = 1) + [1 + E(T)]\, P(T > 1)$$
$$= p + [1 + E(T)](1 - p)$$

Solving this equation for $E(T)$ gives $E(T) = 1/p$. Using this formula in Example 16-14 tells us that the expected number of applicants the personnel officer must interview until a suitable applicant is discovered is $1/.2 = 5$.

Example 16-15. Suppose that each package of a certain brand of bubble gum contains a picture of 1 of 3 baseball stars. Assuming the bubble-gum company has distributed equal numbers of each picture, how many packs of bubble gum can a child expect to buy before obtaining the entire set of 3 pictures?

SOLUTION: The child will obtain his first picture in the first pack he buys. The number of packs he must buy until he obtains the second picture has a geometric distribution with $p = 2/3$ since each time he buys a new pack he has probability $2/3$ of obtaining a different picture independently of previous purchases. After 2 different pictures are obtained, the number of purchases necessary to obtain the third picture has a geometric distribution with $p = 1/3$. (Only 1 of the 3 remaining pictures is different.)

Therefore, using the formula for the expectation of a geometric variable, the expected number of purchases from the first pack until obtaining the second picture is $1/(2/3) = 3/2$. Likewise, after obtaining the second picture, the expected number of purchases until obtaining the third picture is $1/(1/3) = 3$. Therefore, using the addition rule for expectation, the total expected number of purchases is $1 + 3/2 + 3 = 5.5$.

Problems

16-7 A piano player has probability .01 of making a mistake each time he plays a note.

    a. How many notes will he play before making a mistake on the average?

    b. What is the probability that he will play an entire piece with 1,000 notes, making no mistakes.

16-8 In a police lineup of 15 people, 4 were involved in a certain robbery. A witness to the robbery can't identify the robbers, and so he selects 4 people at random from the lineup.

    a. What is the probability that the witness correctly selects only 2 of the robbers?

    b. What is the probability that the witness correctly selects at least 1 of the robbers?

16-9 A gambler playing the game of roulette has $50 left with which he bets $1 on number 00 each spin of the wheel. (A roulette wheel has 38 numbers: 00, 0, 1, 2, . . . , 36; if you bet $1 on a particular number and win, you get $36.)

    a. What is the probability that he goes broke without ever winning?

    b. What is the probability that he wins exactly once before going broke?

16-10 A committee contains 8 men and 12 women. If a subcommittee of size 6 is selected at random, find P(4 men and 2 women are selected).

16-11 You are an oil prospector. When you dig for oil, if the location is dry, you will find oil with probability 1/3; if the location is rocky, you will find oil with probability 1/2; and if the location is wet, you will find oil with probability 1/4. You know from prior experience that a location is dry with probability 3/5, P(rocky) = 1/5, and P(wet) = 1/5. If you keep digging in various locations until you strike oil the first time, how many times should you expect to dig?

16-12 If 5 cards are chosen at random from a deck of cards, what is the probability that they are all black?

Poisson distribution

The distributions we have studied so far in this chapter arise in a fairly obvious way from simply described situations. Another important **dis-**

tribution, whose origin is not so easily explained, is the *Poisson distribution*. The Poisson distribution (named after a famous French mathematician) often serves as a satisfactory model for physical processes involving objects that are randomly distributed in space and for events which occur at random times. Some variables which experience shows usually have a Poisson distribution are the number of disintegrating atoms from a radioactive substance during a specified interval of time, the number of telephone calls entering a busy switchboard during a given period, the number of meteorites found in an area of a desert, and the number of raisins in a particular cookie made from a large batch of batter and raisins.

We now give the formula for the Poisson distribution:

A variable X is said to have a Poisson distribution with mean μ if

$$P(X = k) = \frac{e^{-\mu}\mu^k}{k!} \qquad (16\text{-}7)$$

for each value $k = 0, 1, 2, \ldots$, and some positive number μ.

Here the number e is the base for natural logarithms, having approximately the value $e \approx 2.718$. It turns out that the expectation of X is μ. We have included tables of Poisson probabilities computed for various values of μ and k in the Appendix.

Some insight into situations where the Poisson distribution is applicable is given by the following fact:

Suppose S is a variable having a binomial distribution with parameters n and p, with n large and np^3 small. Then S will have approximately a Poisson distribution with mean $\mu = np$. That is,

$$P(S = k) \approx \frac{e^{-np}(np)^k}{k!} \qquad (16\text{-}8)$$

Besides providing a method for approximating the binomial distribution in certain cases (see Example 16-18), Eq. (16-8) suggests a criterion for determining whether a variable has a Poisson distribution. Suppose a variable X denotes the number of occurrences of a certain event. If X can be envisioned as being the number of successes in a binomial experiment with a large value of n and a small value of p then (16-8) tells us that the distribution of X is roughly a Poisson distribution.

Consider, for example, the situation where telephone calls arrive at a central switchboard of a large city during a certain period of the day, say, the noon hour. Suppose that from past experience it has been observed that on the average 20 calls arrive during this period. Envision the hour as being divided into a large number of small time intervals, say, 1,000. Since the calls arrive "randomly," we expect on the average about $20/1,000 = .02$ calls per interval. Moreover, since the intervals are so short, it is highly unlikely that more than 1 call will arrive in any particular interval. Thus, there will probably be either 1 call or none in each interval, the probability of 1 call during a particular interval being about .02. Since the calls are independent in a large city, we may consider the intervals as comprising trials in a binomial experiment with $n = 1,000$ and $p = .02$. Thus, according to Eq. (16-8), the variable $X =$ number of calls during the noon hour will have approximately a Poisson distribution with mean $\mu = np = 1,000(.02) = 20$, and probabilities given by

$$P(X = k) = \frac{e^{-20}(20)^k}{k!} \qquad \text{for } k = 0, 1, \ldots$$

Example 16-16. A certain radioactive element is known to emit particles at the rate of about 3 particles per second. In a 1-second interval, what is the probability that the number of particles emitted X is less than 6?

SOLUTION: We may break up the time interval just as in the telephone example, and hence it is reasonable to assume X has a Poisson distribution with $\mu = 3$. Therefore

$$P(X < 6) = P(X = 0) + P(X = 1) + P(X = 2) + \cdots + P(X = 5)$$

$$= \frac{e^{-3}(3)^0}{0!} + \frac{e^{-3}(3)^1}{1!} + \frac{e^{-3}(3)^2}{2!} + \cdots + \frac{e^{-3}(3)^5}{5!}$$

$$= .9161$$

Example 16-17. A popular 100-acre area of public beach has been found to contain about 15 watches after a crowded Sunday afternoon. If a person thoroughly searches 1 acre of this beach with a metal detector after such a day, what is the probability of finding at least 1 watch?

SOLUTION: Let $X =$ no. of watches found on the acre that is searched. It is reasonable to assume that the watches are randomly distributed over this area, and hence X has a Poisson distribution with a mean $\mu = 15/100 = .15$ watches per acre. Now,

$$P(X \geq 1) = 1 - P(X = 0)$$

$$= 1 - \frac{e^{-.15}(.15)^0}{(0)!}$$

$$= 1 - .86$$

$$= .14$$

Example 16-18. Suppose that a certain birth defect occurs in 1 out of 10,000 births on the average. Estimate the probability that in a city of 25,000 none of the population will have the birth defect.

SOLUTION: We may think of the births as independent trials, and hence the variable S = number of birth defects will have a binomial distribution with $n = 25,000$ and $p = .0001$. We wish to estimate $P(S = 0)$. Notice that we cannot use the normal approximation to the binomial to estimate this probability because $np = 25,000(.0001) = 2.5$ is less than 5. However, we may use (16-8) to obtain

$$P(S = 0) \approx \frac{e^{-2.5}(2.5)^0}{0!} = e^{-2.5} = .08$$

Problems

16-13 In a certain store, 1 out of every 500 checks is bad on the average. Over a 1-year period, 20,000 checks were cashed. Find the probability that the number of bad checks is less than or equal to 15.

16-14 Assume that 1% of the population is allergic to cat hairs. If 100 people are chosen at random, what is the probability that at least 6 are allergic to cat hairs? (Use the Poisson approximation to the binomial.)

16-15 The number of calls entering each of 10 centrally located telephone switchboards has a Poisson distribution with a mean of 5 calls per minute. In a given minute what is the probability that exactly 7 of the switchboards have no calls coming in?

Gambling

We now apply some of the concepts we have discussed to games of chance.

Example 16-19. The game of craps is played as follows: A player rolls two dice. If the sum of the dots on the first roll is 7 or 11, the player wins; if it is 2, 3, or 12, the player loses. If the sum is another number, the dice are rolled repeatedly either until the player repeats the first sum, in which case he wins, or until he rolls a 7, in which case he loses. What is the probability of winning this game?

SOLUTION: The game can be won either on the first roll or later, and so $P(\text{win}) = P(\text{win on first roll}) + P(\text{win on later roll})$. The first probability is easily computed: $P(\text{win on first roll}) = P(7) + P(11) = 6/36 + 2/36 = 8/36$. (You should verify this by considering the $6 \times 6 = 36$ possible outcomes of the rolling of a pair of dice.) Considering the 6 possible sums appearing on the first roll which continue the game, namely, 4, 5, 6, 8, 9, and 10, we compute the second probability according to $P(\text{win on later roll}) = P(4 \text{ on first roll and win on later roll}) + P(5 \text{ on first roll and win on later roll}) + \cdots + P(10 \text{ on first roll and win on later roll})$. A 4 on the first roll will occur with probability 3/36, and given this, the player will win only if he then rolls a 4 before a 7. Now, $P(\text{rolling 4 before 7}) = 3/9$ since there are 3 ways to get a 4 and 6 ways to get a 7, giving a total of 9 ways to get 4 or 7, and all of the 9 ways are equally likely. Hence, $P(4 \text{ on first roll and win on later roll}) = (3/36)(3/9) = 1/36$, and by symmetry $P(10 \text{ on first roll and win on later roll}) = 1/36$.

Using the same reasoning, we obtain $P(5 \text{ on first roll and win on later roll}) = P(9 \text{ on first roll and win on later roll}) = (4/36)(4/10) = 2/45$ and $P(6 \text{ on first roll and win on later roll}) = P(8 \text{ on first roll and win on later roll}) = (5/36)(5/11) = 25/396$. Thus,

$P(\text{win}) = P(\text{win on first roll}) + P(\text{win on later roll})$

$$= \frac{8}{36} + \left(\frac{1}{36} + \frac{1}{36} + \frac{2}{45} + \frac{2}{45} + \frac{25}{396} + \frac{25}{396} \right) = .493$$

We shall now explain how to find the probabilities of various poker hands. Poker is a game played between at least two persons. In one variation, an ordinary deck of cards is thoroughly shuffled and then dealt out until each player has five cards. Then bets are made, and the hands are compared to see whose is highest, the person with the highest hand being declared winner. The order of hands from lowest to highest is 1 pair, 2 pair, 3 of a kind, straight (for example, 5 consecutive denominations), flush (for example, 5 cards of 1 suit), full house (for example, 1 pair and 3 of a kind), 4 of a kind, and straight flush (5 consecutive denominations of the same suit).

Example 16-20. In a game of poker what is the probability of 1 pair being dealt on the first hand?

SOLUTION: First of all, using the methods of counting we developed previously, there are $\binom{52}{5}$ possible poker hands, all of them being equally likely. There are 13 possible types of pairs (aces, twos, threes, etc.) and $\binom{4}{2}$ ways to choose 2 cards from the 4 cards of each type. Given the type of the pair, there are $\binom{12}{3}$ ways to choose the types of the cards not in the pairs and 4^3 ways to choose the suits from these types. Thus, there are $13\binom{4}{2}\binom{12}{3}4^3$ hands with exactly 1 pair, implying

$$P(1\text{ pair}) = \frac{13\binom{4}{2}\binom{12}{3}4^3}{\binom{52}{5}} = .42256903$$

Example 16-21. What is the probability that a poker hand is a flush?

SOLUTION: There are 4 possible suits for the flush, $\binom{13}{5}$ ways to choose the 5 denominations, and hence $4\binom{13}{5}$ hands making a flush among the $\binom{52}{5}$ possible hands. Hence,

$$P(\text{flush}) = \frac{4\binom{13}{5}}{\binom{52}{5}} = .00198079$$

Example 16-22. What is the probability of a full house being dealt in a poker hand?

SOLUTION: There are 13 possible types of pairs and $\binom{4}{2}$ ways to choose the suits of the pair. Once the pair has been specified, there are 12 possible types for the 3 of a kind and $\binom{4}{3}$ ways to choose their suits. Since there are $\binom{52}{5}$ equally likely hands,

$$P(\text{full house}) = \frac{\binom{4}{2} \, 12\binom{4}{3}}{\binom{52}{5}} = .00144057$$

Summary

Several common probability distributions have been presented. It was shown how the binomial distribution arises in situations where an experiment having only two possible outcomes is independently repeated, for example, when sampling with replacement. If a sample from a population is drawn without replacement and the members are in one of two categories, then the hypergeometric distribution is used. The geometric distribution describes the random behavior of the first time a success occurs in an experiment which consists of repeating independent trials having only two outcomes, success and failure. In describing physical phenomena involving items randomly distributed in space and events occurring at random times, the Poisson distribution is often used.

In many situations where an event can be easily described in terms of the outcomes of an experiment, there are too many outcomes to simply list them and add their probabilities. In such cases it is often possible to use the methods of counting permutations and combinations. These methods may be used to derive the formulas for the distributions we have studied.

Important terms

Binomial distribution

Combination

Geometric distribution

Hypergeometric distribution

Permutation

Poisson distribution

Additional problems

16-16 A box contains 10 balls numbered 1 to 10.

    a. How many different collections of 4 balls can be selected?

    b. If you select 5 balls at random without replacement, what is $P(3$ of the 5 balls are odd)?

16-17 If a room contains 30 people, what is the probability that at least 2 of them have the same birthdays?

16-18 Suppose you are playing poker and you are dealt 2 kings. If you discard the other 3 cards and draw 3 new cards, what is the probability that you draw at least 1 more king?

16-19 A movie theater has 40 rows with 30 seats in each row. If you are among 600 people who are randomly seated in the theater, what is the probability that someone is sitting immediately in front of you?

16-20 A door-to-door salesman has a probability of .2 that he will make a sale at each house he approaches.

    a. If he goes to 10 houses, what is the probability he will sell 6 items?

    b. What is the expected number of houses he will visit before making his first sale?

16-21 If 10 indistinguishable balls are randomly placed in 4 boxes numbered from 1 to 4, find the probability that exactly 6 balls are placed in box number 3.

16-22 In a certain court 4 prosecuting attorneys take turns prosecuting cases. Out of a total of 6 cases in which the defendant was acquitted, 4 were cases handled by the youngest attorney. Was he justified in attributing his lack of success to chance?

16-23 A list of committee members contains the first names of both husbands and wives for 20 couples. If 10 people are chosen at random from the list, what is the probability that no couple is included?

16-24 An insurance company has insured 20,000 families against theft. Assuming that the probability of theft to a family during a 1-year period is .0004, find the probability that the company will have to pay off at least one claim during the year.

16-25 If there are 15 points on the circumference of a circle, how many triangles may be formed by connecting 3 points?

16-26 When you enter a certain hotel, the probability that you will find the elevator on the ground floor is 1/2.

    a. If you enter the hotel 20 times, what is the probability that it will be on the ground floor exactly 7 times?

    b. How many times will you have to enter the hotel, on the average, between times when you find the elevator on the ground floor?

16-27 The uptown and downtown trains each come into the station once every hour. You like to go uptown about half the time and downtown about half the time, and so you walk into the station at random times and take whichever train comes first. To your dismay, you find that you are going uptown 9 times as often as downtown. Explain.

16-28 a. Devise a binomial experiment.

    b. Devise a variable having a hypergeometric distribution.

    c. Devise a variable having a Poisson distribution.

16-29 A machine that produces cans has a probability of .05 of producing a defective can. What is the probability that the twentieth can produced is the fifth defective can?

*16-30 When a famous statistics professor got married, he made a resolution to have babies until the number of boys and girls was equal. For example, if the second child were a boy after the first was a girl, the professor would have no more. Supposing that both sexes have the same probability of being born, obtain the probability that the professor will end up with n children (n is an even number).

True-false test

Circle the correct letter.

T F   1. A coin having probability 1/10 of heads is tossed 3 times. Then the probability of obtaining at least 1 head is 3/10.

T F   2. A group contains 20 women and 10 men. If 5 people are selected at random (without replacement), then S = number of women in the sample has a hypergeometric distribution.

T F   3. An underground nuclear test has probability .000001 of having an accident causing dangerous radiation leakage. If this test is repeated over and over again, eventually an accident will certainly result.

T F   4. The binomial distribution is a good approximation to the hypergeometric distribution when the sample size is small relative to the population size.

T F   5. The number of ways of obtaining 4 heads in 10 tosses of a coin is the same as the number of ways of selecting 6 people from a group of 10 people.

T F   6. If you draw 5 cards from a deck, you are 5 times as likely to draw an ace than if you draw only 1 card.

T F   7. If an experiment having 10 possible outcomes is repeated independently 7 times, then the number of times a specific outcome occurs has a binomial distribution.

T F   8. The reason binomial experiments occur so frequently is because our society is overly concerned with "success" and "failure."

T F   9. If 20 indistinguishable guppies are placed into 6 aquariums, then the number of guppies put in the fifth aquarium is a binomial variable with $n = 20$ and $p = 1/6$.

T F   10. Suppose 20 cards numbered from 1 to 20 are placed face down on a table in random order and you have to guess the numbers. If you guess the numbers in increasing order from 1 to 20, then on a lucky day you could guess 19 correctly.

Conditional probability and decision theory

17

We shall now discuss the concept of *conditional probability* and see how it can be used to solve certain problems of statistical inference. In these problems one is asked to determine the best decision or course of action when faced with a number of alternatives. This branch of statistics is often called *decision theory*.

Conditional probability

We shall introduce the concept of conditional probability and derive a computational formula by considering the following example.

Example 17-1. A dart is thrown at random onto a circular dartboard. The probability that the dart will land in a particular region is the relative area taken up by that region. In other words,

$$P(\text{region}) = \frac{\text{Area of region}}{\text{Total area}}$$

We have split the dartboard into three regions *A*, *B*, and *C*, as shown in Fig. 17-1. Region *A* takes up 2/3 of the dartboard, and region *B* takes up 1/3 of the dartboard. Region *C* takes up 1/2 of the dartboard, 1/4 of which is in region *A* and 1/4 of which is in region *B*. We can write this in probabilistic form:

$$P(A) = \frac{2}{3} \qquad P(B) = \frac{1}{3} \qquad P(C) = \frac{1}{2}$$

$$P(A \text{ and } C) = \frac{1}{4} \qquad P(B \text{ and } C) = \frac{1}{4}$$

Suppose we are told that the dart landed in region *A*. What would be the probability that it also landed in region *C?*

SOLUTION: We shall denote this as $P(C \mid A)$, which reads "the probability of *C* given *A*." This type of probability is called a *conditional probability* because it depends on, or is conditional on, the occurrence of an event in the experiment. In this case, we are given that the dart landed in region *A*, and based on this fact, we wish to know the probability that the dart also landed in *C*.

Since the probability that the dart will land in a particular region is the relative area taken up by that region, $P(C \mid A)$ should be the proportion of the area of region *A* taken up by region *C*. In other words,

$$P(C \mid A) = \frac{\text{Area of } (C \text{ and } A)}{\text{Area of } A}$$

We do not know these areas, but we can compute this probability using the probabilities we already know:

$$P(C \mid A) = \frac{\text{Area of } (C \text{ and } A)}{\text{Area of } A}$$

$$= \frac{\text{Area of } (C \text{ and } A)/\text{total area}}{\text{Area of } A/\text{total area}}$$

$$= \frac{P(C \text{ and } A)}{P(A)}$$

$$= \frac{1/4}{2/3}$$

$$= \frac{3}{8}$$

Example 17-1 yields a useful formula for computing conditional probabilities which we state in general:

Conditional probability formula. For any events A, B,

$$P(B \mid A) = \frac{P(B \text{ and } A)}{P(A)} \qquad (17\text{-}1)$$

As with regular probabilities, an addition rule also holds for conditional probabilities, providing the conditioning is based on the occurrence of a fixed event. Using this, we obtain

For any events A, B,
$$P(\text{not } B \mid A) = 1 - P(B \mid A) \qquad (17\text{-}2)$$

If we multiply both sides of formula (17-1) by $P(A)$, we obtain a general multiplication rule.

General multiplication rule. For any two events A, B,
$$P(A \text{ and } B) = P(B \mid A)\, P(A) \qquad (17\text{-}3)$$

The multiplication rule for independent events, which we studied in Chap. 3, states that if A and B are independent, then $P(A$ and

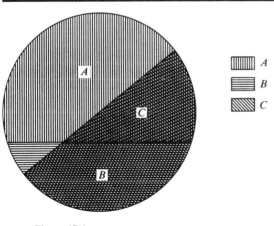

Figure 17-1

$B) = P(A) \, P(B)$. We can compare this with formula (17-3) to obtain the following:

If A and B are independent, $P(B \mid A) = P(B)$. In other words, if two events are independent, knowledge of whether or not one event has occurred does not affect one's opinion as to whether the other event has occurred.

In the next example, we further illustrate the conditional probability formula, introducing the concept of "updating" information based on the results of a random experiment.

Example 17-2. City A contains a proportion of .8 who distrust politicians, and city B contains a proportion of .4 who distrust politicians. One of these cities is chosen at random, and two people are then randomly selected from that city. You are told whether or not the people selected distrust politicians, and you must then guess which city was selected. How should you make your decision?

SOLUTION: If you had no knowledge about the city selected other than the fact that it was selected at random, and therefore that $P(A) = P(B) = .5$, you would be indifferent as to which city to guess. We call these probabilities *prior probabilities* because they are known to you prior to the experiment. However, you are told whether or not people selected distrust politicians, and you can therefore *update* your prior probabilities based on this information by computing the conditional probabilities of the cities, given the results of the experiment.

If we let D stand for *distrusts politicians* and N stand for *doesn't distrust politicians,* the probabilities of interest are as follows:

$P(A \mid DD) \qquad P(B \mid DD) \qquad P(A \mid DN) \qquad P(B \mid DN)$

$P(A \mid ND) \qquad P(B \mid ND) \qquad P(A \mid NN) \qquad P(B \mid NN)$

Using the conditional probability formulas (17-1) and (17-2) and the tree diagram in Fig. 17-2, we compute these conditional probabilities as follows:

$$P(A \mid DD) = \frac{P(A \text{ and } DD)}{P(DD)} = \frac{.32}{.40} = .8$$

$$P(B \mid DD) = 1 - P(A \mid DD) = .2$$

$$P(A \mid DN) = \frac{P(A \text{ and } DN)}{P(DN)} = \frac{.08}{.20} = .4$$

$$P(B \mid DN) = 1 - P(A \mid DN) = .6$$

Likewise,

$$P(A \mid ND) = .4 \quad P(B \mid ND) = .6 \quad P(A \mid NN) = .1$$

$$P(B \mid NN) = .9$$

Thus, if you are told *DD*, that both selected people distrust politicians, you should guess that city *A* was the city selected since $P(A \mid DD) = .8$ while $P(B \mid DD) = .2$. If you observe *DN* or *ND*, you should guess city *B* since $P(B \mid DN) = P(B \mid ND) = .6$ while $P(A \mid DN) = P(A \mid ND) = .4$. Finally, if you observe *NN*, you should also guess city *B* since $P(B \mid NN) = .9$ while $P(A \mid NN) = .1$.

The conditional probabilities we used in Example 17-2 have updated prior probabilities using the results of the experiment and are called *posterior probabilities*. We used the criterion of choosing the city with highest posterior probability.

Example 17-3. You have been placed in one of three rooms and you don't know which one. However, you do know that with probability 2/5 you have been placed in room *A*, with probability 2/5 you have been placed in room *B*, and with probability 1/5 you have been placed in room *C* (these are the prior probabilities). You also know that room

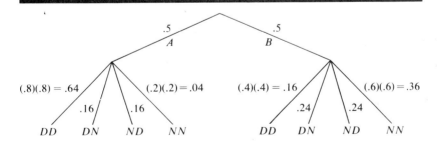

Figure 17-2 $P(DD) = (.5)(.64) + (.5)(.16) = .40; P(DN) = P(ND) = (.5)(.16) + (.5)(.24) = .20; P(NN) = (.5)(.04) + (.5)(.36) = .20$

A contains a coin with P(heads) = 1/2, room *B* contains a coin with P(heads) = 5/12, and room *C* contains a coin with P(heads) = 5/8. To help you decide where you are, you toss the coin in and it comes up heads. Based on this, which room do you think you're in?

SOLUTION: The relevant probabilities to consider are the posterior probabilities: P(room *A* | heads), P(room *B* | heads), and P(room *C* | heads). An appropriate decision would be to decide upon the room with highest posterior probability. Using formula (17-1), we write these probabilities as follows:

$$P(\text{room } A \mid \text{heads}) = \frac{P(\text{room } A \text{ and heads})}{P(\text{heads})}$$

$$P(\text{room } B \mid \text{heads}) = \frac{P(\text{room } B \text{ and heads})}{P(\text{heads})}$$

$$P(\text{room } C \mid \text{heads}) = \frac{P(\text{room } C \text{ and heads})}{P(\text{heads})}$$

To obtain the necessary probabilities on the right side of the preceding equations, we construct the tree diagram for this experiment in Fig. 17-3. Using the tree diagram in the proper way we obtain the desired probabilities:

$$P(\text{room } A \text{ and heads}) = \frac{2}{5} \times \frac{1}{2} = \frac{1}{5}$$

$$P(\text{room } B \text{ and heads}) = \frac{2}{5} \times \frac{5}{12} = \frac{1}{6}$$

$$P(\text{room } C \text{ and heads}) = \frac{1}{5} \times \frac{5}{8} = \frac{1}{8}$$

$$P(\text{heads}) = \frac{1}{5} + \frac{1}{6} + \frac{1}{8} = \frac{59}{120}$$

Now we apply formula (17-1) and obtain

$$P(\text{room } A \mid \text{heads}) = \frac{1/5}{59/120} = \frac{24}{59}$$

$$P(\text{room } B \mid \text{heads}) = \frac{1/6}{59/120} = \frac{20}{59}$$

$$P(\text{room } C \mid \text{heads}) = \frac{1/8}{59/120} = \frac{15}{59}$$

Thus, if heads is the outcome of the toss, it is most likely that we are in room A since $P(\text{room } A\,|\,\text{heads}) = 24/59$, the highest posterior probability.

Example 17-4. It is known that 60% of marijuana smokers are cigarette smokers, and it is also known that 30% of non-marijuana smokers are cigarette smokers. At a certain university 50% of the students are marijuana smokers. If a student is selected at random and it is determined that he smokes cigarettes, what is the probability he smokes marijuana? In other words, find $P(\text{smokes marijuana}\,|\,\text{smokes cigarettes})$.

SOLUTION: Formula (17-1) yields

$P(\text{smokes marijuana} \mid \text{smokes cigarettes})$

$$= \frac{P(\text{smokes marijuana and smokes cigarettes})}{P(\text{smokes cigarettes})}$$

Since .6 of marijuana smokers are cigarette smokers and since .5 of the students are marijuana smokers, we use formula (17-3) and deduce that $.6 \times .5 = .3$ of the students are both marijuana smokers and cigarette smokers. Therefore, $P(\text{smokes marijuana and smokes cigarettes}) = .3$. Likewise, we obtain $P(\text{doesn't smoke marijuana and smokes cigarettes}) = .3 \times .5 = .15$. Thus, $P(\text{smokes cigarettes}) = .3 + .15 = .45$ and $P(\text{smokes marijuana}\,|\,\text{smokes cigarettes}) = .3/.45 = .67$.

Example 17-5. Recall Example 3-14. When you travel on a certain freeway during rush hour, you have a .7 chance of getting off at the correct exit (exit A). If you get off at exit A, you have a .9 chance of not

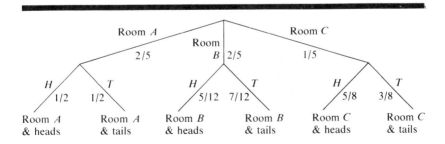

Figure 17-3

getting lost. If, however, you miss exit *A* and have to get off at exit *B*, your chances of getting lost are .8. Given that you get lost, what is the probability that you got off at exit *B?* In other words, compute *P*(got off at *B*|you're lost).

SOLUTION: We use formula (17-1):

$$P(\text{got off at } B|\text{you're lost}) = \frac{P(\text{got off at } B \text{ and you're lost})}{P(\text{you're lost})}$$

Turning to Example 3-14 and the tree diagram in Fig. 3-5, we see: *P*(*B* and lost) = .3 × .8 = .24; *P*(lost) = 1 − .69 = .31. Thus *P*(got off at *B*|you're lost) = .24/.31 = .77.

Problems

17-1 You believe that a suspect in a certain criminal case is lying with probability .8. You decide to give him a polygraph test knowing that 90% of liars fail the polygraph test while 5% of nonliars fail the test. Suppose the suspect fails the test. Find *P*(he is a liar|he fails test).

17-2 In Detroit 3/4 of the people favor capital punishment, while in San Francisco 1/2 favor capital punishment. One of these cities is selected at random and then 2 people are selected from the chosen city. Find

    a. *P*(both selected people favor capital punishment)

    b. *P*(Detroit was the selected city|both people selected favor capital punishment)

17-3 You have a 1/3 chance of seeing a poisonous snake in forest A, a 1/4 chance in forest B, and you have a 1/8 chance in forest C. You are in one of these forests, and you're not sure which one; but you believe that with probability 1/12 you're in forest A, with probability 1/6 you're in forest B, and with probability 3/4 you're in forest C. Suddenly, you see a poisonous snake. Which forest do you think you're in?

Decision theory

The next example illustrates a basic type of problem in *decision theory*. We are confronted with a situation in which various courses of action are available to the experimenter and in which various random factors are involved. In order to decide which course of action is optimal in the sense of maximizing an expected reward of some type, conditional probabilities are computed and the methods we have been studying are fully utilized.

Example 17-6. An oil prospector wishes to decide whether or not to drill for oil in a certain locale. He knows that there are three states possible in the site under consideration: no oil, some oil, and "bonanza," which we denote by θ_1, θ_2, θ_3, respectively. The worth of these states to the prospector, taking into account the $50,000 cost of drilling a well, is given in the following net gains table:

Net gains to prospector if he
drills (in thousands of dollars)

θ_1	-50
θ_2	50
θ_3	100

Of course, if the prospector decides not to drill, he neither gains nor loses anything, and so the worth of not drilling is $0.

In the past, the prospector's drillings have yielded state θ_1 a proportion of .5 of the time, state θ_2 a proportion of .3 of the time, and state θ_3 a proportion of .2 of the time. Before drilling, at a cost of $5,000, the prospector can take a seismic sounding which gives two possible readings $+$ and $-$. Past experience with seismic soundings has indicated a relationship between soundings and the states, which is shown in Eqs. (17-4):

$$P(+|\theta_1) = .2 \qquad P(-|\theta_1) = .8$$

$$P(+|\theta_2) = .4 \qquad P(-|\theta_2) = .6 \qquad\qquad (17\text{-}4)$$

$$P(+|\theta_3) = .9 \qquad P(-|\theta_3) = .1$$

If the prospector wishes to maximize his expected gain, what should he do?

SOLUTION: First, we list the prior probabilities for the possible states based on the prospector's past experience:

$$P(\theta_1) = .5$$

$$P(\theta_2) = .3$$

$$P(\theta_3) = .2$$

With these probabilities, we compute the expected gain to the prospector if he decides to drill without taking a seismic sounding:

$$E(\text{gain without sounding}) = (-50)(.5) + (50)(.3)$$

$$+ (100)(.2) = 10 \qquad (17\text{-}5)$$

In other words, if the prospector decides not to take a seismic sounding he should drill for oil since he could expect to gain $10,000 whereas not drilling would yield him nothing.

In order to decide whether or not it is advantageous to take a seismic sounding, we must compare (17-5) with the prospector's expected gain if he takes a sounding before deciding whether or not to drill. To do this, we update the prospector's prior probabilities and obtain posterior probabilities based on the results of the sounding. We then compute the prospector's expected gain from drilling based on the posterior probabilities, which will tell the prospector how to act after taking the sounding and whether or not he should take the sounding at all.

The posterior probabilities of the states given the results of the sounding are derived using the conditional probability formula and the tree diagram in Fig. 17-4:

$$P(\theta_1 \,|\, +) = .25 \qquad P(\theta_2 \,|\, +) = .30 \qquad P(\theta_3 \,|\, +) = .45 \qquad (17\text{-}6)$$

$$P(\theta_1 \,|\, -) = .67 \qquad P(\theta_2 \,|\, -) = .30 \qquad P(\theta_3 \,|\, -) = .03 \qquad (17\text{-}7)$$

Using the net gains table and the probabilities in (17-6) and (17-7), we now compute the prospector's expected gain from drilling, given that the sounding resulted in $+$, and then the expected gain, given that the sounding resulted in $-$.

$$E(\text{gain} \,|\, +) = (-50)(.25) + (50)(.3) + (100)(.45) = 47.5 \qquad (17\text{-}8)$$

$$E(\text{gain} \,|\, -) = (-50)(.67) + (50)(.3) + (100)(.03) = -15.5 \qquad (17\text{-}9)$$

From (17-8), we see that if the sounding indicates $+$, the prospector should drill, this action giving him an expected gain of $47,500. On the other hand, (17-9) shows that if the sounding yields $-$, the prospector should not drill since if he did, he could expect to lose $15,500, which is clearly worse than not drilling at all.

We must now evaluate the overall expected gain to the

prospector from taking a sounding. Looking back to Fig. 17-4, we see $P(+) = .4$ and $P(-) = .6$. Therefore, if the prospector decides to take a sounding and uses the strategy "Drill if $+$; don't drill if $-$," his overall expected gain from sounding is given by

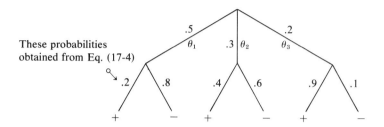

These probabilities obtained from Eq. (17-4)

Figure 17-4 $P(\theta_1, +) = (.5)(.2) = .10; P(\theta_2, +) = (.3)(.4) = .12;$
$P(\theta_3, +) = (.2)(.9) = .18; P(+) = .10 + .12 + .18 = .40;$
$P(\theta_1, -) = (.5)(.8) = .40; P(\theta_2, -) = (.3)(.6) = .18;$
$P(\theta_3, -) = (.2)(.1) = .02; P(-) = .40 + .18 + .02 = .60$

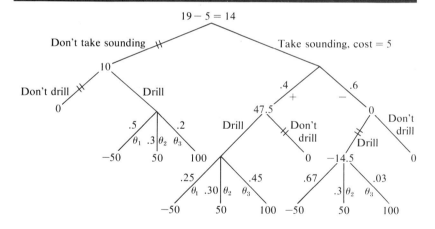

Figure 17-5 Decision tree for oil prospector (gains listed in thousands of dollars). Optimal strategy: Take sounding, at cost $5,000. If sounding yields $+$, drill. If sounding yields $-$, don't drill. Expected gain $= \$14,000.$

$E(\text{gain from sounding}) = (47.5)(.4) + (0)(.6) = 19 \qquad (17\text{-}10)$

We subtract from this the \$5,000 cost of taking a sounding and obtain

$E(\text{net gain from sounding}) = \$19,000 - \$5,000 = \$14,000 \quad (17\text{-}11)$

Since (17-5) showed that the prospector's expected gain without sounding was only \$10,000, we conclude that he should take a sounding and use the optimal strategy:

Take a seismic sounding for \$5,000. Then, drill if the sounding yields + but do not drill if the sounding yields −.

We can display this problem and its solution in a diagram similar to a tree diagram called a *decision tree*. A decision tree lists the various stages and options of the experiment in tree diagram form, with the relevant probabilities and expected gains located at the appropriate nodes and branches. The undesirable actions are denoted by two lines marked through them, making clear the desired path of optimal actions. In Fig. 17-5, we have listed the decision tree for the oil drilling problem.

Example 17-6 is a typical problem in decision theory. We have used the prior probabilities in an essential manner. If no such prior probabilities are available, (in this case, if the prospector had no past records or other prior knowledge of the likelihood of the states), a different approach to the problem would have to be used.

Summary

We have developed the concept of conditional probability and have applied it to problems in decision theory. We have seen that the basic procedure in such problems can be listed as follows:

1 Before taking a sample or trying to obtain information, record what knowledge is available in the form of prior probabilities.

2 Consider all alternatives open to you and assess your expected gains, or costs for each action using the appropriate posterior probabilities.

3 Choose as the optimal strategy the course of action which maximizes your expected gain.

Important terms

Conditional probability

Decision theory

Expected gain

Posterior probability

Prior probability

Additional problems

17-4 A New Year's Eve party-goer who has had one too many is staggering down the street. He suddenly starts lurching sideways back and forth, going one step toward the gutter with probability 1/2 and one step toward a brick wall with probability 1/2. If he starts one step from the gutter and three steps from the wall what is P(he lands in the gutter before he crashes into the wall)?

17-5 In a certain criminal trial, the prosecuter argues that only .000001 of the population fit the description of the criminal given at the scene of the crime and since the suspect on trial fits this description, he is undoubtedly guilty. Suppose the probability of a citizen committing this particular crime is .001. Suppose also that .000001 of the people who commit this crime fit the suspect's description. How should the defense reply to the prosecutor's claim?

17-6 A coin with $P(H) = p$ is tossed 20 times, and $S = 13$ heads are observed.

   a. Given this information, what is the probability heads occurred on the eighth toss; for example, what is $P(H$ on eighth$|S = 13)$?

   b. What is $P(7$ heads occur in last 10 tosses$|S = 13)$?

17-7 Your lottery number is 3,631. The winning number is announced over the radio just as your radio blows a tube; however, you are able to hear that there is a 6 in the winning number. Assuming that the lottery numbers range from 0000 to 9,999, what is P(you win$|$there is a 6 in the winning number)?

17-8 You are looking for a job, and you know that on any given day, independent of other days, *P*(getting a job) = .2. What is *P*(you get a job on the fifteenth day|you don't get a job in the first 10 days)?

17-9   *a.* Make up an experiment with two events *A*, *B* such that $P(A) = 1/3$, $P(B) = 1/4$, and $P(A|B) = 1/2$.

     b. For the experiment in (*a*), what is *P*(*A* and *B*)?

17-10 On a given day the state of the stock market is either up, down, or average. Past experience leads you to believe that $P(\text{up}) = 1/4$, $P(\text{down}) = 1/4$, and $P(\text{average}) = 1/2$. If you invest on a given day, on the next day your stock will climb according to the following probabilities:

$$P(\text{climb}|\text{up}) = \frac{1}{3} \quad P(\text{climb}|\text{down}) = \frac{1}{6} \quad P(\text{climb}|\text{average}) = \frac{1}{4}$$

  a. What is *P*(down|climb)? *P*(up|not climb)?

  b. Suppose you observe whether or not a stock climbed the day after it was bought and you have to guess the state of the market on the day of purchase, winning $1 if you guess correctly and $0 otherwise. What is your best strategy and expected gain if you use this strategy?

17-11 An ice-cream maker has found that the demand for his home-made ice cream depends on the weather. Summer days are classified as θ_1 = hot and θ_2 = moderate, each occurring with about equal frequency. On a given day the ice-cream maker can take actions a_1: prepare a huge amount of ice cream, or a_2: prepare a moderate amount of ice cream, his gains being reflected in the following table (a negative gain is a loss):

	a_1	a_2
θ_1 (hot)	20	-10
θ_2 (moderate)	-15	30

For a cost of $4, he can obtain a weather report which has the following probabilities:

$P(\text{report predicts hot}|\theta_1) = .7$

$P(\text{predicts moderate}|\theta_2) = .6$

What is the ice-cream maker's best strategy?

17-12 You have to decide whether to invest in oil stocks or farm commodities. The market can be in two possible states:

θ_1: favorable to oil stocks

θ_2: favorable to farm commodities

Here is a gains table for your actions:

	a_1 (oil)	a_2 (farm)
θ_1	2	-8
θ_2	-6	1

To help you decide what to do, you observe if the price of gold went up (+) or down (−) on the previous day, knowing the following conditional probabilities:

	$P(+\mid\theta)$	$P(-\mid\theta)$
θ_1	1/3	2/3
θ_2	3/4	1/4

You also have prior knowledge about θ as follows:

	Prior probability
θ_1	1/3
θ_2	2/3

What is your best strategy?

17-13 Acme Nuts and Bolts is an old company that is thinking of replacing its antiquated capital equipment. There are four possible actions the firm can take:

a_1: Replace no equipment.

a_2: Replace some equipment at cost $30,000.

a_3: Replace all equipment with used equipment at cost $50,000.

a_4: Replace all equipment with new equipment at cost $90,000.

The management would like to know the state of the economy

before deciding what to do, and it feels that there are three possible states:

θ_1: Recession

θ_2: Moderate growth

θ_3: Expansionary period with strong demand for products

The prior probabilities for these states are given as follows:

$P(\theta_1) = .2$ $P(\theta_2) = .6$ $P(\theta_3) = .2$

The company also computes the following gains table (in thousands of dollars):

	a_1	a_2	a_3	a_4
θ_1	0	20	20	20
θ_2	0	30	50	50
θ_3	0	30	60	150

Finally, for a cost of $15,000, the company can hire a consulting firm which has access to certain economic indicators. The reliability of these indicators is measured by the following probabilities:

Indicators

z_1: down	$P(z_1\|\theta_1) = .75$	$P(z_1\|\theta_2) = .25$	$P(z_1\|\theta_3) = 0$
z_2: mixed	$P(z_2\|\theta_1) = .25$	$P(z_2\|\theta_2) = .67$	$P(z_2\|\theta_3) = .25$
z_3: up	$P(z_3\|\theta_1) = 0$	$P(z_3\|\theta_2) = .08$	$P(z_3\|\theta_3) = .75$

What should the company do?

17-14 A certain coin is either fair (θ_1) or has $P(\text{heads}) = 1/3$ (θ_2). You toss the coin twice and then have to guess which type the coin is, your gains being determined by the following gains table:

	State	
Action	θ_1	θ_2
Guess θ_1	0	-1
Guess θ_2	-2	1

Suppose you are given the following prior distribution on θ: $P(\theta_1) = 2/5$; $P(\theta_2) = 3/5$. What is the best decision rule and what is your average gain if you use it?

True-false test

Circle the correct letter.

T F   1. If you toss a fair coin repeatedly and observe 20 successive heads, then the chance of tails on the twenty-first toss is greater than 1/2.

T F   2. Conditional probability is symmetric in the sense $P(A|B) = P(B|A)$.

T F   3. If A and B are mutually exclusive, then $P(A|B) = 0$.

T F   4. $P(A$ and B and $C) = P(A)P(B|A)P(C|B$ and $A)$

T F   5. Conditional probability is so named because it can be used only under certain conditions.

T F   6. If $P(A) = 1/2$, then $P(A|A) = 1/2$.

T F   7. If $P(A) = 1/2$ and $P(B) = 1/2$, then $P(A|B)$ is no greater than 1/2.

T F   8. If $P(A|B) = .9$ and $P(A|C) = .9$ and you know that B or C must occur, then $P(A) \geq .9$.

T F   9. If $P(A|B) = .9$ and $P(B|C) = .9$, then $P(A|C) \geq .9$.

T F   10. $P(A|\text{not } B) + P(A|B) = 1$

A method of estimating sensitive parameters

18

An implicit aspect of every statistical procedure considered so far has been the assumption that sample members will avail themselves to any request made by the researcher. Needless to say, in many situations involving personal areas of people's lives this assumption may not be satisfied. It is only natural that one would be reluctant to reveal his innermost secrets to a strange researcher. In this way, the "human element" itself is often the biggest barrier preventing the collection of data about human beings.

A simple case of this dilemma occurs when the researcher wishes to estimate the proportion of people in a population within a certain category, say, category A, but when asked, a person will not readily admit whether or not he or she is in category A. This difficulty might be circumvented in some situations by simply collecting data through secret ballots. Each subject gives a response in a sealed envelope, and a tally is made only after all responses are obtained, thus preserving the anonymity of individual subjects. In this case, the usual estimate, the sample proportion of members in category A, could be used to estimate the corresponding population proportion. However, if responses are collected on a door-to-door basis or in personal interviews, or if the topic of investigation is so sensitive and so personal that some individuals would not reveal the category to which they belong even in a secret ballot (fearing that their ballots might be identified), then the ballot technique cannot be used. For such situations, an alternative technique is available.

The method we present is based on the fact that we need know only the *number* of sample members in category A, not *which members* are in this category. In fact, for an estimate of the population proportion of people in category A, we don't even need to know exactly how many sample members are in this category, but only the approximate number. Let us consider an example.

Suppose that in a large city a researcher wishes to estimate the proportion of married people who have committed adultery, and suppose that 100 married persons are selected at random. There is reason to believe that many sample members would not wish to answer the question "Have you committed adultery?" and therefore the usual procedure based on sample answers to this question could not be used. Instead, the following method is put into action: Each person in the sample is presented with an *unfair* coin having probability 1/4 of heads and 3/4 of tails. If no such coin is available, then a box containing 4 similar slips of paper with 1 of them marked "head" and 3 of them marked "tail" may be used instead. In fact, any mechanism that simulates a coin with probability 1/4 of heads will do. To make the discussion simpler, assume we have such a coin.

An individual is asked to toss the coin once without letting the researcher observe the outcome. A response of + or − is then given to the researcher according to the following rule (Fig. 18-1):

Response rule

Answer to the question "Have you committed adultery?"	Result of coin toss	Response to researcher
Yes	Head	+
No	Head	−
Yes	Tail	−
No	Tail	+

Figure 18-1

For example, if a person answers no to the question and the coin lands heads, then he responds with a −; or if the answer is no and the coin lands tails, the response to the researcher is +. Notice that the researcher receives only the response + or − from each subject; he is not told either the answer to the question or the result of the coin toss. Since a response of + could have resulted from either a yes or a no answer to the question, depending on the outcome of the coin toss, the researcher cannot tell whether a person has answered yes or no just by knowing whether the response is + or −. Accordingly, even if the answer to the question is yes for a sample member, he may confidently participate without fear of being found out. As we shall now explain, based on the sample statistic $S =$ number of + responses, the researcher can make an intelligent estimate of p, the population proportion of married people who have committed adultery.

Notice that the response from any person is independent of the response from any other person. Moreover, since each sample member is selected at random, the probability of a + response is the same for each person (each person uses the same response rule). Thus, we have a binomial experiment, and the variable S has a binomial distribution with parameters $n = 100$ and success probability equal to the probability of a + response. Let us denote by $p+$ the probability of a + response.

Since 100 is a reasonably large sample size, the law of large numbers tells us that the sample proportion of + responses $S/100$ is likely to be close to the probability of a + response $p+$. That is,

$$p+ \approx \frac{S}{100} \tag{18-1}$$

Now, $p+$ and p, the proportion of married people who have committed adultery, are directly related by a simple formula, which we shall derive in a moment. If we substitute the value of $p+$ expressed in terms of p into Eq. (18-1), we may solve the resulting formula for p, thus obtaining an expression for p in terms of $S/100$, at least approximately.

Let us now express $p+$ in terms of p. First, notice that a response of + results from either (yes and head) or (no and tail). Thus

$$p+ = P(\text{yes and head}) + P(\text{no and tail})$$

Next, using the fact that the coin toss is independent of whether a person has committed adultery, this equation becomes

$$p+ = P(\text{yes})P(\text{head}) + P(\text{no})P(\text{tail})$$

$$= p\frac{1}{4} + (1-p)\frac{3}{4}$$

or

$$p+ = \frac{3}{4} - \frac{1}{2}\,p$$

Substituting this expression for $p+$ into (18-1) gives

$$\frac{3}{4} - \frac{1}{2}\,p \approx \frac{S}{100}$$

or equivalently

$$p \approx \frac{3}{2} - 2\!\left(\frac{S}{100}\right)$$

Accordingly, as our estimate of p we take the sample statistic $\hat{p}$ defined by

$$\hat{p} = \frac{3}{2} - 2\!\left(\frac{S}{100}\right)$$

More generally, suppose we wish to estimate the proportion p of people in a certain category, say, category A, within some large population. A sample of size n is selected, and each member responds with a $+$ or $-$ according to the response rule specified in Fig. 18-1, using the question "Are you in category A?" in place of "Have you committed adultery?" The number of $+$ responses S is then tallied. The unknown "sensitive" parameter p is estimated by $\hat{p}$ defined according to the equation

$$\hat{p} = \frac{3}{2} - 2\!\left(\frac{S}{n}\right) \tag{18-3}$$

Example 18-1. In the adultery example if there were 35 $+$ responses and a sample of size 100 had been selected, then $S = 35$, $n = 100$, and the estimate of the proportion of adulterers in the population would be

$$\hat{p} = \frac{3}{2} - 2\!\left(\frac{35}{100}\right) = .8$$

The procedure we have just outlined assumes the researcher possesses an unfair coin (or some equivalent mechanism) having probability of head equal to 1/4. This requirement is not really necessary just as long as the coin is unfair. However, Eq. (18-3) must be modified accordingly when an unfair coin having probability of heads different than 1/4 is used. We leave you to carry out the details of this modification as an exercise.

Desirable properties of the estimate $\hat{p}$

As we have stated, the estimate $\hat{p}$ is valuable because it can be obtained from the responses of people when they do not wish the researcher to know their positions on a sensitive question. It is also a "good" estimate in the following sense:

$\hat{p}$ *is an unbiased estimate of p. That is, on the average* $\hat{p} = p;$
$E(\hat{p}) = p.$ (18-4)

$\hat{p}$ *is a consistent estimate of p. This means that the larger n is, the more likely it is that $\hat{p}$ is close to p.* (18-5)

We shall now verify these two facts.

That $\hat{p}$ is unbiased is easy to check using the techniques of Chap. 14.

$$E(\hat{p}) = E\left(\frac{3}{2} - 2\frac{S}{n}\right) = E\left(\frac{3}{2}\right) - E\left(2\frac{S}{n}\right) = \frac{3}{2} - 2E\left(\frac{S}{n}\right)$$

Now recall $E(S/n) = p+$ since S is binomial with parameters n and $p+$. Using the relation in (18-2), we obtain

$$E(\hat{p}) = \frac{3}{2} - 2E\left(\frac{S}{n}\right) = \frac{3}{2} - 2p+$$

$$= \frac{3}{2} - 2\left[p\left(\frac{1}{4}\right) + (1-p)\left(\frac{3}{4}\right)\right]$$

$$= \frac{3}{2} - 2\left(\frac{3}{4} - \frac{1}{2}p\right) = p$$

as (18-4) states. Property (18-5) follows from the law of large numbers. The sample proportion of $+$ responses S/n gets close to the probability of a $+$ response $p+$ as n gets large. This forces $\hat{p}$ to be close to p, as can be seen by substituting $p+$ in place of S/n in Eq. (18-3) and changing $=$ to $\approx$.

We may easily compute the variance of $\hat{p}$ using the methods discussed in Chap. 15, and find

$$\text{Var } \hat{p} = \frac{3/4 + p(1 - p)}{n}$$

It is interesting to compare the variance of the estimate $\hat{p}$ with the variance of the sample proportion which would be used if answers to the question "Are you in category A?" are given directly. In the latter case, the sample proportion = (number of yes answers)/(sample size) is used to estimate p. Since the number of yes answers is a binomial variable with parameters n and p, the variance of this estimate is $p(1 - p)/n$, as we have shown in Chap. 15. Comparing this value with the variance of the estimate $[3/4 + p(1 - p)]/n$, we see that an estimate with lower variance is obtained when answers are given directly. This is not surprising since the coin toss required in the present method employs an additional element of randomness that is not present when answers are given directly.

CONFIDENCE INTERVALS

The central limit theorem for sample proportions implies that for large n, the estimate $\hat{p}$ is approximately normally distributed with mean p and standard deviation $\sqrt{[3/4 + p(1 - p)]/n}$. This allows us to compute a confidence interval for p using the approach of Chap. 5. For example, it can be shown that an approximate 95% confidence interval for p is given by $\hat{p} \pm 1.96\sqrt{1/n}$. Of course, the width of the confidence interval is smaller as the sample size n is larger.

Example 18-2. The Alcoholics Anonymous chapter in a large city wishes to determine the proportion of members who have had at least one drink during the past week. Based on a random sample of 200 members, 118 + responses were received, using the method described in Fig. 18.1 with the question "Have you had at least one drink in the past week?" What is the estimate $\hat{p}$ for p, the proportion of members who have had at least one drink in the past week? Find a 95% confidence interval for p.

> SOLUTION: In this example $n = 200$ and $S = 118$; hence $\hat{p} = 3/2 - 2(S/n) = 1.50 - 2(118/200) = .32$. Using the formula $\hat{p} \pm 1.96\sqrt{1/n}$, we obtain the 95% confidence interval for p as $.32 \pm 1.96\sqrt{1/200}$, or $.32 \pm .14$. In other words, we may be 95% confident that the true value of p lies between .18 and .46.

Tables

N	$\sqrt{N}$	$\sqrt{10N}$	N	$\sqrt{N}$	$\sqrt{10N}$	N	$\sqrt{N}$	$\sqrt{10N}$
1.00	1.000	3.162	1.50	1.225	3,873	2.00	1.414	4.472
1.01	1.005	3.178	1.51	1.229	3.886	2.01	1.418	4.483
1.02	1.010	3.194	1.52	1.233	3.899	2.02	1.421	4.494
1.03	1.015	3.209	1.53	1.237	3.912	2.03	1.425	4.506
1.04	1.020	3.225	1.54	1.241	3.924	2.04	1.428	4.517
1.05	1.025	3.240	1.55	1.245	3.937	2.05	1.432	4.528
1.06	1.030	3.256	1.56	1.249	3.950	2.06	1.435	4.539
1.07	1.034	3.271	1.57	1.253	3.962	2.07	1.439	4.550
1.08	1.039	3.286	1.58	1.257	3.975	2.08	1.442	4.571
1.09	1.044	3.302	1.59	1.261	3.987	2.09	1.446	4.572
1.10	1.049	3.317	1.60	1.265	4.000	2.10	1.449	4.583
1.11	1.054	3.332	1.61	1.269	4.012	2.11	1.453	4.593
1.12	1.058	3.347	1.62	1.273	4.025	2.12	1.456	4.604
1.13	1.063	3.362	1.63	1.277	4.037	2.13	1.459	4.615
1.14	1.068	3.376	1.64	1.281	4.050	2.14	1.463	4.626
1.15	1.072	3.391	1.65	1.285	4.062	2.15	1.466	4.637
1.16	1.077	3.406	1.66	1.288	4.074	2.16	1.470	4.648
1.17	1.082	3.421	1.67	1.292	4.087	2.17	1.473	4.658
1.18	1.086	3.435	1.68	1.296	4.099	2.18	1.476	4.669
1.19	1.091	3.450	1.69	1.300	4.111	2.19	1.480	4.680
1.20	1.095	3.464	1.70	1.304	4.123	2.20	1.483	4.690
1.21	1.100	3.479	1.71	1.308	4.135	2.21	1.487	4.701
1.22	1.105	3.493	1.72	1.311	4.147	2.22	1.490	4.712
1.23	1.109	3.507	1.73	1.315	4.159	2.23	1.493	4.722
1.24	1.114	3.521	1.74	1.319	4.171	2.24	1.497	4.733
1.25	1.118	3.536	1.75	1.323	4.183	2.25	1.500	4.743
1.26	1.122	3.550	1.76	1.327	4.195	2.26	1.503	4.754
1.27	1.127	3.564	1.77	1.330	4.207	2.27	1.507	4.764
1.28	1.131	3.578	1.78	1.334	4.219	2.28	1.510	4.775
1.29	1.136	3.592	1.79	1.338	4.231	2.29	1.513	4.785
1.30	1.140	3.606	1.80	1.342	4.243	2.30	1.517	4.796
1.31	1.145	3.619	1.81	1.345	4.254	2.31	1.520	4,806
1.32	1.149	3.633	1.82	1.349	4.266	2.32	1.523	4,817
1.33	1.153	3.647	1.83	1.353	4.278	2.33	1.526	4,827
1.34	1.158	3.661	1.84	1.356	4.290	2.34	1.530	4,837
1.35	1.162	3.674	1.85	1.360	4.301	2.35	1.533	4,848
1.36	1.166	3.688	1.86	1.364	4.313	2.36	1.536	4.858
1.37	1.170	3.701	1.87	1.367	4.324	2.37	1.539	4.868
1.38	1.175	3.715	1.88	1.371	4.336	2.38	1.543	4.879
1.39	1.179	3.728	1.89	1.375	4.347	2.39	1.546	4.889
1.40	1.183	3.742	1.90	1.378	4.359	2.40	1.549	4.899
1.41	1.187	3.755	1.91	1.382	4.370	2.41	1.552	4.909
1.42	1.192	3.768	1.92	1.386	4.382	2.42	1.556	4.919
1.43	1.196	3.782	1.93	1.389	4.393	2.43	1.559	4.930
1.44	1.200	3.795	1.94	1.393	4.405	2.44	1.562	4.940
1.45	1.204	3.808	1.95	1.396	4.416	2.45	1.565	4.950
1.46	1.208	3.821	1.96	1.400	4.427	2.46	1.568	4.960
1.47	1.212	3.834	1.97	1.404	4.438	2.47	1.572	4.970
1.48	1.217	3.847	1.98	1.407	4.450	2.48	1.575	4.980
1.49	1.221	3.860	1.99	1.411	4.461	2.49	1.578	4.990

Square roots

N	$\sqrt{N}$	$\sqrt{10N}$	N	$\sqrt{N}$	$\sqrt{10N}$	N	$\sqrt{N}$	$\sqrt{10N}$
2.50	1.581	5.000	3.00	1.732	5.477	3.50	1.871	5.916
2.51	1.584	5.010	3.01	1.735	5.486	3.51	1.873	5.925
2.52	1.587	5.020	3.02	1.738	5.495	3.52	1.876	5.933
2.53	1.591	5.030	3.03	1.741	5.505	3.53	1.879	5.941
2.54	1.594	5.040	3.04	1.744	5.514	3.54	1.881	5.950
2.55	1.597	5.050	3.05	1.746	5.523	3.55	1.884	5.958
2.56	1.600	5.060	3.06	1.749	5.532	3.56	1.887	5.967
2.57	1.603	5.070	3.07	1.752	5.541	3.57	1.889	5.975
2.58	1.606	5.079	3.08	1.755	5.550	3.58	1.892	5.983
2.59	1.609	5.089	3.09	1.758	5.559	3.59	1.895	5.992
2.60	1.612	5.099	3.10	1.761	5.568	3.60	1.897	6.000
2.61	1.616	5.109	3.11	1.764	5.577	3.61	1.900	6.008
2.62	1.619	5.119	3.12	1.766	5.586	3.62	1.903	6.017
2.63	1.622	5.128	3.13	1.769	5.595	3.63	1.905	6.025
2.64	1.625	5.138	3.14	1.772	5.604	3.64	1.908	6.033
2.65	1.628	5.148	3.15	1.775	5.612	3.65	1.910	6.042
2.66	1.631	5.158	3.16	1.778	5.621	3.66	1.913	6.050
2.67	1.634	5.167	3.17	1.780	5.630	3.67	1.916	6.058
2.68	1.637	5.177	3.18	1.783	5.639	3.68	1.918	6.066
2.69	1.640	5.187	3.19	1.786	5.648	3.69	1.921	6.075
2.70	1.643	5.196	3.20	1.789	5.657	3.70	1.924	6.083
2.71	1.646	5.206	3.21	1.792	5.666	3.71	1.926	6.091
2.72	1.649	5.215	3.22	1.794	5.675	3.72	1.929	6.099
2.73	1.652	5.225	3.23	1.797	5.683	3.73	1.931	6.107
2.74	1.655	5.234	3.24	1.800	5.692	3.74	1.934	6.116
2.75	1.658	5.244	3.25	1.803	5.701	3.75	1.936	6.124
2.76	1.661	5.254	3.26	1.806	5.710	3.76	1.939	6.132
2.77	1.664	5.263	3.27	1.808	5.718	3.77	1.942	6.140
2.78	1.667	5.273	3.28	1.811	5.727	3.78	1.944	6.148
2.79	1.670	5.282	3.29	1.814	5.736	3.79	1.947	6.156
2.80	1.673	5.292	3.30	1.817	5.745	3.80	1.949	6.164
2.81	1.676	5.301	3.31	1.819	5.753	3.81	1.852	6.173
2.82	1.679	5.310	3.32	1.822	5.762	3.82	1.954	6.181
2.83	1.682	5.320	3.33	1.825	5.771	3.83	1.957	6.189
2.84	1.685	5.329	3.34	1.828	5.779	3.84	1.960	6.197
2.85	1.688	5.339	3.35	1.830	5.788	3.85	1.962	6.205
2.86	1.691	5.348	3.36	1.833	5.797	3.86	1.965	6.213
2.87	1.694	5.357	3.37	1.836	5.805	3.87	1.967	6.221
2.88	1.697	5.367	3.38	1.838	5.814	3.88	1.970	6.229
2.89	1.700	5.376	3.39	1.841	5.822	3.89	1.972	6.237
2.90	1.703	5.385	3.40	1.844	5.831	3.90	1.975	6.245
2.91	1.706	5.394	3.41	1.847	5.840	3.91	1.977	6.253
2.92	1.709	5.404	3.42	1.849	5.848	3.92	1.980	6.261
2.93	1.712	5.413	3.43	1.852	5.857	3.93	1.982	6.269
2.94	1.715	5.422	3.44	1.855	5.865	3.94	1.985	6.277
2.95	1.718	5.431	3.45	1.857	5.874	3.95	1.987	6.285
2.96	1.720	5.441	3.46	1.860	5.882	3.96	1.990	6.293
2.97	1.723	5.450	3.47	1.863	5.891	3.97	1.992	6.301
2.98	1.726	5.459	3.48	1.865	5.899	3.98	1.995	6.309
2.99	1.729	5.468	3.49	1.868	5.908	3.99	1.997	6.317

N	$\sqrt{N}$	$\sqrt{10N}$	N	$\sqrt{N}$	$\sqrt{10N}$	N	$\sqrt{N}$	$\sqrt{10N}$
4.00	2.000	6.325	4.50	2.121	6.708	5.00	2.236	7.071
4.01	2.002	6.332	4.51	2.124	6.716	5.01	2.238	7.078
4.02	2.005	6.340	4.52	2.126	6.723	5.02	2.241	7.085
4.03	2.007	6.348	4.53	2.128	6.731	5.03	2.243	7.092
4.04	2.010	6.356	4.54	2.131	6.738	5.04	2.245	7.099
4.05	2.012	6.364	4.55	2.133	6.745	5.05	2.247	7.106
4.06	2.015	6.372	4.56	2.135	6.753	5.06	2.249	7.113
4.07	2.017	6.380	4.57	2.138	6.760	5.07	2.252	7.120
4.08	2.020	6.387	4.58	2.140	6.768	5.08	2.254	7.127
4.09	2.022	6.395	4.59	2.142	6.775	5.09	2.256	7.134
4.10	2.025	6.403	4.60	2.145	6.782	5.10	2.258	7.141
4.11	2.027	6.411	4.61	2.147	6.790	5.11	2.261	7.148
4.12	2.030	6.419	4.62	2.149	6.797	5.12	2.263	7.155
4.13	2.032	6.427	4.63	2.152	6.804	5.13	2.265	7.162
4.14	2.035	6.434	4.64	2.154	6.812	5.14	2.267	7.169
4.15	2.037	6.442	4.65	2.156	6.819	5.15	2.269	7.176
4.16	2.040	6.450	4.66	2.159	6.826	5.16	2.272	7.183
4.17	2.042	6.458	4.67	2.161	6.834	5.17	2.274	7.190
4.18	2.045	6.465	4.68	2.163	6.841	5.18	2.276	7.197
4.19	2.047	6.473	4.69	2.166	6.848	5.19	2.278	7.204
4.20	2.049	6.481	4.70	2.168	6.856	5.20	2.280	7.211
4.21	2.052	6.488	4.71	2.170	6.863	5.21	2.280	7.218
4.22	2.054	6.496	4.72	2.173	6.870	5.22	2.285	7.225
4.23	2.057	6.504	4.73	2.175	6.877	5.23	2.287	7.232
4.24	2.059	6.512	4.74	2.177	6.885	5.24	2.289	7.239
4.25	2.062	6.519	4.75	2.179	6.892	5.25	2.291	7.246
4.26	2.064	6.527	4.76	2.182	6.899	5.26	2.293	7.253
4.27	2.066	6.535	4.77	2.184	6.907	5.27	2.296	7.259
4.28	2.069	6.542	4.78	2.186	6.914	5.28	2.298	7.266
4.29	2.071	6.550	4.79	2.189	6.921	5.29	2.300	7.273
4.30	2.074	6.557	4.80	2.191	6.928	5.30	2.302	7.280
4.31	2.076	6.565	4.81	2.193	6.935	5.31	2.304	7.287
4.32	2.078	6.573	4.82	2.195	6.943	5.32	2.307	7.294
4.33	2.081	6.580	4.83	2.198	6.950	5.33	2.309	7.301
4.34	2.083	6.588	4.84	2.200	6.957	5.34	2.311	7.308
4.35	2.086	6.595	4.85	2.202	6.964	5.35	2.313	7.314
4.36	2.088	6.603	4.86	2.205	6.971	5.36	2.315	7.321
4.37	2.090	6.611	4.87	2.207	6.979	5.37	2.317	7.328
4.38	2.093	6.618	4.88	2.209	6.986	5.38	2.319	7.335
4.39	2.095	6.626	4.89	2.211	6.993	5.39	2.322	7.342
4.40	2.098	6.633	4.90	2.214	7.000	5.40	2.324	7.348
4.41	2.100	6.641	4.91	2.216	7.007	5.41	2.326	7.355
4.42	2.102	6.648	4.92	2.218	7.014	5.42	2.328	7.632
4.43	2.105	6.656	4.93	2.220	7.021	5.43	2.330	7.369
4.44	2.107	6.663	4.94	2.223	7.029	5.44	2.332	7.376
4.45	2.110	6.671	4.95	2.225	7.036	5.45	2.335	7.382
4.46	2.112	6.678	4.96	2.227	7.043	5.46	2.337	7.389
4.47	2.114	6.686	4.97	2.229	7.050	5.47	2.339	7.396
4.48	2.117	6.693	4.98	2.232	7.057	5.48	2.341	7.403
4.49	2.119	6.701	4.99	2.234	7.064	5.49	2.343	7.409

Square roots (continued)

N	$\sqrt{N}$	$\sqrt{10N}$	N	$\sqrt{N}$	$\sqrt{10N}$	N	$\sqrt{N}$	$\sqrt{10N}$
5.50	2.345	7.416	6.00	2.449	7.746	6.50	2.550	8.062
5.51	2.347	7.423	6.01	2.452	7.752	6.51	2.551	8.068
5.52	2.349	7.430	6.02	2.454	7.759	6.52	2.553	8.075
5.53	2.352	7.436	6.03	2.456	7.765	6.53	2.555	8.081
5.54	2.354	7.443	6.04	2.458	7.772	6.54	2.557	8.087
5.55	2.356	7.450	6.05	2.460	7.778	6.55	2.559	8.093
5.56	2.358	7.457	6.06	2.462	7.785	6.56	2.561	8.099
5.57	2.360	7.463	6.07	2.464	7.791	6.57	2.563	8.106
5.58	2.362	7.470	6.08	2.466	7.797	6.58	2.565	8.112
5.59	2.364	7.477	6.09	2.468	7.804	6.59	2.567	8.118
5.60	2.366	7.483	6.10	2.470	7.810	6.60	2.569	8.124
5.61	2.369	7.490	6.11	2.472	7.817	6.61	2.571	8.130
5.62	2.371	7.497	6.12	2.474	7.823	6.62	2.573	8.136
5.63	2.373	7.503	6.13	2.476	7.829	6.63	2.575	8.142
5.64	2.375	7.510	6.14	2.478	7.836	6.64	2.577	8.149
5.65	2.377	7.517	6.15	2.480	7.842	6.65	2.579	8.155
5.66	2.379	7.523	6.16	2.482	7.849	6.66	2.581	8.161
5.67	2.381	7.530	6.17	2.484	7.855	6.67	2.583	8.167
5.68	2.383	7.537	6.18	2.486	7.861	6.68	2.585	8.173
5.69	2.385	7.543	6.19	2.488	7.868	6.69	2.587	8.179
5.70	2.387	7.550	6.20	2.490	7.874	6.70	2.588	8.185
5.71	2.390	7.556	6.21	2.492	7.880	6.71	2.590	8.191
5.72	2.392	7.563	6.22	2.494	7.887	6.72	2.592	8.198
5.73	2.394	7.570	6.23	2.496	7.893	6.73	2.594	8.204
5.74	2.396	7.576	6.24	2.498	7.899	6.74	2.596	8.210
5.75	2.398	7.583	6.25	2.500	7.906	6.75	2.598	8.216
5.76	2.400	7.589	6.26	2.502	7.912	6.76	2.600	8.222
5.77	2.402	7.596	6.27	2.504	7.918	6.77	2.602	8.228
5.78	2.404	7.603	6.28	2.506	7.925	6.78	2.604	8.234
5.79	2.406	7.609	6.29	2.508	7.931	6.79	2.606	8.240
5.80	2.408	7.616	6.30	2.510	7.937	6.80	2.608	8.246
5.81	2.410	7.622	6.31	2.512	7.944	6.81	2.610	8.252
5.82	2.412	7.629	6.32	2.514	7.950	6.82	2.612	8.258
5.83	2.415	7.635	6.33	2.516	7.956	6.83	2.613	8.264
5.84	2.417	7.642	6.34	2.518	7.962	6.84	2.615	8.270
5.85	2.419	7.649	6.35	2.520	7.969	6.85	2.617	8.276
5.86	2.421	7.655	6.36	2.522	7.975	6.86	2.619	8.283
5.87	2.423	7.662	6.37	2.524	7.981	6.87	2.621	8.289
5.88	2.425	7.668	6.38	2.526	7.987	6.88	2.623	8.295
5.89	2.427	7.675	6.39	2.528	7.994	6.89	2.625	8.301
5.90	2.429	7.681	6.40	2.530	8.000	6.90	2.627	8.307
5.91	2.431	7.688	6.41	2.532	8.006	6.91	2.629	8.313
5.92	2.433	7.694	6.42	2.534	8.012	6.92	2.631	8.319
5.93	2.435	7.701	6.43	2.536	8.019	6.93	2.632	8.325
5.94	2.437	7.707	6.44	2.538	8.025	6.94	2.634	8.331
5.95	2.439	7.714	6.45	2.540	8.031	6.95	2.636	8.337
5.96	2.441	7.720	6.46	2.542	8.037	6.96	2.638	8.343
5.97	2.443	7.727	6.47	2.544	8.044	6.97	2.640	8.349
5.98	2.445	7.733	6.48	2.546	8.50	6.98	2.642	8.355
5.99	2.447	7.740	6.49	2.548	8.056	6.99	2.644	8.361

N	$\sqrt{N}$	$\sqrt{10N}$	N	$\sqrt{N}$	$\sqrt{10N}$	N	$\sqrt{N}$	$\sqrt{10N}$
7.00	2.646	8.367	7.50	2.739	8.660	8.00	2.828	8.944
7.01	2.648	8.373	7.51	2.740	8.666	8.01	2.830	8.950
7.02	2.650	8.379	7.52	2.742	8.672	8.02	2.832	8.955
7.03	2.651	8.385	7.53	2.744	8.678	8.03	2.834	8.961
7.04	2.653	8.390	7.54	2.746	8.683	8.04	2.835	8.967
7.05	2.655	8.396	7.55	2.748	8.689	8.05	2.837	8.972
7.06	2.657	8.402	7.56	2.750	8.695	8.06	2.839	8.978
7.07	2.659	8.408	7.57	2.751	8.701	8.07	2.841	8.983
7.08	2.661	8.414	7.58	2.753	8.706	8.08	2.843	8.989
7.09	2.663	8.420	7.59	2.755	8.712	8.09	2.844	8.994
7.10	2.665	8.426	7.60	2.757	8.718	8.10	2.846	9.000
7.11	2.666	8.432	7.61	2.759	8.724	8.11	2.848	9.006
7.12	2.668	8.438	7.62	2.760	8.729	8.12	2.850	9.011
7.13	2.670	8.444	7.63	2.762	8.735	8.13	2.851	9.017
7.14	2.672	8.450	7.64	2.764	8.741	8.14	2.853	9.022
7.15	2.674	8.456	7.65	2.766	8.746	8.15	2.855	9.028
7.16	2.676	8.462	7.66	2.768	8.752	8.16	2.857	9.033
7.17	2.678	8.468	7.67	2.769	8.758	8.17	2.858	9.039
7.18	2.680	8.473	7.68	2.771	8.764	8.18	2.860	9.044
7.19	2.681	8.479	7.69	2.773	8.769	8.19	2.862	9.050
7.20	2.683	8.485	7.70	2.775	8.775	8.20	2.864	9.055
7.21	2.685	8.491	7.71	2.777	8.781	8.21	2.865	9.061
7.22	2.687	8.497	7.72	2.778	8.786	8.22	2.867	9.066
7.23	2.689	8.503	7.73	2.780	8.792	8.23	2.869	9.072
7.24	2.691	8.509	7.74	2.782	8.798	8.24	2.871	9.077
7.25	2.693	8.515	7.75	2.784	8.803	8.25	2.872	9.083
7.26	2.694	8.521	7.76	2.786	8.809	8.26	2.874	9.088
7.27	2.696	8.526	7.77	2.787	8.815	8.27	2.876	9.094
7.28	2.698	8.532	7.78	2.789	8.820	8.28	2.877	9.099
7.29	2.700	8.538	7.79	2.791	8.826	8.29	2.879	9.105
7.30	2.702	8.544	7.80	2.793	8.832	8.30	2.881	9.110
7.31	2.704	8.550	7.81	2.795	8.837	8.31	2.883	9.116
7.32	2.706	8.556	7.82	2.796	8.843	8.32	2.884	9.121
7.33	2.707	8.562	7.83	2.798	8.849	8.33	2.886	9.127
7.34	2.709	8.567	7.84	2.800	8.854	8.34	2.888	9.132
7.35	2.711	8.573	7.85	2.802	8.860	8.35	2.890	9.138
7.36	2.713	8.579	7.86	2.804	8.866	8.36	2.891	9.143
7.37	2.715	8.585	7.87	2.805	8.871	8.37	2.893	9.149
7.38	2.717	8.591	7.88	2.807	8.877	8.38	2.895	9.154
7.39	2.718	8.597	7.89	2.809	8.883	8.39	2.897	9.160
7.40	2.720	8.602	7.90	2.811	8.888	8.40	2.898	9.165
7.41	2.722	8.608	7.91	2.812	8.894	8.41	2.900	9.171
7.42	2.724	8.614	7.92	2.814	8.899	8.42	2.902	9.176
7.43	2.726	8.620	7.93	2.816	8.905	8.43	2.903	9.182
7.44	2.728	8.626	7.94	2.818	8.911	8.44	2.905	9.187
7.45	2.729	8.631	7.95	2.820	8.916	8.45	2.907	9.192
7.46	2.731	8.637	7.96	2.821	8.922	8.46	2.909	9.198
7.47	2.733	8.643	7.97	2.823	8.927	8.47	2.910	9.203
7.48	2.735	8.649	7.98	2.825	8.933	8.48	2.912	9.209
7.49	2.737	8.654	7.99	2.827	8.939	8.49	2.914	9.214

Square roots (continued)

N	$\sqrt{N}$	$\sqrt{10N}$	N	$\sqrt{N}$	$\sqrt{10N}$	N	$\sqrt{N}$	$\sqrt{10N}$
8.50	2.915	9.220	9.00	3.000	9.480	9.50	3.082	9.747
8.51	2.917	9.225	9.01	3.002	9.492	9.51	3.084	9.752
8.52	2.919	9.230	9.02	3.003	9.497	9.52	3.085	9.757
8.53	2.921	9.236	9.03	3.005	9.503	9.53	3.087	9.762
8.54	2.922	9.241	9.04	3.007	9.508	9.54	3.089	9.767
8.55	2.924	9.247	9.05	3.008	9.513	9.55	3.090	9.772
8.56	2.926	9.252	9.06	3.010	9.518	9.56	3.092	9.778
8.57	2.927	9.257	9.07	3.012	9.524	9.57	3.094	9.783
8.58	2.929	9.263	9.08	3.013	9.529	9.58	3.095	9.788
8.59	2.931	9.268	9.09	3.015	9.534	9.59	3.097	9.793
8.60	2.933	9.274	9.10	3.017	9.539	9.60	3.098	9.798
8.61	2.934	9.279	9.11	3.017	9.545	9.61	3.100	9.803
8.62	2.936	9.284	9.12	3.020	9.550	9.62	3.102	9.808
8.63	2.938	9.290	9.13	3.022	9.555	9.63	3.103	9.813
8.64	2.939	9.295	9.14	3.023	9.560	9.64	3.105	9.818
8.65	2.941	9.301	9.15	3.025	9.566	9.65	3.106	9.823
8.66	2.943	9.306	9.16	3.027	9.571	9.66	3.108	9.829
8.67	2.944	9.311	9.17	3.028	9.576	9.67	3.110	9.834
8.68	2.946	9.317	9.18	3.030	9.581	9.68	3.111	9.839
8.69	2.948	9.322	9.19	3.031	9.586	9.69	3.113	9.844
8.70	2.950	9.327	9.20	3.033	9.592	9.70	3.114	9.849
8.71	2.951	9.333	9.21	3.035	9.597	9.71	3.116	9.854
8.72	2.953	9.338	9.22	3.036	9.602	9.72	3.118	9.859
8.73	2.955	9.343	9.23	3.038	9.607	9.73	3.119	9.864
8.74	2.956	9.349	9.24	3.040	9.612	9.74	3.121	9.869
8.75	2.958	9.354	9.25	3.041	9.618	9.75	3.122	9.874
8.76	2.960	9.359	9.26	3.043	9.623	9.76	3.124	9.879
8.77	2.961	9.365	9.27	3.045	9.628	9.77	3.126	9.884
8.78	2.963	9.370	9.28	3.046	9.633	9.78	3.127	9.889
8.79	2.965	9.375	9.29	3.048	9.638	9.79	3.129	9.894
8.80	2.966	9.381	9.30	3.050	9.644	9.80	3.130	9.899
8.81	2.968	9.386	9.31	3.051	9.649	9.81	3.132	9.905
8.82	2.970	9.391	9.32	3.053	9.654	9.82	3.134	9.910
8.83	2.972	9.397	9.33	3.055	9.659	9.83	3.135	9.915
8.84	2.973	9.402	9.34	3.056	9.664	9.84	3.137	9.920
8.85	2.975	9.407	9.35	3.058	9.670	9.85	3.138	9.925
8.86	2.977	9.413	9.36	3.059	9.675	9.86	3.140	9.930
8.87	2.978	9.418	9.37	3.061	9.680	9.87	3.142	9.935
8.88	2.980	9.423	9.38	3.063	9.685	9.88	3.143	9.940
8.89	2.982	9.429	9.39	3.064	9.690	9.89	3.145	9.945
8.90	2.983	9.434	9.40	3.066	9.695	9.90	3.146	9.950
8.91	2.985	9.439	9.41	3.068	9.701	9.91	3.148	9.955
8.92	2.987	9.445	9.42	3.069	9.706	9.92	3.150	9.960
8.93	2.988	9.450	9.43	3.071	9.711	9.93	3.151	9.965
8.94	2.990	9.455	9.44	3.072	9.716	9.94	3.153	9.970
8.95	2.992	9.460	9.45	3.074	9.721	9.95	3.154	9.975
8.96	2.993	9.466	9.46	3.076	9.726	9.96	3.156	9.980
8.97	2.995	9.471	9.47	3.077	9.731	9.97	3.158	9.985
8.98	2.997	9.476	9.48	3.079	9.737	9.98	3.158	9.990
8.99	2.998	9.482	9.49	3.081	9.742	9.99	3.161	9.995

Binomial distribution

The table gives cumulative binomial probabilities $P(S \leq k)$ for a variable S having a binomial distribution with parameters n and p. For example, if $n = 10$ and $p = .40$ then $P(S \leq 3) = .3823$, or if $n = 13$ and $p = .85$ then $P(S \leq 10) = .2704$. Other probabilities of interest may be obtained according to the formulas

$$P(S > k) = 1 - P(S \leq k)$$ $P(S \geq k) = 1 - P(S \leq k-1)$

$$P(S = k) = P(S \leq k) - P(S \leq k - 1)$$

n	k	p = .05	.10	.15	.20	.25	.30	.35	.40	.45
1	0	.9500	.9000	.8500	.8000	.7500	.7000	.6500	.6000	.5500
	1	1.0000	1.0000	1.0000	1.0000	1.0000	1.0000	1.0000	1.0000	1.0000
2	0	.9025	.8100	.7225	.6400	.5625	.4900	.4225	.3600	.3025
	1	.9975	.9900	.9775	.9600	.9375	.9100	.8775	.8400	.7975
	2	1.0000	1.0000	1.0000	1.0000	1.0000	1.0000	1.0000	1.0000	1.0000
3	0	.8574	.7290	.6141	.5129	.4219	.3439	.2746	.2160	.1664
	1	.9928	.9720	.9392	.8960	.8438	.7840	.7182	.6480	.5748
	2	.9999	.9990	.9966	.9929	.9844	.9730	.9571	.9360	.9089
	3	1.0000	1.0000	1.0000	1.0000	1.0000	1.0000	1.0000	1.0000	1.0000
4	0	.8145	.6561	.5200	.4096	.3164	.2401	.1785	.1296	.0915
	1	.9860	.9477	.8905	.8192	.7383	.6517	.5630	.4752	.3910
	2	.9995	.9963	.9880	.9728	.9492	.9163	.8735	.8208	.7585
	3	1.0000	.9999	.9995	.9984	.9961	.9919	.9850	.9743	.9590
	4	1.0000	1.0000	1.0000	1.0000	1.0000	1.0000	1.0000	1.0000	1.0000
5	0	.7738	.5905	.4437	.3277	.2373	.1681	.1160	.0778	.0503
	1	.9774	.9185	.8352	.7373	.6328	.5282	.4284	.3370	.2562
	2	.9988	.9924	.9734	.9421	.8965	.8369	.7648	.6826	.5931
	3	1.0000	.9995	.9978	.9933	.9844	.9692	.9460	.9130	.8688
	4	1.0000	1.0000	.9999	.9997	.9990	.9976	.9947	.9898	.9815
	5	1.0000	1.0000	1.0000	1.0000	1.0000	1.0000	1.0000	1.0000	1.0000
6	0	.7351	.5314	.3771	.2621	.1780	.1176	.0754	.0467	.0277
	1	.9672	.8857	.7765	.6554	.5339	.4202	.3191	.2333	.1636
	2	.9978	.9842	.9527	.9011	.8306	.7443	.6471	.5443	.4415
	3	.9999	.9987	.9941	.9830	.9624	.9295	.8826	.8208	.7447
	4	1.0000	.9999	.9996	.9984	.9954	.9891	.9777	.9590	.9308
	5	1.0000	1.0000	1.0000	.9999	.9998	.9993	.9982	.9959	.9917
	6	1.0000	1.0000	1.0000	1.0000	1.0000	1.0000	1.0000	1.0000	1.0000
7	0	.6983	.4783	.3206	.2097	.1335	.0824	.0490	.0280	.0152
	1	.9556	.8503	.7166	.5767	.4449	.3294	.2338	.1586	.1024
	2	.9962	.9743	.9262	.8520	.7564	.6471	.5323	.4199	.3164
	3	.9998	.9973	.9879	.9667	.9294	.8740	.8002	.7102	.6083
	4	1.0000	.9998	.9988	.9953	.9871	.9812	.9444	.9037	.8643
	5	1.0000	1.0000	.9999	.9996	.9987	.9962	.9910	.9812	.9643
	6	1.0000	1.0000	1.0000	1.0000	.9999	.9998	.9994	.9984	.9963
	7	1.0000	1.0000	1.0000	1.0000	1.0000	1.0000	1.0000	1.0000	1.0000

Binomial distribution

n	k	p = .50	.55	.60	.65	.70	.75	.80	.85	.90	.95
1	0	.5000	.4500	.4000	.3500	.3000	.2500	.2000	.1500	.1000	.0500
	1	1.0000	1.0000	1.0000	1.0000	1.0000	1.0000	1.0000	1.0000	1.0000	1.0000
2	0	.2500	.2025	.1600	.1225	.0900	.0625	.0400	.0225	.0100	.0025
	1	.7500	.6975	.6400	.5775	.5100	.4373	.3600	.2775	.1900	.0975
	2	1.0000	1.0000	1.0000	1.0000	1.0000	1.0000	1.0000	1.0000	1.0000	1.0000
3	0	.1250	.0911	.0640	.0429	.0270	.0156	.0080	.0034	.0010	.0001
	1	.5000	.4252	.3520	.2818	.2160	.1562	.1040	.0608	.0280	.0072
	2	.8750	.8336	.7840	.7254	.6570	.5781	.4880	.3959	.2710	.1426
	3	1.0000	1.0000	1.0000	1.0000	1.0000	1.0000	1.0000	1.0000	1.0000	1.0000
4	0	.0625	.0410	.0256	.0150	.0081	.0039	.0016	.0005	.0001	.0000
	1	.3125	.2415	.1792	.1265	.0837	.0508	.0272	.0120	.0037	.0005
	2	.6875	.6090	.5248	.4370	.3483	.2617	.1808	.1095	.0523	.0140
	3	.9375	.9085	.8705	.8215	.7599	.6836	.5904	.4780	.3439	.1855
	4	1.0000	1.0000	1.0000	1.0000	1.0000	1.0000	1.0000	1.0000	1.0000	1.0000
5	0	.0312	.0185	.0102	.0053	.0024	.0010	.0003	.0001	.0000	.0000
	1	.1875	1312	.0870	.0540	.0308	.0156	.0067	.0022	.0005	.0000
	2	.5000	.4069	.3174	.2352	.1631	.1035	.0579	.0266	.0086	.0012
	3	.8125	.7438	.6630	.5716	.4718	.3672	.2627	.1648	.0815	.0226
	4	.9688	.9497	.9222	.8840	.8319	.7627	.6723	.5563	.4095	.2262
	5	1.0000	1.0000	1.0000	1.0000	1.0000	1.0000	1.0000	1.0000	1.0000	1.0000
6	0	.0156	.0083	.0041	.0018	.0007	.0002	.0001	.0000	.0000	.0000
	1	.1094	.0692	.0410	.0223	.0109	.0046	.0016	.0004	.0001	.0000
	2	.3438	.2553	.1792	.1174	.0705	.0376	.0170	.0059	.0013	.0001
	3	.6562	.5585	.4557	.3529	.2557	.1694	.0989	.0473	.0158	.0022
	4	.8906	.8364	.7667	.6809	.5798	.4661	.3446	.2235	.1143	.0328
	5	.9844	.9723	.9533	.9246	.8824	.8220	.7379	.6229	.4686	.2649
	6	1.0000	1.0000	1.0000	1.0000	1.0000	1.0000	1.0000	1.0000	1.0000	1.0000
7	0	.0078	.0037	.0016	.0006	.0002	.0001	.0000	.0000	.0000	.0000
	1	.0625	.0357	.0188	.0090	.0038	.0013	.0004	.0001	.0000	.0000
	2	.2266	.1529	.0963	.0556	.0288	.0129	.0047	.0012	.0002	.0000
	3	.5000	.3917	.2898	.1998	.1260	.0706	.0333	.0121	.0027	.0002
	4	.7734	.6836	.5801	.4677	.3529	.2436	.1480	.0738	.0257	.0038
	5	.9375	.8976	.8414	.7662	.6706	.5551	.4233	.2834	.1497	.0444
	6	.9922	.9848	.9720	.9510	.9176	.8665	.7903	.6794	.5217	.3917
	7	1.0000	1.0000	1.0000	1.0000	1.0000	1.0000	1.0000	1.0000	1.0000	1.0000

n	k	p = .05	.10	.15	.20	.25	.30	.35	.40	.45
8	0	.6634	.4305	.2725	.1678	.1001	.0576	.0319	.0168	.0084
	1	.9428	.8131	.6572	.5033	.3671	.2553	.1691	.1064	.0632
	2	.9942	.9619	.8948	.7969	.6785	.5518	.4278	.3154	.2201
	3	.9996	.9950	.9786	.9437	.8862	.8059	.7064	.5941	.4770
	4	1.0000	.9996	.9971	.9896	.9727	.9420	.8939	.8263	.7396
	5	1.0000	1.0000	.9998	.9988	.9958	.9887	.9747	.9502	.9115
	6	1.0000	1.0000	1.0000	.9999	.9996	.9987	.9964	.9915	.9819
	7	1.0000	1.0000	1.0000	1.0000	1.0000	.9999	.9988	.9993	.9983
	8	1.0000	1.0000	1.0000	1.0000	1.0000	1.0000	1.0000	1.0000	1.0000
9	0	.6302	.3874	.2316	.1342	.0751	.0404	.0207	.0101	.0046
	1	.9288	.7748	.5995	.4362	.3003	.1960	.1211	.0705	.0385
	2	.9916	.9470	.8591	.7382	.6007	.4628	.3373	.2318	.1495
	3	.9994	.9917	.9661	.9144	.8343	.7297	.6089	.4826	.3614
	4	1.0000	.9991	.9944	.9804	.9511	.9012	.8283	.7334	.6214
	5	1.0000	.9999	.9994	.9969	.9900	.9747	.9464	.9006	.9342
	6	1.0000	1.0000	1.0000	.9997	.9987	.9957	.9888	.9750	.9502
	7	1.0000	1.0000	1.0000	1.0000	.9999	.9996	.9986	.9962	.9909
	8	1.0000	1.0000	1.0000	1.0000	1.0000	1.0000	.9999	.9997	.9992
	9	1.0000	1.0000	1.0000	1.0000	1.0000	1.0000	1.0000	1.0000	1.0000
10	0	.5987	.3487	.1969	.1074	.0563	.0282	.0135	.0060	.0025
	1	.9139	.7361	.5443	.3758	.2440	.1493	.0860	.0464	.0233
	2	.9885	.9298	.8202	.6778	.5256	.3828	.2616	.1673	.0996
	3	.9990	.9872	.9500	.8791	.7759	.6496	.5138	.3823	.2660
	4	.9999	.9984	.9901	.9672	.9219	.8497	.7515	.6331	.5044
	5	1.0000	.9999	.9986	.9936	.9803	.9527	.9051	.8338	.7384
	6	1.0000	1.0000	.9999	.9991	.9965	.9894	.9740	.9452	.8980
	7	1.0000	1.0000	1.0000	.9999	.9996	.9984	.9952	.9877	.9726
	8	1.0000	1.0000	1.0000	1.0000	1.0000	.9999	.9995	.9983	.9955
	9	1.0000	1.0000	1.0000	1.0000	1.0000	1.0000	1.0000	.9999	.9997
	10	1.0000	1.0000	1.0000	1.0000	1.0000	1.0000	1.0000	1.0000	1.0000
11	0	.5688	.3138	.1673	.0859	.0422	.0198	.0088	.0036	.0014
	1	.8981	.6974	.4922	.3221	.1971	.1130	.0606	.0302	.0139
	2	.9848	.9104	.7788	.6174	.4552	.3127	.2001	.1189	.0652
	3	.9984	.9815	.9306	.8389	.7133	.5696	.4256	.2963	.1911
	4	.9999	.9972	.9841	.9496	.8854	.7897	.6683	.5328	.3971
	5	1.0000	.9997	.9973	.9883	.9657	.9218	.8513	.7535	.6331
	6	1.0000	1.0000	.9997	.9980	.9924	.9784	.9499	.9006	.8262
	7	1.0000	1.0000	1.0000	.9998	.9988	.9957	.9878	.9707	.9390
	8	1.0000	1.0000	1.0000	1.0000	.9999	.9994	.9980	.9941	.9852
	9	1.0000	1.0000	1.0000	1.0000	1.0000	1.0000	.9998	.9993	.9978
	10	1.0000	1.0000	1.0000	1.0000	1.0000	1.0000	1.0000	1.0000	.9998
	11	1.0000	1.0000	1.0000	1.0000	1.0000	1.0000	1.0000	1.0000	1.0000

Binomial distribution (continued)

n	k	p = .50	.55	.60	.65	.70	.75	.80	.85	.90	.95
8	0	.0030	.0017	.0007	.9992	.0001	.0000	.0000	.0000	.0000	.0000
	1	.0352	.0181	.0085	.0086	.0013	.0004	.0001	.0000	.0000	.0000
	2	.1445	.0885	.0498	.0253	.0113	.0042	.0012	.0002	.0000	.0000
	3	.3633	.2604	.1737	.1061	.0580	.0273	.0104	.0029	.0004	.0000
	4	.6367	.5230	.4059	.2936	.1941	.1138	.0563	.0214	.0050	.0004
	5	.8555	.7799	.6846	.5722	.4482	.3215	.2031	.1052	.0381	.0058
	6	.9648	.9368	.8936	.8309	.7447	.6329	.4967	.3428	.1869	.0572
	7	.9961	.9916	.9832	.9681	.9424	.8999	.8322	.7275	.5695	.3366
	8	1.0000	1.0000	1.0000	1.0000	1.0000	1.0000	1.0000	1.0000	1.0000	1.0000
9	0	.0020	.0008	.0003	.0001	.0000	.0000	.0000	.0000	.0000	.0000
	1	.0195	.0091	.0038	.0014	.0004	.0001	.0000	.0000	.0000	.0000
	2	.0898	.0498	.0250	.0112	.0043	.0013	.0003	.0000	.0000	.0000
	3	.2539	.1658	.0994	.0536	.0253	.0100	.0031	.0006	.0001	.0000
	4	.5000	.3786	.2666	.1717	.0988	.0489	.0196	.0056	.0009	.0000
	5	.7461	.6386	.5174	.3911	.2703	.1657	.0856	.0339	.0083	.0006
	6	.9102	.8505	.7682	.6627	.5372	.3993	.2618	.1409	.0530	.0084
	7	.9805	.9615	.9295	.8789	.8040	.6997	.5638	.4005	.2252	.0712
	8	.9980	.9954	.9899	.9793	.9596	.9249	.8658	.7684	.6126	.3698
	9	1.0000	1.0000	1.0000	1.0000	1.0000	1.0000	1.0000	1.0000	1.0000	1.0000
10	0	.0010	.0003	.0001	.0000	.0000	.0000	.0000	.0000	.0000	.0000
	1	.0107	.0045	.0017	.0005	.0001	.0000	.0000	.0000	.0000	.0000
	2	.0547	.0274	.0123	.0048	.0016	.0004	.0001	.0000	.0000	.0000
	3	.1719	.1020	.0548	.0260	.0106	.0035	.0009	.0001	.0000	.0000
	4	.3770	.2616	.1662	.0949	.0473	.0197	.0064	.0014	.0001	.0000
	5	.6230	.4956	.3669	.2485	.1503	.0781	.0328	.0099	.0016	.0001
	6	.8281	.7340	.6177	.4862	.3504	.2241	.1209	.0500	.0128	.0010
	7	.9453	.9004	.8327	.7184	.6172	.4744	.3222	.1798	.0702	.0115
	8	.9893	.9767	.9536	.9140	.8507	.7560	.6242	.4557	.2639	.0861
	9	.9990	.9975	.9940	.9865	.9718	.9437	.8926	.8031	.6513	.4013
	10	1.0000	1.0000	1.0000	1.0000	1.0000	1.0000	1.0000	1.0000	1.0000	1.0000
11	0	.0005	.0002	.0000	.0000	.0000	.0000	.0000	.0000	.0000	.0000
	1	.0059	.0022	.0007	.0002	.0000	.0000	.0000	.0000	.0000	.0000
	2	.0327	.0148	.0059	.0020	.0006	.0001	.0000	.0000	.0000	.0000
	3	.1133	.0610	.0293	.0122	.0043	.0012	.0002	.0000	.0000	.0000
	4	.2744	.1738	.0994	.0501	.0216	.0076	.0020	.0003	.0000	.0000
	5	.5000	.3669	.2465	.1487	.0782	.0343	.0117	.0027	.0003	.0000
	6	.7256	.6029	.4672	.3317	.2103	.1146	.0504	.0159	.0028	.0001
	7	.8867	.8089	.7037	.5744	.4304	.2867	.1611	.0694	.0185	.0016
	8	.9673	.9348	.8811	.7999	.6873	.5448	.3826	.2212	.0896	.0152
	9	.9941	.9861	.9698	.9394	.8870	.8029	.6779	.5078	.3026	.1019
	10	.9995	.9986	.9964	.9912	.9802	.0578	.9141	.8327	.6862	.4312
	11	1.0000	1.0000	1.0000	1.0000	1.0000	1.0000	1.0000	1.0000	1.0000	1.0000

n	k	p = .05	.10	.15	.20	.25	.30	.35	.40	.45
12	0	.5404	.2824	.1422	.0687	.0317	.0138	.0057	.0022	.0008
	1	.8816	.6590	.4435	.2749	.1584	.0850	.0424	.0424	.0083
	2	.9804	.8891	.7358	.5583	.3907	.2528	.1513	.0834	.0421
	3	.9978	.9744	.9078	.7946	.6488	.4925	.3467	.2253	.1345
	4	.9998	.9956	.9761	.9274	.8424	.7237	.5833	.4382	.3044
	5	1.0000	.9995	.9954	.9806	.9456	.8822	.7873	.6652	.5269
	6	1.0000	.9999	.9993	.9961	.9857	.9614	.9154	.8418	.7393
	7	1.0000	1.0000	.9999	.9994	.9972	.9905	.9745	.9427	.8883
	8	1.0000	1.0000	1.0000	.9999	.9996	.9983	.9944	.9847	.9644
	9	1.0000	1.0000	1.0000	1.0000	1.0000	.9998	.9992	.9972	.9921
	10	1.0000	1.0000	1.0000	1.0000	1.0000	1.0000	.9999	.9997	.9989
	11	1.0000	1.0000	1.0000	1.0000	1.0000	1.0000	1.0000	1.0000	.9999
	12	1.0000	1.0000	1.0000	1.0000	1.0000	1.0000	1.0000	1.0000	1.0000
13	0	.5133	.2542	.1209	.0550	.0238	.0097	.0037	.0013	.0004
	1	.8646	.6213	.3983	.2336	.1267	.0637	.9296	.0126	.0049
	2	.9755	.8661	.7296	.5017	.3326	.2025	.1132	.0579	.0269
	3	.9969	.9658	.9033	.7473	.5843	.4206	.2783	.1686	.0929
	4	.9997	.9935	.9740	.9009	.7940	.6543	.5005	.3530	.2279
	5	1.0000	.9991	.9947	.9700	.9198	.8346	.7159	.5744	.4268
	6	1.0000	.9999	.9987	.9930	.9757	.9376	.8705	.7712	.6437
	7	1.0000	1.0000	.9998	.9988	.9944	.9818	.9538	.9023	.8212
	8	1.0000	1.0000	1.0000	.9998	.9990	.9960	.9874	.9679	.9302
	9	1.0000	1.0000	1.0000	1.0000	.9999	.9993	.9975	.9922	.9797
	10	1.0000	1.0000	1.0000	1.0000	1.0000	.9999	.9997	.9987	.9959
	11	1.0000	1.0000	1.0000	1.0000	1.0000	10000	1.0000	.9999	.9995
	12	1.0000	1.0000	1.0000	1.0000	1.0000	1.0000	1.0000	1.0000	1.0000
	13	1.0000	1.0000	1.0000	1.0000	1.0000	1.0000	1.0000	1.0000	1.0000
14	0	.4877	.2288	.1028	.0440	.0178	.0068	.0024	.0008	.0002
	1	.8470	.5846	.3567	.1979	.1010	.0475	.0205	.0081	.0029
	2	.9699	.8416	.6479	.4481	.2811	.1608	.0839	.0398	.0170
	3	.9958	.9559	.8535	.6982	.5213	.3552	.2205	.1243	.0632
	4	.9996	.9908	.9533	.8702	.7415	.5842	.4227	.2793	.1672
	5	1.0000	.9985	.9885	.9561	.8883	.7805	.6405	.4859	.3373
	6	1.0000	.9998	.9978	.9884	.9617	.9067	.8164	.6925	.5461
	7	1.0000	1.0000	.9997	.9976	.9897	.9685	.9247	.8499	.7414
	8	1.0000	1.0000	1.0000	.9996	.9978	.9917	.9757	.9417	.8811
	9	1.0000	1.0000	1.0000	1.0000	.9997	.9983	.9940	.9825	.9574
	10	1.0000	1.0000	1.0000	1.0000	1.0000	.9998	.9989	.9961	.9886
	11	1.0000	1.0000	1.0000	1.0000	1.0000	1.0000	.9999	.9994	.9978
	12	1.0000	1.0000	1.0000	1.0000	1.0000	1.0000	1.0000	.9999	.9997
	13	1.0000	1.0000	1.0000	1.0000	1.0000	1.0000	1.0000	1.0000	1.0000
	14	1.0000	1.0000	1.0000	1.0000	1.0000	1.0000	1.0000	1.0000	1.0000

Binomial distribution (continued)

n	k	p = .50	.55	.60	.65	.70	.75	.80	.85	.90	.95
12	0	.0002	.0001	.0000	.0000	.0000	.0000	.0000	.0000	.0000	.0000
	1	.0032	.0011	.0003	.0001	.0000	.0000	.0000	.0000	.0000	.0000
	2	.0193	.0079	.0028	.0008	.0002	.0000	.0000	.0000	.0000	.0000
	3	.0730	.0356	.0153	.0056	.0017	.0004	.0001	.0000	.0000	.0000
	4	.1938	.1117	.0573	.0255	.0095	.0028	.0006	.0001	.0000	.0000
	5	.3872	.2607	.1582	.0846	.0386	.0143	.0039	.0007	.0001	.0000
	6	.6128	.4731	.3348	.2127	.1178	.0544	.0194	.0046	.0005	.0000
	7	.8062	.6956	.5618	.4167	.2763	.1576	.0726	.0239	.0043	.0002
	8	.9270	.8655	.7747	.6533	.5075	.3512	.2054	.0922	.0256	.0022
	9	.9807	.9579	.9166	.8487	.7472	.6093	.4417	.2642	.1109	.0196
	10	.9968	.9917	.9804	.9576	.9150	.8416	.7251	.5565	.3410	.1184
	11	.9998	.9992	.9978	.9943	.9862	.9683	.9313	.8578	.7176	.4596
	12	1.0000	1.0000	1.0000	1.0000	1.0000	1.0000	1.0000	1.0000	1.0000	1.0000
13	0	.0001	.0000	.0000	.0000	.0000	.0000	.0000	.0000	.0000	.0000
	1	.0017	.0005	.0001	.0000	.0000	.0000	.0000	.0000	.0000	.0000
	2	.0112	.0041	.0013	.0003	.0001	.0000	.0000	.0000	.0000	.0000
	3	.0461	.0203	.0078	.0025	.0007	.0001	.0000	.0000	.0000	.0000
	4	.1334	.0698	.0321	.0126	.0040	.0010	.0002	.0000	.0000	.0000
	5	.2905	.1788	.0977	.0462	.0182	.0056	.0012	.0002	.0000	.0000
	6	.5000	.3563	.2288	.1295	.0624	.0243	.0070	.0013	.0001	.0000
	7	.7095	.5732	.4256	.2841	.1654	.0802	.0300	.0053	.0009	.0000
	8	.8666	.7721	.6470	.4995	.3457	.2060	.0991	.0260	.0065	.0003
	9	.9539	.9071	.8314	.7217	.5794	.4157	.2527	.0967	.0342	.0031
	10	.9888	.9731	.9421	.8868	.7975	.6674	.4983	.2704	.1339	.0245
	11	.9983	.9951	.9874	.9704	.9363	.8733	.7664	.6017	.3787	.1354
	12	.9999	.9996	.9987	.9963	.9903	.9762	.9450	.8791	.7458	.4867
	13	1.0000	1.0000	1.0000	1.0000	1.0000	1.0000	1.0000	1.0000	1.0000	1.0000
14	0	.0000	.0000	.0000	.0000	.0000	.0000	.0000	.0000	.0000	.0000
	1	.0009	.0003	.0001	.0000	.0000	.0000	.0000	.0000	.0000	.0000
	2	.0065	.0022	.0006	.0001	.0000	.0000	.0000	.0000	.0000	.0000
	3	.0287	.0114	.0039	.0011	.0002	.0000	.0000	.0000	.0000	.0000
	4	.0898	.0462	.0175	.0060	.0017	.0003	.0000	.0000	.0000	.0000
	5	.2120	.1189	.0583	.0243	.0083	.0022	.0004	.0000	.0000	.0000
	6	.3953	.2586	.1501	.0753	.0315	.0108	.0024	.0003	.0000	.0000
	7	.6047	.4539	.3075	.1836	.0933	.0383	.0116	.0022	.0002	.0000
	8	.7880	.6627	.5141	.3595	.2195	.1117	.0439	.0115	.0015	.0000
	9	.9102	.8328	.7207	.5773	.4158	.2585	.1298	.0467	.0092	.0004
	10	.9713	.9368	.8757	.7795	.6448	.4787	.3018	.1465	.0441	.0042
	11	.9935	.9830	.9602	.9161	.8392	.7189	.5519	.3521	.1584	.0301
	12	.9991	.9971	.9919	.9795	.9525	.8990	.8021	.6433	.4154	.1530
	13	.9999	.9998	.9992	.9976	.9932	.9822	.9560	.8972	.7712	.5123
	14	1.0000	1.0000	1.0000	1.0000	1.0000	1.0000	1.0000	1.0000	1.0000	1.0000

n	k	p = .05	.10	.15	.20	.25	.30	.35	.40	.45
15	0	.4633	.2059	.0874	.0352	.0134	.0047	.0016	.0005	.0001
	1	.8290	.5490	.3186	.1671	.0802	.0353	.0142	.0052	.0017
	2	.9638	.8159	.6042	.3980	.2361	.1268	.0617	.0271	.0107
	3	.9945	.9444	.8227	.6482	.4613	.2969	.1727	.0905	.1424
	4	.9994	.9873	.9383	.8358	.6865	.5155	.3519	.2173	.1204
	5	.9999	.9978	.9832	.9389	.8516	.7216	.5643	.4032	.2608
	6	1.0000	.9997	.9964	.9819	.9434	.8689	.7548	.6098	.4522
	7	1.0000	1.0000	.9994	.9958	.9827	.9500	.8868	.7869	.6535
	8	1.0000	1.0000	.9999	.9992	.9958	.9848	.9578	.9050	.8121
	9	1.0000	1.0000	1.0000	.9999	.9992	.9963	.9876	.9662	.9231
	10	1.0000	1.0000	1.0000	1.0000	.9999	.9993	.9972	.9907	.9745
	11	1.0000	1.0000	1.0000	1.0000	1.0000	.9999	.9995	.9981	.9937
	12	1.0000	1.0000	1.0000	1.0000	1.0000	1.0000	.9999	.9997	.9989
	13	1.0000	1.0000	1.0000	1.0000	1.0000	1.0000	1.0000	1.0000	.9999
	14	1.0000	1.0000	1.0000	1.0000	1.0000	1.0000	1.0000	1.0000	1.0000
	15	1.0000	1.0000	1.0000	1.0000	1.0000	1.0000	1.0000	1.0000	1.0000
16	0	.4401	.1853	.0743	.0281	.0100	.0033	.0010	.0003	.0001
	1	.8108	.5147	.2839	.1407	.0635	.0261	.0098	.0088	.0010
	2	.9571	.7892	.5614	.3518	.1971	.0904	.0451	.0183	.0066
	3	.9930	.9316	.7899	.5981	.4050	.2459	.1339	.0651	.0281
	4	.9991	.9830	.9209	.7982	.6302	.4499	.2892	.1666	.0853
	5	.9999	.9967	.9765	.9183	.8103	.6598	.4900	.3288	.1970
	6	1.0000	.9995	.9944	.9733	.9204	.8247	.6881	.5272	.3660
	7	1.0000	.9999	.9989	.9930	.9729	.9256	.8406	.7161	.5629
	8	1.0000	1.0000	.9998	.9985	.9925	.9743	.9329	.8577	.7441
	9	1.0000	1.0000	1.0000	.9998	.9984	.9938	.9809	.9514	.8759
	10	1.0000	1.0000	1.0000	1.0000	.9997	.9984	.9938	.9809	.9514
	11	1.0000	1.0000	1.0000	1.0000	1.0000	.9997	.9987	.9951	.9851
	12	1.0000	1.0000	1.0000	1.0000	1.0000	1.0000	.9998	.9991	.9965
	13	1.0000	1.0000	1.0000	1.0000	1.0000	1.0000	1.0000	.9999	.9965
	14	1.0000	1.0000	1.0000	1.0000	1.0000	1.0000	1.0000	1.0000	.9999
	15	1.0000	1.0000	1.0000	1.0000	1.0000	1.0000	1.0000	1.0000	1.0000
	16	1.0000	1.0000	1.0000	1.0000	1.0000	1.0000	1.0000	1.0000	1.0000

Binomial distribution (continued)

n	k	p = .50	.55	.60	.65	.70	.75	.80	.85	.90	.95
15	0	.0000	.0000	.0000	.0000	.0000	.0000	.0000	.0000	.0000	.0000
	1	.0005	.0001	.0000	.0000	.0000	.0000	.0000	.0000	.0000	.0000
	2	.0037	.0011	.0003	.0001	.0000	.0000	.0000	.0000	.0000	.0000
	3	.0176	.0063	.0019	.0005	.0001	.0000	.0000	.0000	.0000	.0000
	4	.0592	.0255	.0093	.0028	.0007	.0001	.0000	.0000	.0000	.0000
	5	.1509	.1769	.0228	.0124	.0037	.0008	.0001	.0000	.0000	.0000
	6	.3036	.1818	.0950	.0422	.0152	.0042	.0008	.0001	.0000	.0000
	7	.5000	.3465	.2131	.1132	.0500	.0173	.0042	.0006	.0000	.0000
	8	.6964	.5478	.3902	.2452	.1311	.0566	.0181	.0036	.0003	.0000
	9	.8491	.7392	.5968	.4357	.2784	.1484	.0611	.0168	.0022	.0001
	10	.9408	.8796	.7827	.6481	.4845	.3135	.1642	.0617	.0127	.0006
	11	.9824	.9576	.9095	.8273	.7031	.5387	.3518	.1773	.0556	.0055
	12	.9963	.9893	.9729	.9383	.8732	.7639	.6020	.3958	.1841	.0362
	13	.9995	.9983	.9948	.9858	.9647	.9198	.8329	.6814	.4510	.1710
	14	1.0000	.9999	.9995	.9984	.9953	.9866	.9648	.9126	.7941	.5367
	15	1.0000	1.0000	1.0000	1.0000	1.0000	1.0000	1.0000	1.0000	1.0000	1.0000
16	0	.0000	.0000	.0000	.0000	.0000	.0000	.0000	.0000	.0000	.0000
	1	.0003	.0001	.0000	.0000	.0000	.0000	.0000	.0000	.0000	.0000
	2	.0021	.0006	.0001	.0000	.0000	.0000	.0000	.0000	.0000	.0000
	3	.0106	.0035	.0009	.0002	.0000	.0000	.0000	.0000	.0000	.0000
	4	.0384	.0149	.0049	.0013	.0003	.0000	.0000	.0000	.0000	.0000
	5	.1051	.0486	.0191	.0062	.0016	.0003	.0000	.0000	.0000	.0000
	6	.2272	.1241	.0583	.0229	.0071	.0016	.0002	.0000	.0000	.0000
	7	.4018	.2559	.1423	.0671	.0257	.0075	.0015	.0002	.0000	.0000
	8	.5982	.4371	.2839	.1594	.0744	.0271	.0070	.0011	.0001	.0000
	9	.7228	.6340	.4728	.3119	.1753	.0796	.0267	.0056	.0005	.0000
	10	.8949	.8024	.6712	.5100	.3402	.1897	.0817	.0235	.0033	.0001
	11	.9616	.9147	.8334	.7108	.5501	.3698	.2018	.0791	.0170	.0009
	12	.9894	9719	.9349	.8661	.7541	.5950	.4019	.2101	.0684	.0070
	13	.9979	.9934	.9817	.9549	.9006	.8729	.6482	.4386	.2108	.0429
	14	.9997	.9990	.9967	.9902	.9739	.9365	.8593	.7176	.4853	.1892
	15	1.0000	.9999	.9997	.9990	.9967	.9900	.9719	.9257	.8147	.5599
	16	1.0000	1.0000	1.0000	1.0000	1.0000	1.0000	1.0000	1.0000	1.0000	1.0000

n	k	p = .05	.10	.15	.20	.25	.30	.35	.40	.45
17	0	.4181	.1668	.0631	.0225	.0075	.0023	.0007	.0002	.0000
	1	.7922	.4818	.2525	.1182	.0501	.0193	.0067	.0021	.0006
	2	.9497	.7618	.5198	.3096	.1637	.0774	.0327	.0123	.0041
	3	.9912	.9174	.7556	.5489	.3530	.2019	.1028	.0464	.0184
	4	.9985	.9718	.8794	.7164	.5187	.3327	.1886	.0942	.0411
	5	.9998	.9936	.9581	.8671	.7175	.5344	.3550	.2088	.1077
	6	1.0000	.9988	.9882	.9487	.8610	.7217	.5491	.3743	.2258
	7	1.0000	.9998	.9973	.9837	.9431	.8593	.7283	.5634	.3915
	8	1.0000	1.0000	.9995	.9957	.9807	.9404	.8609	.7368	.5778
	9	1.0000	1.0000	1.0000	.9995	.9969	.9873	.9611	.9081	.8166
	10	1.0000	1.0000	1.0000	.9999	.9994	.9968	.9880	.9652	.9174
	11	1.0000	1.0000	1.0000	1.0000	.9999	.9993	.9970	.9894	.9699
	12	1.0000	1.0000	1.0000	1.0000	1.0000	.9999	.9994	.9975	.9914
	13	1.0000	1.0000	1.0000	1.0000	1.0000	1.0000	.9999	.9995	.9981
	14	1.0000	1.0000	1.0000	1.0000	1.0000	1.0000	1.0000	.9999	.9997
	15	1.0000	1.0000	1.0000	1.0000	1.0000	1.0000	1.0000	1.0000	1.0000
	16	1.0000	1.0000	1.0000	1.0000	1.0000	1.0000	1.0000	1.0000	1.0000
	17	1.0000	1.0000	1.0000	1.0000	1.0000	1.0000	1.0000	1.0000	1.0000
18	0	3972	.1501	.0536	.0180	.0056	.0016	.0004	.0001	.0000
	1	.7735	.4503	.2241	.0991	.0395	.0142	.0046	.0013	.0003
	2	.9419	.7338	.4797	.2713	.1353	.0600	.0236	.0082	.0025
	3	.9891	.9018	.7202	.5010	.3057	.1646	.0783	.0328	.0120
	4	.9985	.9718	.8794	.7164	.5187	.3327	.1886	.0942	.0411
	5	.9998	.9936	.9581	.8671	.7175	.5344	.3550	.2088	.1077
	6	1.0000	.9988	.9882	.9487	.8610	.7217	.5491	.3743	.2258
	7	1.0000	.9998	.9973	.9837	.9431	.8593	.7283	.5634	.3915
	8	1.0000	1.0000	.9995	.9957	.9807	.9404	.8609	.7368	.5778
	9	1.0000	1.0000	.9999	.9991	.9946	.9790	.9403	.8653	.7473
	10	1.0000	1.0000	1.0000	.9998	.9988	.9939	.9788	.9424	.8720
	11	1.0000	1.0000	1.0000	1.0000	.9998	.9986	.9938	.9797	.9463
	12	1.0000	1.0000	1.0000	1.0000	1.0000	.9997	.9986	.9942	.9817
	13	1.0000	1.0000	1.0000	1.0000	1.0000	1.0000	.9997	.9987	.9951
	14	1.0000	1.0000	1.0000	1.0000	1.0000	1.0000	1.0000	.9998	.9990
	15	1.0000	1.0000	1.0000	1.0000	1.0000	1.0000	1.0000	1.0000	.9999
	16	1.0000	1.0000	1.0000	1.0000	1.0000	1.0000	1.0000	1.0000	1.0000
	17	1.0000	1.0000	1.0000	1.0000	1.0000	1.0000	1.0000	1.0000	1.0000
	18	1.0000	1.0000	1.0000	1.0000	1.0000	1.0000	1.0000	1.0000	1.0000

Binomial distribution (continued)

n	k	p = .50	.55	.60	.65	.70	.75	.80	.85	.90	.95
17	0	.0000	.0000	.0000	.0000	.0000	.0000	.0000	.0000	.0000	.0000
	1	.0001	.0000	.0000	.0000	.0000	.0000	.0000	.0000	.0000	.0000
	2	.0012	.0003	.0001	.0000	.0000	.0000	.0000	.0000	.0000	.0000
	3	.0064	.0019	.0005	.0001	.0000	.0000	.0000	.0000	.0000	.0000
	4	.0245	.0086	.0025	.0006	.0001	.0000	.0000	.0000	.0000	.0000
	5	.0717	.0301	.0106	.0030	.0007	.0001	.0000	.0000	.0000	.0000
	6	.1662	.0826	.0348	.0120	.0032	.0006	.0001	.0000	.0000	.0000
	7	.3145	.1834	.0919	.0383	.0127	.0031	.0005	.0000	.0000	.0000
	8	.5000	.3374	.1989	.0994	.0403	.0124	.0026	.0003	.0000	.0000
	9	.6855	.5257	.3595	.2128	.1046	.0402	.0109	.0017	.0001	.0000
	10	.8338	.7098	.5522	.3812	.2248	.1071	.0377	.0083	.0008	.0000
	11	.9283	.8529	.7361	.5803	.4032	.2347	.1057	.0319	.0047	.0001
	12	.9755	.9404	.8740	.7652	.6113	.4261	.2418	.0987	.0221	.0012
	13	.9936	.9816	.9536	.8972	.7981	.6470	.4511	.2444	.0826	.0088
	14	.9988	.9959	.9877	.9673	.9226	.8363	.6904	.4802	.2382	.0503
	15	.9999	.9994	.9979	.9933	.9807	.9499	.8818	.7475	.5182	.2078
	16	1.0000	1.0000	.9998	.9997	.9977	.9925	.9775	.9369	.8332	.5819
	17	1.0000	1.0000	1.0000	1.0000	1.0000	1.0000	1.0000	1.0000	1.0000	1.0000
18	0	.0000	.0000	.0000	.0000	.0000	.0000	.0000	.0000	.0000	.0000
	1	.0001	.0000	.0000	.0000	.0000	.0000	.0000	.0000	.0000	.0000
	2	.0007	.0001	.0000	.0000	.0000	.0000	.0000	.0000	.0000	.0000
	3	.0038	.0010	.0002	.0000	.0000	.0000	.0000	.0000	.0000	.0000
	4	.0154	.0049	.0013	.0003	.0000	.0000	.0000	.0000	.0000	.0000
	5	.0481	.0183	.0058	.0014	.0003	.0000	.0000	.0000	.0000	.0000
	6	.1189	.0537	.0203	.0062	.0014	.0002	.0000	.0000	.0000	.0000
	7	.2403	.1280	.0576	.0212	.0061	.0012	.0002	.0000	.0000	.0000
	8	.4073	.2527	.1347	.0597	.0210	.0054	.0009	.0001	.0000	.0000
	9	.5927	.4222	.2632	.2717	.1407	.0569	.0163	.0027	.0002	.0000
	10	.7597	.6085	.4366	.2717	.1407	.0569	.0163	.0027	.0002	.0000
	11	.8811	.7742	.6457	.4509	.2783	.1390	.0513	.0118	.0012	.0000
	12	.9519	.8923	.7912	.6450	.4656	.2825	.1329	.0419	.0064	.0002
	13	.9846	.9589	.9058	.8114	.6673	.4813	.2836	.1206	.0282	.0015
	14	.9962	.9880	.9672	.9217	.8354	.6943	.4990	.2798	.0982	.0109
	15	.9993	.9975	.9918	.9764	.9400	.8647	.7287	.5203	.2662	.0581
	16	.9999	.9997	.9987	.9954	.9858	.9605	.9009	.7759	.5497	.2265
	17	1.0000	1.0000	.9999	.9996	.9984	.9944	.9820	.9464	.8499	.6028
	18	1.0000	1.0000	1.0000	1.0000	1.0000	1.0000	1.0000	1.0000	1.0000	1.0000

n	k	p = .05	.10	.15	.20	.25	.30	.35	.40	.45
19	0	.3774	.1351	.0456	.0144	.0042	.0011	.0003	.0001	.0000
	1	.7547	.4203	.1985	.0829	.0310	.0008	.0002	.0008	.0002
	2	.9335	.7054	.4413	.2369	.1113	.0462	.0170	.0055	.0015
	3	.9869	.8850	.6841	.4551	.2631	.1332	.0591	.0230	.0077
	4	.9869	.8850	.6841	.4551	.2631	.1332	.0591	.0230	.0077
	5	.9998	.9914	.9463	.8369	.6678	.4739	.2968	.1629	.0777
	6	1.0000	.9983	.9837	.9324	.8251	.6655	.4912	.3081	.1727
	7	1.0000	.9997	.9959	.9767	.9225	.8180	.6656	.4878	.3169
	8	1.0000	1.0000	.9992	.9933	.9713	.9161	.8145	.6675	.4940
	9	1.0000	1.0000	.9999	.9984	.9911	.9674	.9125	.8139	.6710
	10	1.0000	1.0000	1.0000	.9997	.9977	.9895	.9653	.9115	.8159
	11	1.0000	1.0000	1.0000	1.0000	.9995	.9972	.9886	.9648	.9129
	12	1.0000	1.0000	1.0000	1.0000	.9999	.9994	.9969	.9884	.9658
	13	1.0000	1.0000	1.0000	1.0000	1.0000	.9999	.9993	.9969	.9891
	14	1.0000	1.0000	1.0000	1.0000	1.0000	1.0000	.9999	.9994	.9972
	15	1.0000	1.0000	1.0000	1.0000	1.0000	1.0000	1.0000	.9999	.9995
	16	1.0000	1.0000	1.0000	1.0000	1.0000	1.0000	1.0000	1.0000	.9999
	17	1.0000	1.0000	1.0000	1.0000	1.0000	1.0000	1.0000	1.0000	1.0000
	18	1.0000	1.0000	1.0000	1.0000	1.0000	1.0000	1.0000	1.0000	1.0000
	19	1.0000	1.0000	1.0000	1.0000	1.0000	1.0000	1.0000	1.0000	1.0000
20	0	.3585	.1261	.0388	.0115	.0032	.0008	.0000	.0000	.0000
	1	.7358	.3917	.1756	.0692	.0243	.0076	.0021	.0005	.0001
	2	.9245	.6769	.4049	.2061	.0913	.0355	.0121	.0036	.0009
	3	.9841	.8670	.6477	.4114	.2252	.1071	.0444	.0160	.0049
	4	.9974	.9568	.8298	.6296	.4148	.2375	.1182	.0510	.0189
	5	.9997	.9887	.9327	.8042	.6172	.4164	.2454	.1256	.0553
	6	1.0000	.9976	.9781	.9133	.7858	.6080	.4166	.2500	.1299
	7	1.0000	.9996	.9941	.9679	.8982	.7723	.6010	.4159	.2520
	8	1.0000	.9999	.9987	.9900	.9591	.8867	.7624	.5956	.4143
	9	1.0000	1.0000	.9998	.9974	.9861	.9520	.8782	.7553	.5914
	10	1.0000	1.0000	1.0000	.9994	.9961	.9829	.9468	.8725	.7507
	11	1.0000	1.0000	1.0000	.9999	.9991	.9949	.9804	.9435	.8692
	12	1.0000	1.0000	1.0000	1.0000	.9998	.9987	.9940	.9790	.9420
	13	1.0000	1.0000	1.0000	1.0000	1.0000	.9997	.9985	.9935	.9786
	14	1.0000	1.0000	1.0000	1.0000	1.0000	1.0000	.9997	.9984	.9936
	15	1.0000	1.0000	1.0000	1.0000	1.0000	1.0000	1.0000	.9997	.9985
	16	1.0000	1.0000	1.0000	1.0000	1.0000	1.0000	1.0000	1.0000	.9997
	17	1.0000	1.0000	1.0000	1.0000	1.0000	1.0000	1.0000	1.0000	1.0000
	18	1.0000	1.0000	1.0000	1.0000	1.0000	1.0000	1.0000	1.0000	1.0000
	19	1.0000	1.0000	1.0000	1.0000	1.0000	1.0000	1.0000	1.0000	1.0000
	20	1.0000	1.0000	1.0000	1.0000	1.0000	1.0000	1.0000	1.0000	1.0000

Binomial distribution (continued)

n	k	p = .50	.55	.60	.65	.70	.75	.80	.85	.90	.95
19	0	.0000	.0000	.0000	.0000	.0000	.0000	.0000	.0000	.0000	.0000
	1	.0000	.0000	.0000	.0000	.0000	.0000	.0000	.0000	.0000	.0000
	2	.0004	.0001	.0000	.0000	.0000	.0000	.0000	.0000	.0000	.0000
	3	.0022	.0005	.0001	.0000	.0000	.0000	.0000	.0000	.0000	.0000
	4	.0096	.0028	.0006	.0001	.0000	.0000	.0000	.0000	.0000	.0000
	5	.0318	.0109	.0031	.0007	.0001	.0000	.0000	.0000	.0000	.0000
	6	.0835	.0342	.0116	.0031	.0006	.0001	.0000	.0000	.0000	.0000
	7	.1796	.0871	.0352	.0114	.0028	.0005	.0000	.0000	.0000	.0000
	8	.3238	.1841	.0885	.0347	.0105	.0023	.0003	.0000	.0000	.0000
	9	.5000	.3290	.1861	.0875	.0326	.0287	.0067	.0008	.0000	.0000
	10	.6762	.5060	.3325	.1855	.0839	.0287	.0067	.0008	.0000	.0000
	11	.8204	.6831	.5122	.3344	.1820	.0775	.0233	.0041	.0008	.0000
	12	.9165	.8273	.6919	.5188	.3345	.1749	.0676	.0163	.0017	.0000
	13	.9682	.9223	.8371	.7032	.5261	.3322	.1631	.0537	.0086	.0002
	14	.9904	.9720	.9304	.8500	.7178	.5346	.3267	.1444	.0352	.0020
	15	.9978	.9923	.9770	.9409	.8668	.7369	.5449	.3159	.1150	.0132
	16	.9996	.9985	.9945	.9830	.9538	.8887	.7631	.5587	.2946	.0665
	17	1.0000	.9998	.9992	.9969	.9896	.9690	.9171	.8015	.5797	.2453
	18	1.0000	1.0000	.9999	.9997	.9989	.9958	.9856	.9544	.8649	.6226
	19	1.0000	1.0000	1.0000	1.0000	1.0000	1.0000	1.0000	1.0000	1.0000	1.0000
20	0	.0000	.0000	.0000	.0000	.0000	.0000	.0000	.0000	.0000	.0000
	1	.0000	.0000	.0000	.0000	.0000	.0000	.0000	.0000	.0000	.0000
	2	.0002	.0000	.0000	.0000	.0000	.0000	.0000	.0000	.0000	.0000
	3	.0013	.0003	.0000	.0000	.0000	.0000	.0000	.0000	.0000	.0000
	4	.0059	.0015	.0003	.0000	.0000	.0000	.0000	.0000	.0000	.0000
	5	.0207	.0064	.0016	.0003	.0000	.0000	.0000	.0000	.0000	.0000
	6	.0577	.0214	.0065	.0015	.0003	.0000	.0000	.0000	.0000	.0000
	7	.1316	.0580	.0210	.0060	.0013	.0002	.0000	.0000	.0000	.0000
	8	.2517	.1308	.0565	.0196	.0051	.0009	.0001	.0000	.0000	.0000
	9	.4119	.2493	.1275	.0532	.0171	.0039	.0006	.0000	.0000	.0000
	10	.5881	.4086	.2447	.1218	.0480	.0139	.0026	.0002	.0000	.0000
	11	.7483	.5857	.4044	.2376	.1133	.0409	.0100	.0013	.0001	.0000
	12	.8684	.7480	.5841	.3990	.2277	.1018	.0321	.0059	.0004	.0000
	13	.9423	.8701	.7500	.5834	.3920	.2142	.0867	.0219	.0024	.0000
	14	.9793	.9447	.8744	.7546	.5836	.3828	.1958	.0673	.0113	.0003
	15	.9941	.9811	.9490	.8818	.7625	.5852	.3704	.1702	.0432	.0026
	16	.9987	.9951	.9840	.9556	.8929	.7748	.5886	.3523	.1330	.0159
	17	.9998	.9991	.9964	.9879	.9645	.9087	.7939	.5951	.3231	.0755
	18	1.0000	.9999	.9995	.9979	.9924	.9757	.9308	.8244	.6083	.2642
	19	1.0000	1.0000	1.0000	.9998	.9992	.9968	.9885	.9612	.8784	.6415
	20	1.0000	1.0000	1.0000	1.0000	1.0000	1.0000	1.0000	1.0000	1.0000	1.0000

Normal distribution

The table gives the area to the left of z under a standard normal curve, for various values of z.

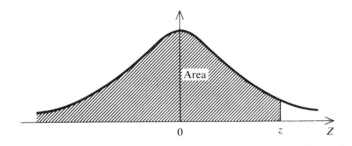

To obtain probabilities use the fact that

Area to the left of $z = P(Z \leq z)$

where Z is a variable with a standard normal distribution.

z	Area		z	Area		z	Area
−4	.00003		−2.74	.0031		−2.29	.0110
−3.9	.00005		−2.73	.0032		−2.28	.0113
−3.8	.0001		−2.72	.0033		−2.27	.0116
−3.7	.0001		−2.71	.0034		−2.26	.0119
−3.6	.0002		−2.70	.0035		−2.25	.0122
−3.5	.0002		−2.69	.0036		−2.24	.0125
−3.4	.0003		−2.68	.0037		−2.23	.0129
−3.3	.0005		−2.67	.0038		−2.22	.0132
−3.2	.0007		−2.66	.0039		−2.21	.0136
−3.1	.0010		−2.65	.0040		−2.20	.0139
−3.09	.0010		−2.64	.0041		−2.19	.0143
−3.08	.0010		−2.63	.0043		−2.18	.0146
−3.07	.0011		−2.62	.0044		−2.17	.0150
−3.06	.0011		−2.61	.0045		−2.16	.0154
−3.05	.0011		−2.60	.0047		−2.15	.0158
−3.04	.0012		−2.59	.0048		−2.14	.0162
−3.03	.0012		−2.58	.0049		−2.13	.0166
−3.02	.0013		−2.57	.0051		−2.12	.0170
−3.01	.0013		−2.56	.0052		−2.11	.0174
−3.00	.0013		−2.55	.0054		−2.10	.0179
−2.99	.0014		−2.54	.0055		−2.09	.0183
−2.98	.0014		−2.53	.0057		−2.08	.0188
−2.97	.0015		−2.52	.0059		−2.07	.0182
−2.96	.0015		−2.51	.0060		−2.06	.0197
−2.95	.0016		−2.50	.0062		−2.05	.0202
−2.94	.0016		−2.49	.0064		−2.04	.0207
−2.93	.0017		−2.48	.0066		−2.03	.0212
−2.92	.0017		−2.47	.0068		−2.02	.0217
−2.91	.0018		−2.46	.0069		−2.01	.0222
−2.90	.0019		−2.45	.0071		−2.00	.0228
−2.89	.0019		−2.44	.0073		−1.99	.0233
−2.88	.0020		−2.43	.0075		−1.98	.0239
−2.87	.0021		−2.42	.0078		−1.97	.0244
−2.86	.0021		−2.41	.0080		−1.96	.0250
−2.85	.0022		−2.40	.0082		−1.95	.0256
−2.84	.0023		−2.39	.0084		−1.94	0262
−2.83	.0023		−2.38	.0087		−1.93	.0268
−2.82	.0024		−2.37	.0089		1.92	.0274
−2.81	.0025		−2.36	.0091		−1.91	.0281
−2.80	.0026		−2.35	.0094		−1.90	.0287
−2.79	.0026		−2.34	.0096		−1.89	.0294
−2.78	.0027		−2.33	.0099		−1.88	.030
−2.77	.0028		−2.32	.0102		−1.87	.0307
−2.76	.0029		−2.31	.0104		−1.86	.0314
−2.75	.0030		−2.30	.0107		−1.85	.0322

z	Area		z	Area		z	Area
−1.84	.0329		−1.39	.0823		− .94	.1736
−1.83	.0336		−1.38	.0838		− .93	.1762
−1.82	.0344		−1.37	.0853		− .92	.1788
−1.81	.0352		−1.36	.0869		− .91	.1814
−1.80	.0359		−1.35	.0885		− .90	.1841
−1.79	.0367		−1.34	.0901		− .89	.1867
−1.78	.0375		−1.33	.0918		− .88	.1894
−1.77	.0384		−1.32	.0934		− .87	.1922
−1.76	.0392		−1.31	.0951		− .86	.1949
−1.75	.0401		−1.30	.0968		− .85	.1977
−1.74	.0409		−1.29	.0985		− .84	.2005
−1.73	.0418		−1.28	.1003		− .83	.2033
−1.72	.0427		−1.27	.1020		− .82	.2061
−1.71	.0436		−1.26	.1038		− .81	.2090
−1.70	.0446		−1.25	.1056		− .80	.2119
−1.69	.0455		−1.24	.1075		− .79	.2148
−1.68	.0465		−1.23	.1093		− .78	.2177
−1.67	.0475		−1.22	.1112		− .77	.2296
−1.66	.0485		−1.21	.1131		− .76	.2236
−1.65	.0495		−1.20	.1151		− .75	.2266
−1.64	.0505		−1.19	.1170		− .74	.2296
−1.63	.0516		−1.18	.1190		− .73	.2327
−1.62	.0526		−1.17	.1210		− .72	.2358
−1.61	.0537		−1.16	.1230		− .71	.2389
−1.60	.0548		−1.15	.1251		− .70	.2420
−1.59	.0559		−1.14	.1271		− .69	.2451
−1.58	.0571		−1.13	.1292		− .68	.2483
−1.57	.0582		−1.12	.1314		− .67	.2514
−1.56	.0594		−1.11	.1335		− .66	.2546
−1.55	.0606		−1.10	.1357		− .65	.2578
−1.54	.0618		−1.09	.1379		− .64	.2611
−1.53	.0630		−1.08	.1401		− .63	.2643
−1.52	.0643		−1.07	.1423		− .62	.2676
−1.51	.0655		−1.06	.1446		− .61	.2709
−1.50	.0668		−1.05	.1469		− .60	.2743
−1.49	.0681		−1.04	.1492		− .59	.2776
−1.48	.0694		−1.03	.1515		− .58	.2810
−1.47	.0708		−1.02	.1539		− .57	.2843
−1.46	.0722		−1.01	.1562		− .56	.2877
−1.45	.0735		−1.00	.1587		− .55	.2912
−1.44	.0749		− .99	.1611		− .54	.2946
−1.43	.0764		− .98	.1635		− .53	.2981
−1.42	.0778		− .97	.1660		− .52	.3015
−1.41	.0793		− .96	.1685		− .51	.3050
−1.40	.0808		− .95	.1711		− .50	.3085

Normal distribution

z	Area		z	Area		z	Area
− .49	.3121		− .04	.4840		.41	.6591
− .48	.3156		− .03	.4880		.42	.6628
− .47	.3192		− .02	.4920		.43	.6664
− .46	.3228		− .01	.4960		.44	.6700
− .45	.3264		.00	.5000		.45	.6736
− .44	.3300		.01	.5040		.46	.6772
− .43	.3336		.02	.5080		.47	.6808
− .42	.3372		.03	.5120		.48	.6844
− .41	.3409		.04	.5160		.49	.6849
− .40	.3446		.05	.5199		.50	.6915
− .39	.3483		.06	.5239		.51	.6950
− .38	.3520		.07	.5279		.52	.6985
− .37	.3557		.08	.5319		.53	.7019
− .36	.3594		.09	.5359		.54	.7054
− .35	.3632		.10	.5398		.55	.7088
− .34	.3669		.11	.5438		.56	.7123
− .33	.3707		.12	.5478		.57	.7157
− .32	.3745		.13	.5517		.58	.7190
− .31	.3783		.14	.5557		.59	.7224
− .30	.3821		.15	.5596		.60	.7257
− .29	.3859		.16	.5636		.61	.7291
− .28	.3897		.17	.5675		.62	.7324
− .27	.3936		.18	.5714		.63	.7357
− .26	.3974		.19	.5753		.64	.7389
− .25	.4013		.20	.5793		.65	.7422
− .24	.4052		.21	.5832		.66	.7454
− .23	.4090		.22	.5871		.67	.7486
− .22	.4129		.23	.5910		.68	.7517
− .21	.4168		.24	.5948		.69	.7549
− .20	.4207		.25	.5987		.70	.7580
− .19	.4247		.26	.6026		.71	.7611
− .18	.4286		.27	.6064		.72	.7642
− .17	.4325		.28	.6103		.73	.7673
− .16	.4364		.29	.6141		.74	.7704
− .15	.4404		.30	.6179		.75	.7734
− .14	.4443		.31	.6217		.76	.7764
− .13	.4483		.32	.6255		.77	.7794
− .12	.4522		.33	.6293		.78	.7823
− .11	.4562		.34	.6331		.79	.7852
− .10	.4602		.35	.6368		.80	.7881
− .09	.4641		.36	.6406		.81	.7910
− .08	.4681		.37	.6443		.82	.7939
− .07	.4721		.38	.6480		.83	.7967
− .06	.4761		.39	.6517		.84	.7995
− .05	.4801		.40	.6554		.85	.8023

z	Area		z	Area		z	Area
.86	.8051		1.31	.9049		1.76	.9608
.87	.8078		1.32	.9066		1.77	.9616
.88	.8106		1.33	.9082		1.78	.9625
.89	.8133		1.34	.9099		1.79	.9633
.90	.8159		1.35	.9115		1.80	.9641
.91	.8186		1.36	.9131		1.81	.9649
.92	.8212		1.37	.9147		1.82	.9656
.93	.8238		1.38	.9162		1.83	.9664
.94	.8264		1.39	.9177		1.84	.9671
.95	.8289		1.40	.9192		1.85	.9678
.96	.8315		1.41	.9207		1.86	.9686
.97	.8340		1.42	.9222		1.87	.9693
.98	.8365		1.43	.9236		1.88	.9699
.99	.8389		1.44	.9251		1.89	.9706
1.00	.8413		1.45	.9265		1.90	.9713
1.01	.8438		1.46	.9278		1.91	.9719
1.02	.8461		1.47	.9292		1.92	.9726
1.03	.8485		1.48	.9306		1.93	.9732
1.04	.8508		1.49	.9319		1.94	.9738
1.05	.8531		1.50	.9332		1.95	.9744
1.06	.8554		1.51	.9345		1.96	.9750
1.07	.8577		1.52	.9357		1.97	.9756
1.08	.8599		1.53	.9370		1.98	.9761
1.09	.8621		1.54	.9382		1.99	.9767
1.10	.8643		1.55	.9394		2.00	.9772
1.11	.8665		1.56	.9406		2.01	.9778
1.12	.8686		1.57	.9418		2.02	.9783
1.13	.8708		1.58	.9429		2.03	.9788
1.14	.8729		1.59	.9441		2.04	.9793
1.15	.8749		1.60	.9452		2.05	.9798
1.16	.8770		1.61	.9463		2.06	.9803
1.17	.8790		1.62	.9474		2.07	.9808
1.18	.8810		1.63	.9484		2.08	.9812
1.19	.8830		1.64	.9495		2.09	.9817
1.20	.8849		1.65	.9505		2.10	.9821
1.21	.8869		1.66	.9515		2.11	.9826
1.22	.8888		1.67	.9525		2.12	.9830
1.23	.8907		1.68	.9535		2.13	.9834
1.24	.8925		1.69	.9545		2.14	.9838
1.25	.8944		1.70	.9554		2.15	.9842
1.26	.8962		1.71	.9564		2.16	.9846
1.27	.8980		1.72	.9573		2.17	.9850
1.28	.8997		1.73	.9582		2.18	.9854
1.29	.9015		1.74	.9591		2.19	.9857
1.30	.9032		1.75	.9599		2.20	.9861

Normal distribution (continued)

z	Area		z	Area		z	Area
2.21	.9864		2.66	.9961		3.2	.9993
2.22	.9868		2.67	.9962		3.3	.9995
2.23	.9871		2.68	.9963		3.4	.9997
2.24	.9875		2.69	.9964		3.5	.9998
2.25	.9878		2.70	.9965		3.6	.9998
						3.7	.9999
2.26	.9881		2.71	.9966		3.8	.9999
2.27	.9884		2.72	.9967			
2.28	.9887		2.73	.9968		3.9	.99995
2.29	.9890		2.74	.9969		4.0	.99997
2.30	.9893		2.75	.9970			
2.31	.9896		2.76	.9971			
2.32	.9898		2.77	.9972			
2.33	.9901		2.78	.9973			
2.34	.9904		2.79	.9974			
2.35	.9906		2.80	.9974			
2.36	.9909		2.81	.9975			
2.37	.9911		2.82	.9976			
2.38	.9913		2.83	.9977			
2.39	.9916		2.84	.9977			
2.40	.9918		2.85	.9978			
2.41	.9920		2.86	.9979			
2.42	.9922		2.87	.9979			
2.43	.9925		2.88	.9980			
2.44	.9927		2.89	.9981			
2.45	.9929		2.90	.9981			
2.46	.9931		2.91	.9982			
2.47	.9932		2.92	.9982			
2.48	.9934		2.93	.9983			
2.49	.9936		2.94	.9984			
2.50	.9938		2.95	.9984			
2.51	.9940		2.96	.9985			
2.52	.9941		2.97	.9985			
2.53	.9943		2.98	.9986			
2.54	.9945		2.99	.9986			
2.55	.9946		3.00	.9987			
2.56	.9948		3.01	.9987			
2.57	.9949		3.02	.9987			
2.58	.9951		3.03	.9988			
2.59	.9952		3.04	.9988			
2.60	.9953		3.05	.9989			
2.61	.9955		3.06	.9989			
2.62	.9956		3.07	.9989			
2.63	.9957		3.08	.9990			
2.64	.9959		3.09	.9990			
2.65	.9960		3.10	.9990			

Student's t distribution

The table gives the critical values of *t* needed for the *t* test, for various degrees of freedom and levels of significance.

Example 1. Suppose the test is one-sided and rejects for large values of *t*, with level of significance .05. Then the critical value is in the column headed .95, on the row corresponding to the appropriate degrees of freedom.

Example 2. Suppose the test is two-sided, and has level of significance .01. Then the critical values are in the columns headed .005 and .995, on the row corresponding to the appropriate degrees of freedom. If there are 10 degrees of freedom, the test would reject if the observed value of *t* is less than −3.17 or greater than 3.17.

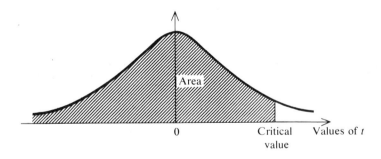

Degrees of freedom	Area to the left of the critical value							
	.005	.01	.025	.05	.95	.975	.99	.995
1	−63.66	−31.82	−12.71	−6.31	6.31	12.71	31.82	63.66
2	− 9.92	− 6.96	− 4.30	−2.92	2.92	4.30	6.96	9.92
3	− 5.84	− 4.54	− 3.18	−2.35	2.35	3.18	4.54	5.84
4	− 4.60	− 3.75	− 2.78	−2.13	2.13	2.78	3.75	4.60
5	− 4.03	− 3.36	− 2.57	−2.02	2.02	2.57	3.36	4.03
6	− 3.71	− 3.14	− 2.45	−1.94	1.94	2.45	3.14	3.71
7	− 3.50	− 3.00	− 2.36	−1.90	1.90	2.36	3.00	3.50
8	− 3.36	− 2.90	− 2.31	−1.86	1.86	2.31	2.90	3.36
9	− 3.25	− 2.82	− 2.26	−1.83	1.83	2.26	2.82	3.25
10	− 3.17	− 2.76	− 2.23	−1.81	1.81	2.23	2.76	3.17
11	− 3.11	− 2.72	− 2.20	−1.80	1.80	2.20	2.72	3.11
12	− 3.06	− 2.68	− 2.18	−1.78	1.78	2.18	2.68	3.06
13	− 3.01	− 2.65	− 2.16	−1.77	1.77	2.16	2.65	3.01
14	− 2.98	− 2.62	− 2.14	−1.76	1.76	2.14	2.62	2.98
15	− 2.95	− 2.60	− 2.13	−1.75	1.75	2.13	2.60	2.95
16	− 2.92	− 2.58	− 2.12	−1.75	1.75	2.12	2.58	2.92
17	− 2.90	− 2.57	− 2.11	−1.74	1.74	2.11	2.57	2.90
18	− 2.88	− 2.55	− 2.10	−1.73	1.73	2.10	2.55	2.88
19	− 2.86	− 2.54	− 2.09	−1.73	1.73	2.09	2.54	2.86
20	− 2.84	− 2.53	− 2.09	−1.72	1.72	2.09	2.53	2.84
21	− 2.83	− 2.52	− 2.08	−1.72	1.72	2.08	2.52	2.83
22	− 2.82	− 2.51	− 2.07	−1.72	1.72	2.07	2.51	2.82
23	− 2.81	− 2.50	− 2.07	−1.71	1.71	2.07	2.50	2.81
24	− 2.80	− 2.49	− 2.06	−1.71	1.71	2.06	2.49	2.80
25	− 2.79	− 2.48	− 2.06	−1.71	1.71	2.06	2.48	2.79
26	− 2.78	− 2.48	− 2.06	−1.71	1.71	2.06	2.48	2.78
27	− 2.77	− 2.47	− 2.05	−1.70	1.70	2.05	2.47	2.77
28	− 2.76	− 2.47	− 2.05	−1.70	1.70	2.05	2.47	2.76
29	− 2.76	− 2.46	− 2.04	−1.70	1.70	2.04	2.46	2.76
30	− 2.75	− 2.46	− 2.04	−1.70	1.70	2.04	2.46	2.75
40	− 2.70	− 2.42	− 2.02	−1.68	1.68	2.02	2.42	2.70
50	− 2.68	− 2.40	− 2.01	−1.68	1.68	2.01	2.40	2.68
60	− 2.66	− 2.39	− 2.00	−1.67	1.67	2.00	2.39	2.66
80	− 2.64	− 2.37	− 1.99	−1.66	1.66	1.99	2.37	2.64
100	− 2.63	− 2.36	− 1.98	−1.66	1.66	1.98	2.36	2.63

Critical values for chi-square tests

The table gives the critical values for a chi-square test, for various degrees of freedom and levels of significance α.

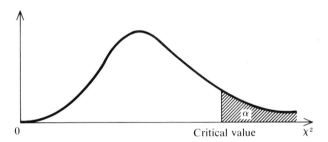

0   Critical value χ^2

Degrees of freedom	Level of significance $= \alpha$	
	.05	.01
1	3.84	6.63
2	5.99	9.21
3	7.82	11.34
4	9.49	13.28
5	11.07	15.09
6	12.59	16.81
7	14.07	18.48
8	15.51	20.09
9	16.92	21.67
10	18.31	23.21
11	19.68	24.72
12	21.03	26.22
13	22.36	27.69
14	23.68	29.14
15	25.00	30.58
16	26.30	32.00
17	27.59	33.41
18	28.87	34.80
19	30.14	36.19
20	31.41	37.57
21	32.67	38.93
22	33.92	40.29
23	35.17	41.64
24	36.42	42.98
25	37.65	44.31
26	38.88	45.64
27	40.11	46.96
28	41.34	48.28
29	42.56	49.59
30	43.77	50.89

F distribution

This table gives the critical value for an F test, for various degrees of freedom and levels of significance $\alpha = .01$, and $\alpha = .05$. γ_1 denotes the numerator degrees of freedom, and γ_2 denotes the denominator degrees of freedom.

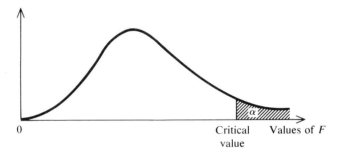

$\nu_2 \backslash \nu_1$	1	2	3	4	5	6	7	8	9	10	15	20	30	40	60	∞
1	161	200	216	225	230	234	237	239	241	242	246	248	250	251	252	254
2	18.5	19.0	19.2	19.2	19.3	19.3	19.4	19.4	19.4	19.4	19.4	19.4	19.5	19.5	19.5	19.5
3	10.1	9.55	9.28	9.12	9.01	8.94	8.89	8.85	8.81	8.79	8.70	8.66	8.62	8.59	8.57	8.53
4	7.71	6.94	6.59	6.39	6.26	6.16	6.09	6.04	6.00	5.96	5.86	5.80	5.75	5.72	5.69	5.63
5	6.61	5.79	5.41	5.19	5.05	4.95	4.88	4.82	4.77	4.74	4.62	4.56	4.50	4.46	4.43	4.37
6	5.99	5.14	4.78	4.53	4.39	4.28	4.21	4.15	4.10	4.06	3.94	3.87	3.81	3.77	3.74	3.67
7	5.59	4.74	4.35	4.12	3.97	3.87	3.79	3.73	3.68	3.64	3.51	3.44	3.38	3.34	3.30	3.23
8	5.32	4.46	4.07	3.84	3.69	3.58	3.50	3.44	3.39	3.35	3.22	3.15	3.08	3.04	3.01	2.93
9	5.12	4.26	3.86	3.63	3.48	3.37	3.29	3.23	3.18	3.14	3.01	2.94	2.86	2.83	2.79	2.71
10	4.96	4.10	3.71	3.48	3.33	3.22	3.14	3.07	3.02	2.98	2.85	2.77	2.70	2.66	2.62	2.54
11	4.84	3.98	3.59	3.36	3.20	3.09	3.01	2.95	2.90	2.85	2.72	2.65	2.57	2.53	2.49	2.40
12	4.75	3.89	3.49	3.26	3.11	3.00	2.91	2.85	2.80	2.75	2.62	2.54	2.47	2.43	2.38	2.30
13	4.67	3.81	3.41	3.18	3.03	2.92	2.83	2.77	2.71	2.67	2.53	2.46	2.38	2.34	2.30	2.21
14	4.60	3.74	3.34	3.11	2.96	2.85	2.76	2.70	2.65	2.60	2.46	2.39	2.31	2.27	2.22	2.13
15	4.54	3.68	3.29	3.06	2.90	2.79	2.71	2.64	2.59	2.54	2.40	2.33	2.25	2.20	2.16	2.07
16	4.49	3.63	3.24	3.01	2.85	2.74	2.66	2.59	2.54	2.49	2.35	2.28	2.19	2.15	2.11	2.01
17	4.45	3.59	3.20	2.96	2.81	2.70	2.61	2.55	2.49	2.45	2.31	2.23	2.15	2.10	2.06	1.96
18	4.41	3.55	3.16	2.93	2.77	2.66	2.58	2.51	2.46	2.41	2.27	2.19	2.11	2.06	2.02	1.92
19	4.38	3.52	3.13	2.90	2.74	2.63	2.54	2.48	2.42	2.38	2.23	2.16	2.07	2.03	1.98	1.88
20	4.35	3.49	3.10	2.87	2.71	2.60	2.51	2.45	2.39	2.35	2.20	2.12	2.04	1.99	1.95	1.84
21	4.32	3.47	3.07	2.84	2.68	2.57	2.49	2.42	2.37	2.32	2.18	2.10	2.01	1.96	1.92	1.81
22	4.30	3.44	3.05	2.82	2.66	2.55	2.46	2.40	2.34	2.30	2.15	2.07	1.98	1.94	1.89	1.78
23	4.28	3.42	3.03	2.80	2.64	2.53	2.44	2.37	2.32	2.27	2.13	2.05	1.96	1.91	1.86	1.76
24	4.26	3.40	3.01	2.78	2.62	2.51	2.42	2.36	2.30	2.25	2.11	2.03	1.94	1.89	1.84	1.73
25	4.24	3.39	2.99	2.76	2.60	2.49	2.40	2.34	2.28	2.24	2.09	2.01	1.92	1.87	1.82	1.71
30	4.17	3.32	2.92	2.69	2.53	2.42	2.33	2.27	2.21	2.16	2.01	1.93	1.84	1.79	1.74	1.62
40	4.08	3.23	2.84	2.61	2.45	2.34	2.25	2.18	2.12	2.08	1.92	1.84	1.74	1.69	1.64	1.51
60	4.00	3.15	2.76	2.53	2.37	2.25	2.17	2.10	2.04	1.99	1.84	1.75	1.65	1.59	1.53	1.39
120	3.92	3.07	2.68	2.45	2.29	2.18	2.09	2.02	1.96	1.91	1.75	1.66	1.55	1.50	1.43	1.25
∞	3.84	3.00	2.60	2.37	2.21	2.10	2.01	1.94	1.88	1.83	1.67	1.57	1.46	1.39	1.32	1.00

Critical values for F for $\alpha = .05$

ν_1 / ν_2	1	2	3	4	5	6	7	8	9	10	15	20	30	40	60	∞
1	4052	5000	5403	5625	5764	5859	5928	5982	6023	6056	6157	6209	6261	6287	6313	6366
2	98.5	99.0	99.2	99.2	99.3	99.3	99.4	99.4	99.4	99.4	99.4	99.4	99.5	99.5	99.5	99.5
3	34.1	30.8	29.5	28.7	28.2	27.9	27.7	27.5	27.3	27.2	26.9	26.7	26.5	26.4	26.3	26.1
4	21.2	18.0	16.7	16.0	15.5	15.2	15.0	14.8	14.7	14.5	14.2	14.0	13.8	13.7	13.7	13.5
5	16.3	13.3	12.1	11.4	11.0	10.7	10.5	10.3	10.2	10.1	9.72	9.55	9.38	9.27	9.20	9.02
6	13.7	10.9	9.78	9.15	8.75	8.47	8.26	8.10	7.98	7.87	7.56	7.40	7.23	7.14	7.06	6.88
7	12.2	9.55	8.45	7.85	7.46	7.19	6.99	6.84	6.72	6.62	6.31	6.16	5.99	5.91	5.82	5.65
8	11.3	8.65	7.59	7.01	6.63	6.37	6.18	6.03	5.91	5.81	5.52	5.36	5.20	5.12	5.03	4.86
9	10.6	8.02	6.99	6.42	6.06	5.80	5.61	5.47	5.35	5.26	4.96	4.81	4.65	4.57	4.48	4.31
10	10.0	7.56	6.55	5.99	5.64	5.39	5.20	5.06	4.94	4.85	4.56	4.41	4.25	4.17	4.08	3.91
11	9.65	7.21	6.22	5.67	5.32	5.07	4.89	4.74	4.63	4.54	4.25	4.10	3.94	3.86	3.78	3.60
12	9.33	6.93	5.95	5.41	5.06	4.82	4.64	4.50	4.39	4.30	4.01	3.86	3.70	3.62	3.54	3.36
13	9.07	6.70	5.74	5.21	4.86	4.62	4.44	4.30	4.19	4.10	3.82	3.66	3.51	3.43	3.34	3.17
14	8.86	6.51	5.56	5.04	4.70	4.46	4.28	4.14	4.03	3.94	3.66	3.51	3.35	3.27	3.18	3.00
15	8.68	6.36	5.42	4.89	4.56	4.32	4.14	4.00	3.89	3.80	3.52	3.37	3.21	3.13	3.05	2.87
16	8.53	6.23	5.29	4.77	4.44	4.20	4.03	3.89	3.78	3.69	3.41	3.26	3.10	3.02	2.93	2.75
17	8.40	6.11	5.19	4.67	4.34	4.10	3.93	3.79	3.68	3.59	3.31	3.16	3.00	2.92	2.83	2.65
18	8.29	6.01	5.09	4.58	4.25	4.01	3.84	3.71	3.60	3.51	3.23	3.08	2.92	2.84	2.75	2.57
19	8.19	5.93	5.01	4.50	4.17	3.94	3.77	3.63	3.52	3.43	3.15	3.00	2.84	2.76	2.67	2.49
20	8.10	5.85	4.94	4.43	4.10	3.87	3.70	3.56	3.46	3.37	3.09	2.94	2.78	2.69	2.61	2.42
21	8.02	5.78	4.87	4.37	4.04	3.81	3.64	3.51	3.40	3.31	3.03	2.88	2.72	2.64	2.55	2.36
22	7.95	5.72	4.82	4.31	3.99	3.76	3.59	3.45	3.35	3.26	2.98	2.83	2.67	2.58	2.50	2.31
23	7.88	5.66	4.78	4.26	3.94	3.71	3.54	3.41	3.30	3.21	2.93	2.78	2.62	2.54	2.45	2.26
24	7.82	5.61	4.72	4.22	3.90	3.67	3.50	3.36	3.26	3.17	2.89	2.74	2.58	2.49	2.40	2.21
25	7.77	5.57	4.68	4.18	3.86	3.63	3.46	3.32	3.22	3.13	2.85	2.70	2.53	2.45	2.36	2.17
30	7.56	5.39	4.50	4.02	3.70	3.47	3.30	3.17	3.07	2.98	2.70	2.55	2.39	2.30	2.21	2.01
40	7.31	5.18	4.31	3.83	3.51	3.29	3.12	2.99	2.89	2.80	2.52	2.37	2.20	2.11	2.02	1.80
60	7.08	4.98	4.13	3.65	3.34	3.12	2.95	2.82	2.72	2.63	2.35	2.20	2.03	1.94	1.84	1.60
120	6.85	4.79	3.95	3.48	3.17	2.96	2.79	2.66	2.56	2.47	2.10	2.03	1.86	1.76	1.66	1.38
∞	6.63	4.61	3.78	3.32	3.02	2.80	2.64	2.51	2.41	2.32	2.04	1.88	1.70	1.59	1.47	1.00

Critical values for F for α = .01

Wilcoxon table

This table gives the significance probabilities for the Wilcoxon signed-rank test for paired comparisons, for various selected values of the test statistic W = sum of all signed ranks. The significance probabilities included in the table are the ones closest to the commonly used levels of significance $\alpha = .10$, $\alpha = .05$, and $\alpha = .01$. Thus the table may be used to obtain the appropriate critical value of W for a given value of α, the level of significance.

The critical values c in the table correspond to the critical value for a one-sided test which rejects for large values of W. If the test is one-sided, and rejects for small (negative) values of W, then the critical value is $-c$, where c is the value in the table for which $P(W \geq c)$ = desired level of significance. If the test is two-sided, then the critical value c is determined by finding the value in the table for which $P(W \geq c) = 1/2 \, \alpha$, where α is the desired level of significance. In this case the test is to reject H_0 if $W \leq -c$ or $W \geq c$.

Examples

(a) The test is one-sided and rejects for large values of W. Suppose $\alpha = .05$ and $n = 8$. Then the critical value is $c = 24$, since $P(W \geq c) = .055$, and .055 is closest to the desired level $\alpha = .05$. Thus, the test rejects H_0 if $W \geq 24$, and accepts otherwise.

(b) The test is one-sided and rejects for small (negative) values of W. Suppose $\alpha = .10$ and $n = 12$. The critical value is -34, since $P(W \geq 34) = .102$, and .102 is the value closest to .10. Thus the test rejects H_0 if $W \leq -34$.

(c) The test is two-sided. Suppose $\alpha = .05$ and $n = 20$. Then the critical values are 106 and -106, since $P(W \geq 106) = .024$, and .024 is the value closest to .025 ($= 1/2\alpha$). Thus the test rejects H_0 if $W \leq -106$ or $W \geq 106$.

n	c	$P(W \geq c)$	n	c	$P(W \geq c)$	n	c	$P(W \geq c)$	n	c	$P(W \geq c)$
1	1	.500	8	32	.012	12	58	.010	16	88	.011
				28	.027		50	.026		76	.025
2	3	.250		24	.055		44	.046		64	.052
				20	.098		34	.102		52	.096
3	6	.125									
			9	39	.010	13	65	.011	17	97	.010
4	10	.062		33	.027		57	.024		83	.025
	8	.125		29	.049		49	.047		71	.049
				23	.102		39	.095		55	.103
5	15	.031									
	13	.062	10	45	.010	14	73	.010	18	105	.010
	11	.094		39	.024		63	.025		91	.024
				33	.053		53	.052		77	.049
6	21	.016		27	.097		43	.097		61	.098
	19	.031									
	17	.047	11	52	.009	15	80	.011	19	114	.010
	13	.109		44	.027		70	.024		98	.025
				38	.051		60	.047		82	.052
7	28	.008		30	.013		46	.104		66	.098
	24	.023									
	20	.055							20	124	.010
	16	.109								106	.024
										90	.049
										70	.101

Mann-Whitney tables

This table gives the significance probabilities for the two-tailed Mann-Whitney test, for various selected values of the test statistic R = sum of ranks from the sample from population 1. The significance probabilities included in the table are the ones closest to the commonly used levels of significance $\alpha = .10$, $\alpha = .05$, and $\alpha = .01$. Thus the table may be used to obtain the appropriate critical value of R, for a given level of significance α, and sample sizes n_1 and n_2 drawn from populations 1 and 2, respectively. The population to be designated as population 1 is the one with the smallest sample size, so $n_1 \leq n_2$.

For a two-tailed test, H_0 is rejected if R is too small or two large. The table gives $P(R \leq a \text{ or } R \geq b)$, for selected pairs (a,b).

Example. Samples of sizes 6 and 7 are drawn from populations 1 and 2, respectively, and $\alpha = .05$. Then $n_1 = 6$, $n_2 = 7$, and the test would reject H_0 if $R \leq 28$ or $R \geq 56$. The exact level of significance for this test is .0512.

	Significance Probability nearest to					
	$\alpha = .10$		$\alpha = .05$		$\alpha = .01$	
$n_1 = 3$						
n_2						
3	(6,15)	.1000				
4	(7,17)	.1142	(6,18)	.0572		
5	(7,20)	.0714	(6,21)	.0358		
6	(8,22)	.0952	(7,23)	.0476		
7	(9,24)	.1166	(8,25)	.0666	(6,27)	.0166
8	(9,27)	.0848	(8,28)	.0482	(6,30)	.0121
$n_1 = 4$						
n_2						
4	(12,24)	.1142	(11,25)	.0572		
5	(13,27)	.1112	(12,28)	.0634	(10,30)	.0158
6	(14,30)	.1142	(13,31)	.0666	(10,34)	.0096
7	(15,33)	.1190	(13,35)	.0424	(11,37)	.0122
8	(16,36)	.1090	(14,38)	.0484	(11,41)	.0080
$n_1 = 5$						
n_2						
5	(19,36)	.0952	(18,37)	.0556	(15,40)	.0080
6	(20,40)	.0822	(19,41)	.0520	(16,44)	.0086
7	(22,43)	.1060	(20,45)	.0480	(17,48)	.0102
8	(23,47)	.0932	(21,49)	.0450	(18,52)	.0108
$n_1 = 6$						
n_2						
6	(28,50)	.0930	(26,52)	.0412	(23,55)	.0086
7	(30,54)	.1014	(28,56)	.0512	(24,60)	.0082
8	(32,58)	.1078	(29,61)	.0426	(25,65)	.0080
$n_1 = 7$						
n_2						
7	(39,66)	.0974	(37,68)	.0530	(33,72)	.0110
8	(41,71)	.0938	(39,73)	.0540	(34,78)	.0094
$n_1 = 8$						
n_2						
8	(52,84)	.1048	(49,87)	.0498	(44,92)	.0104

Critical values for the correlation coefficient r

	$\alpha = .05$		$\alpha = .01$	
n	*one tail*	*two tail*	*one tail*	*two tail*
3	.99	1.00	1.00	1.00
4	.90	.95	.98	.99
5	.81	.88	.93	.96
6	.73	.81	.88	.92
7	.67	.75	.83	.87
8	.62	.71	.79	.83
9	.58	.67	.75	.80
10	.54	.63	.72	.76
11	.52	.60	.69	.73
12	.50	.58	.66	.71
13	.48	.53	.63	.68
14	.46	.53	.61	.66
15	.44	.51	.59	.64
16	.42	.50	.57	.61
17	.41	.48	.56	.61
18	.40	.47	.54	.59
19	.39	.46	.53	.58
20	.38	.44	.52	.56
21	.37	.43	.50	.55
22	.36	.42	.49	.54
23	.35	.41	.48	.53
24	.34	.40	.47	.52
25	.34	.40	.46	.51
26	.33	.39	.45	.50
27	.32	.38	.45	.49
28	.32	.37	.44	.48
29	.31	.37	.43	.47
30	.31	.36	.42	.46

For tests of *r* when $n > 30$, an approximation using the *t* table must be done.

Poisson distribution

X denotes a variable with a Poisson distribution having expectation μ. The table gives the values of $P(X = k) = (e^{-\mu}\mu^k/k!)$ for various values of k and μ.

> **Examples.**   For $\mu = 1.0$,
>
> $P(X = 1) = .368$
>
> $P(X = 3) = .061$
>
> $P(X \le 2) = P(X = 0) + P(X = 1) + P(X = 2)$
>
> $\qquad\qquad = .368 + .368 + .184$
>
> $\qquad\qquad = .920$

k	$\mu = .1$	$\mu = .5$	$\mu = 1.0$	$\mu = 2.0$
0	.905	.607	.368	.135
1	.090	.303	.368	.271
2	.005	.076	.184	.271
3		.013	.061	.180
4		.002	.015	.090
5			.003	.036
6			.001	.012
7				.003
8				.001

k	$\mu = 3.0$	$\mu = 4.0$	$\mu = 5.0$	$\mu = 6.0$
0	.050	.018	.007	.002
1	.149	.073	.034	.015
2	.224	.147	.084	.045
3	.224	.195	.140	.089
4	.168	.195	.175	134
5	.101	.156	.175	.161
6	.050	.104	.146	.161
7	.022	.060	.104	.138
8	.008	.030	.065	.103
9	.003	.013	.036	.069
10	.001	.005	.018	.041
11		.002	.008	.023
12		.001	.003	.011
13			.001	.005
14				.002
15				.001

k	$\mu = 7.0$	$\mu = 8.0$	$\mu = 9.0$	$\mu = 10.0$
0	.001			
1	.006	.003	.001	
2	.022	.011	.005	.002
3	.052	.029	.015	.008
4	.091	.057	.034	.019
5	.128	.092	.061	.038
6	.149	.122	.091	.063
7	.149	.140	.117	.090
8	.130	.140	.132	.113
9	.101	.124	.132	.125
10	.071	.099	.119	.125
11	.045	.072	.097	.114
12	.026	.048	.073	.095
13	.014	.030	.050	.073
14	.007	.017	.032	.052
15	.003	.009	.019	.035
16	.001	.005	.011	.022
17	.001	.002	.006	.013
18		.001	.003	.007
19			.001	.004
20			.001	.002
21				.001

Answers to Odd-numbered problems

Chapter 2

2-1

Boundaries	Frequency
20.5–30.5	1
30.5–40.5	1
40.5–50.5	2
50.5–60.5	2
60.5–70.5	9
70.5–80.5	8
80.5–90.5	11
90.5–100.5	6

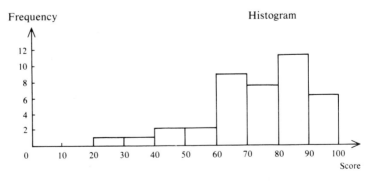

2-3

Boundaries	Frequency	Cumulative frequency
20.5–30.5	1	1
30.5–40.5	1	2
40.5–40.5	2	4
50.5–60.5	2	6
60.5–70.5	9	15
70.5–80.5	8	23
80.5–90.5	11	34
90.5–100.5	6	40

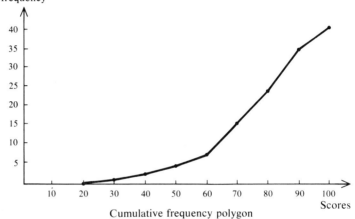

Cumulative frequency polygon

Percentile rank of 90 is 85
Percentile rank of 98 is 95
Percentile rank of 87 is 73
Percentile rank of 68 is 35

2-5

	mean	median	mode
a	12.167	3	1, 3
b	4	4	1, 3, 5, 7
c	2.857	0	0

2-7 11.967

2-9 $\bar{X} = 2646.0$, median $= 2505$

2-11 2.449

2-13   5.547

2-15   *a*

Boundaries	Frequency	Cumulative frequency
2.5–7.5	1	1
7.5–12.5	11	12
12.5–17.5	6	18
17.5–22.5	10	28
22.5–27.5	4	32
27.5–32.5	4	36
32.5–37.5	4	40
37.5–42.5	4	44
42.5–47.5	2	46
47.5–52.5	4	50

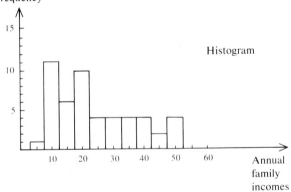

Histogram

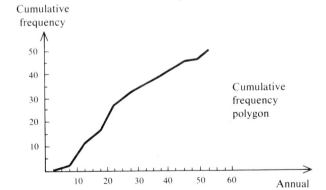

Cumulative frequency polygon

b $\bar{X} = 24.12$, $s = 12.795$

Percentile rank of \$20,000 is 50;
Percentile rank of \$38,000 is 82

d \$20,000; \$20,500

2-17

Boundaries	Frequencies	Cumulative frequencies
2.75–7.25	7	7
7.25–11.75	9	16
11.75–16.25	8	24
16.25–20.75	7	31
20.75–25.25	6	37
25.25–29.75	3	40

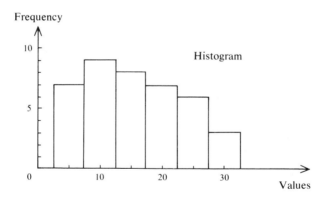

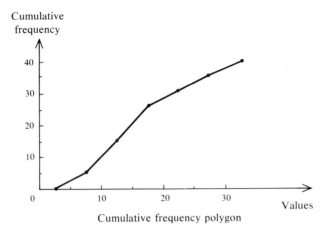

Cumulative frequency polygon

2-19 *a* 72.5; 92.5; 85; 35; *b* 7.3; 13.7; 18.9

TRUE-FALSE

1F	*2F*	*3T*	*4F*	*5T*
6T	*7F*	*8T*	*9F*	*10F*

Chapter 3

3-1 *1/4, 3/4*

3-3 *a* 1/2; *b* 1/10; *c* 1/100; *d* 1/10; *e* 199/1,000; *f* 729/1,000

3-5 .288

3-7 *a* 8/27; *b* 32/243; *c* 224/6,561

3-9 *a* 3/7; *b* 1/2

3-11 *a* 1/2; *b* 1/2; *c* 1/2; *d* 31/42

3-13 36/125, 117/125

3-15 *a* 1/3; *b* 5/6; *c* 2/9

3-17 *a* .001 = 1/1,024; *b* .999 = 1,023/1,024;
c .001 = 1/1,024; *d* .246 = 252/1,024

3-19   19/27

3-21   15/216

3-23   1/6

3-25   3/8

3-27   *a*   3/5;   *b*   154/4,845

3-29   *a*   21;   *b*   1/81;   *c*   50/81

3-31   1

TRUE-FALSE

1 T	*2 F*	*3 T*	*4 F*	*5 T*	*6F*	*7 F*
8 F	*9 T*	*10 T*				

Chapter 4

4-1   *a*   .6331;   *b*   .3669;   *c*   .2007

4-3   *a*   .2669;   *b*   .9444

4-5   *a*   .0228;   *b*   .6912;   *c*   .1359   *d*   .5319

4-7   .8664

4-9   − .30

4-11   8.68

4-13   .5596

4-15   .9854

4-17   *a*   .7625;   *b*   .3

4-19   *a*   .4207;   .1151;   .1151;   .6449;   .6554;   .9772
        b   .1587;   .3085;   .6247
        c   .1587;   .3085;   .6247

4-21   *a*   .8185;   *b*   0

4-23   *a*   .6554;   *b*   .7078

4-25 .0388

4-27 *a* .0359; *b* .2061

4-29 0

4-31 *a* 1; *b* 1

4-33 $m = 127.92, s = 41.67$

TRUE-FALSE

1 T	*2 F*	*3 T*	*4 T*	*5 T*	*6 T*	*7 F*
8 F	*9 F*	*10 T*				

Chapter 5

5-1 .77804–.78196

5-3 7.871–8.529
7.684–8.716

5-5 .0584–.1416

5-7 .1216–.2784

5-9 246

5-11 95 percent confidence interval is 75.614–78.386. This interval does not contain 80, which is strong evidence that men score higher than women.

5-13 95 percent confidence interval is .9173–.9827, which does not contain 1.0, so results are not consistent with claim.

5-15 *a.* 6.852–7.582 *b.* .7660–.9562
6.646–7.788

5-17 .3920–.5080

5-19 9,604

5-21 *a* .1166–.2434
b You know the area beyond 10,000 is .18. A confidence interval is possible.

TRUE-FALSE

1 T	*2 T*	*3 T*	*4 T*	*5 T*	*6 F*	*7 F*
8 T	*9 T*	*10 F*				

Chapter 6

6-1 H_0: $p = 1/2$; H_a: $p > 1/2$; accept H_0

6-3 H_0: $p = 1/4$; H: $p > 1/4$; $\alpha = .05$, reject H_0;
$\alpha = .01$ so accept H_0

6-5 H_0: $p = .90$ H_a: $p < .90$; accept H_0

6-7 H_0: $p = .20$; H_a: $p > .20$; reject H_0; Yes

6-9 H_0: $p = .15$; H_a: $p > .15$; accept H_0

TRUE-FALSE

1 F	*2 T*	*3 F*	*4 F*	*5 F*	*6 T*	*7 T*
8 F	*9 T*	*10 F*				

Chapter 7

7-1 $Z = -1.18$ don't reject H_0

7-3 $Z = -1.41$ don't reject H_0

7-5 $t = -1.58$ don't reject H_0

7-7 $Z = -2.23$ reject H_0

7-9 $t = 2.86$ reject H_0

TRUE-FALSE

1 F	*2 T*	*3 T*	*4 F*	*5 F*	*6 T*	*7 F*
8 F	*9 F*	*10 T*				

Chapter 8

8-1 H_0: $\mu_A - \mu_B = 0$; H_a: $\mu_A - \mu_B \neq 0$; reject H_0

8-3   Yes

8-5   Reject H_0; there is a significant difference in the effect of the two drugs.

8-7   Accept H_0, there is not sufficient evidence to conclude that one method is better than the other.

8-9   Reject H_0, there is a difference in the off-schedule amounts.

TRUE-FALSE

1 F	2 T	3 T	4 F	5 F	6 F	7 T
8 F	9 F	10 T				

Chapter 9

9-1 $p = .0547$, reject H_0

9-3 $p = .1938$, accept H_0

9-5   One-sided test, $W = 35$, reject H_0

9-9   Two-sided test, $R = 43.5$, reject H_0

TRUE-FALSE

1 T	2 F	3 F	4 F	5 T	6 F	7 T
8 T	9 F	10 T				

Chapter 10

10-1 Results are consistent with ranger's belief.

10-3 Data indicate change in ethnic-group composition.

10-5 Type of street light and number of muggings are not independent.

10-7 There is a relationship between a person's age and his recovery time from a sprained ankle.

TRUE-FALSE

1 F	*2 T*	*3 F*	*4 F*	*5 T*	*6 T*	*7 T*
8 T	*9 F*	*10 T*				

Chapter 11

11-1 $\chi^2 = 8.1$ reject H_0

11-3 $\chi^2 = 14.4$ reject H_0

11-5 $\chi^2 = 2.81$ don't reject H_0

11-7 $\chi^2 = 5.66$ don't reject H_0

TRUE-FALSE

1 F	*2 F*	*3 T*	*4 T*	*5 F*

Chapter 12

12-1   Insignificant difference

12-3   There is no difference in average yields.

12-5   There is no difference in location.

TRUE-FALSE

1 T	*2 F*	*3 F*	*4 F*	*5 F*	*6 F*	*7 F*
8 T	*9 F*	*10 T*				

Chapter 13

13-1

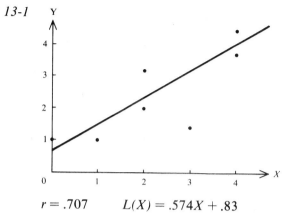

$r = .707$ $L(X) = .574X + .83$

X and Y are positively correlated.

13-3

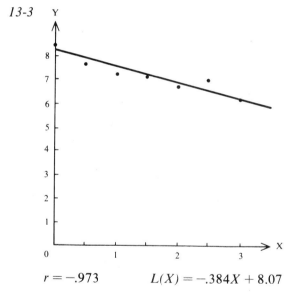

$r = -.973$ $L(X) = -.384X + 8.07$

13-5 $L(x) = .952x + 3.12,\ r = .645$

TRUE-FALSE

1 F	*2 F*	*3 T*	*4 F*	*5 T*	*6 F*	*7 F*
8 F	*9 T*	*10 T*				

Chapter 14

14-1   1.75

14-3   1.2

14-5   1.6 hours

14-7   4.095

14-9 $1.605; $16.05

14-11 $522.92

14-13 $\dfrac{16,807}{7,776} \approx 2.16$

14-15   3 years

14-19   6 hours

14-21   Infinite amount

TRUE-FALSE

1 F	*2 F*	*3 T*	*4 F*	*5 F*	*6 F*	*7 F*
8 F	*9 T*	*10 F*				

Chapter 15

15-1   *a*   1.5625
         b   8.6875
         c   2.2

15-3   13.5

15-5   .951

15-7   At least 3/4

15-9   *a* 1/4
         b .0456

TRUE-FALSE

1 T	*2 T*	*3 F*	*4 F*	*5 F*	*6 T*	*7 T*
8 F	*9 T*	*10 F*				

Chapter 16

16-1 40!/10!

16-3 20!/5!

16-5 $a \begin{pmatrix} 15 \\ 8 \end{pmatrix} (1/3)^8 (2/3)^7$
        b 5

16-7 a 100
        b .99^{1,000}

16-9 $a \left(\dfrac{37}{38}\right)^{50}$ b $50\left(\dfrac{1}{38}\right)\left(\dfrac{37}{38}\right)^{85}$

16-11 20/7

16-13 $\displaystyle\sum_{k=0}^{15}\left(e^{-40}40^k/k!\right)$

16-15 $\begin{pmatrix} 10 \\ 7 \end{pmatrix}\left(e^{-5}\right)^7\left(1-e^{-5}\right)^3$

16-17 $1 - [(365)\,(364)\,(363)\ldots(336)/(365)^{30}] = .71$

16-19 $\left(\dfrac{1,170}{1,200}\right) \times \left(\dfrac{599}{1,199}\right)$

16-21 .0162

16-23 .223

16-25 455

16-27 Downtown train arrives 6 minutes after uptown train leaves.

16-29 $\begin{pmatrix} 19 \\ 4 \end{pmatrix}\left(.05\right)^5\left(.95\right)^{15}$

TRUE-FALSE

1 F	*2 T*	*3 T*	*4 T*	*5 T*	*6 F*	*7 T*
8 F	*9 T*	*10 F*				

Chapter 17

17-1   .986

17-3   C

17-5 P(He's guilty) = .001

17-7   1/3439

17-9   *a* Suppose there are the numbers 1–96 in a hat. Let A be the event you draw one of the numbers 1–32, and B be the event you draw one of the numbers 21-44.
b 1/8

17-11   Choose action a_2, without weather report.

17-13   Hire consulting firm to evaluate indicators and use the following decision rule:
If z_1, do a_3 or a_4;
If z_2, do a_4;
If z_3, do a_4.
Using this procedure, your expected gain is $77,000.

TRUE-FALSE

1 F	2 F	3 T	4 T	5 F	6 F	7 F
8 T	9 F	10 F				

Index